STUDENT'S SOLUTIONS MANUAL

RENATO MIROLLO

Boston College

SALVATORE SCIANDRA

Niagara County Community College

FINITE MATHEMATICS & ITS APPLICATIONS

TENTH EDITION

Larry J. Goldstein

Goldstein Educational Technologies

David I. Schneider

University of Maryland

Martha J. Siegel

Towson University

Prentice Hall
is an imprint of

PEARSON

The author and publisher of this book have used their best efforts in preparing this book. These efforts include the development, research, and testing of the theories and programs to determine their effectiveness. The author and publisher make no warranty of any kind, expressed or implied, with regard to these programs or the documentation contained in this book. The author and publisher shall not be liable in any event for incidental or consequential damages in connection with, or arising out of, the furnishing, performance, or use of these programs.

Reproduced by Pearson Prentice Hall from electronic files supplied by the author.

ISBN-13: 978-0-321-59898-1
ISBN-10: 0-321-59898-9

2 3 4 5 6 BRR 12 11 10

Prentice Hall
is an imprint of

www.pearsonhighered.com

Contents

Chapter 1: Linear Equations and Straight Lines   1 – 1

Chapter 2: Matrices   2 – 1

Chapter 3: Linear Programming, A Geometric Approach   3 – 1

Chapter 4: The Simplex Method   4 – 1

Chapter 5: Sets and Counting   5 – 1

Chapter 6: Probability   6 – 1

Chapter 7: Probability and Statistics   7 – 1

Chapter 8: Markov Processes   8 – 1

Chapter 9: The Theory of Games   9 – 1

Chapter 10: The Mathematics of Finance   10 – 1

Chapter 11: Difference Equations and Mathematical Models 11 – 1

Chapter 12: Logic   12 – 1

Explorations in Finite Mathematics   EFM 1

Chapter 1

1. Right 2, up 3

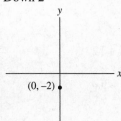

3. Down 2

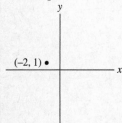

5. Left 2, up 1

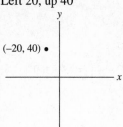

7. Left 20, up 40

9. e

11. $-2(1) + \dfrac{1}{3}(3) = -2 + 1 = -1$ so the point is on the line.

13. $-2x + \dfrac{1}{3}y = -1$ Substitute the x and y coordinates of the point into the equation:

$$\left(\dfrac{1}{2}, 3\right) \rightarrow -2\left(\dfrac{1}{2}\right) + \dfrac{1}{3}(3) = -1 \rightarrow -1 + 1 = -1$$

is a false statement. So the point is not on the line.

15. $m = 5, b = 8$

17. $y = 0x + 3; m = 0, b = 3$

19. $14x + 7y = 21$

$7y = -14x + 21$

$y = -2x + 3$

21. $3x = 5$

$x = \dfrac{5}{3}$

23. $0 = -4x + 8$

$4x = 8$

$x = 2$

x-intercept: $(2, 0)$

$y = -4(0) + 8$

$y = 8$

y-intercept: $(0, 8)$

25. When $y = 0, x = 7$

x-intercept: $(7, 0)$

$0 = 7$

no solution

y-intercept: none

27. $0 = \dfrac{1}{3}x - 1$

$x = 3$

x-intercept: $(3, 0)$

$y = \dfrac{1}{3}(0) - 1$

$y = -1$

y-intercept: $(0, -1)$

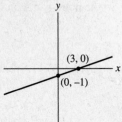

29. $0 = \dfrac{5}{2}$

no solution

x-intercept: none

When $x = 0,\ y = \dfrac{5}{2}$

y-intercept: $\left(0, \dfrac{5}{2}\right)$

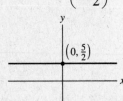

31. $3x + 4(0) = 24$

$x = 8$

x-intercept: $(8, 0)$

$3(0) + 4y = 24$

$y = 6$

y-intercept: $(0, 6)$

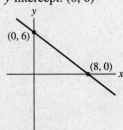

33. $x = -\dfrac{5}{2}$

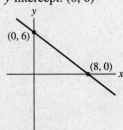

35. $2x + 3y = 6$

$3y = -2x + 6$

$y = -\dfrac{2}{3}x + 2$

a. $4x + 6y = 12$

$6y = -4x + 12$

$y = -\dfrac{2}{3}x + 2$

Yes

b. Yes

c. $x = 3 - \dfrac{3}{2}y$

$\dfrac{3}{2}y = -x + 3$

$y = -\dfrac{2}{3}x + 2$

$y = -\dfrac{2}{3}x + 2$

Yes

d. $6 - 2x - y = 0$

$y = 6 - 2x = -2x + 6$

No

e. $y = 2 - \dfrac{2}{3}x = -\dfrac{2}{3}x + 2$

Yes

f. $x + y = 1$

$y = -x + 1$

No

37. a. $x + y = 3$

$y = -x + 3$

$m = -1, b = 3$

L_3

b. $2x - y = -2$

$-y = -2x - 2$

$y = 2x + 2$

$m = 2, b = 2$

L_1

c. $x = 3y + 3$

$3y = x - 3$

$y = \dfrac{1}{3}x - 1$

$m = \dfrac{1}{3}, b = -1$

L_2

39. $y = 30x + 72$

a. When x = 0, y = 72. This is the temperature of the water at time = 0 before the kettle is turned on.

b. $y = 30(3) + 72$

$y = 162°F$

c. Water boils when y = 212 so we have $212 = 30x + 72$. Solving for x gives x = 4.67 minutes or 4 minutes 40 seconds.

41. a. x-intercept: $\left(-33\dfrac{1}{3}, 0\right)$

y-intercept: (0, 2.5)

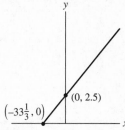

b. In 1960, 2.5 trillion cigarettes were sold.

c. $4 = .075x + 2.5$

$x = 20$

$1960 + 20 = 1980$

d. $2020 - 1960 = 60$

$y = .075(60) + 2.5$

$y = 7$

7 trillion

43. a. x-intercept: (−11.3, 0)

y-intercept: (0, 678)

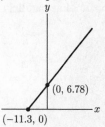

b. In 1997 the car insurance rate for a small car was $678.

c. $2000 - 1997 = 3$

$y = 60(3) + 678$

$y = 858$

$858

d. $1578 = 60x + 678$

$x = 15$

$1997 + 15 = 2012$

The year 2012

45. a. In 2000, 10% of college freshmen smoked.

b. $2005 - 2000 = 5$

$y = \left(-\dfrac{26}{35}\right)(5) + 10$

$y \approx 6.3$

6.3% of college freshmen smoked in 2005.

c. $4.8 = -\dfrac{26}{35}x + 10$

$x = 7$

$2000 + 7 = 2007$

In 2007, the percent of college freshmen that smoked was 4.8.

47. $y = mx + b$

$0.9 = m(0) + b$

$b = 0.9$

$0 = m(0.6) + 0.9$

$m = -1.5$

$y = -1.5x + 0.9$

49. On the *x*-axis, $y = 0$.

51. $y = b$ is an equation of a line parallel to the x-axis.

53. $2x - y = -3$

55. $1 \cdot x + 0 \cdot y = -3$

57. $\dfrac{2}{3}x + y = -5$

$2x + 3y = -15$

59. Since (a,0) and (0,b) are points on the line the slope of the line is (b-0)/(0-a) = -b/a. Since the y intercept is (0,b), the equation of the line is $y = -(b/a)x + b$ or $ay = -bx + ab$. In general form, the equation is bx + ay = ab.

61. One possible equation is $y = x - 9$.

63. One possible equation is $y = x + 7$.

65. One possible equation is $y = x + 2$.

67. One possible equation is $y = x + 9$.

69. a. $y = -3x + 6$

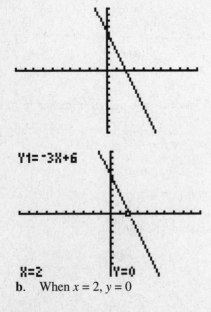

b. When $x = 2$, $y = 0$

c. The intercepts are at the points (2, 0) and (0, 6)

71. a. $3y - 2x = 9$

$3y = 2x + 9$

$y = \dfrac{2}{3}x + 3$

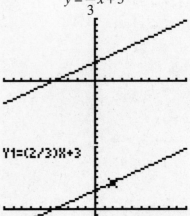

b. When $x = 2$, $y = 4.33$ or $13 / 3$.

c. The intercepts are at the points (–4.5, 0) and (0, 3).

73. 2y + x = 100. When y = 0, x = 100. and when x = 0, y = 50. An appropriate window might be [-10, 110] and [-10,60]. Other answers are possible.

Exercises 1.2

1. False

3. True

5. $2x - 5 \geq 3$

$2x \geq 8$

$x \geq 4$

7. $-5x + 13 \leq -2$

$-5x \leq -15$

$x \geq 3$

9. $2x + y \leq 5$

$y \leq -2x + 5$

11. $5x - \dfrac{1}{3}y \leq 6$

$-\dfrac{1}{3}y \leq -5x + 6$

$y \geq 15x - 18$

13. $4x \geq -3$

$$x \geq -\frac{3}{4}$$

15. $3(2) + 5(1) \leq 12$

$6 + 5 \leq 12$

$11 \leq 12$

Yes

17. $0 \geq -2(3) + 7$

$0 \geq -6 + 7$

$0 \geq 1$

No

19. $5 \leq 3(3) - 4$

$5 \leq 9 - 4$

$5 \leq 5$

Yes

21. $7 \geq 5$

Yes

23.

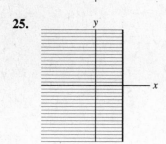

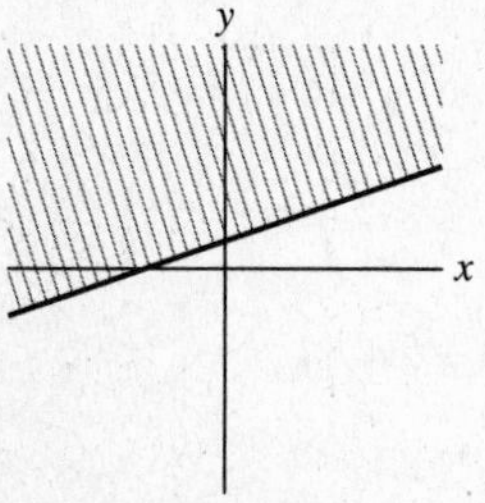

25.

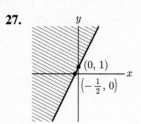

27.

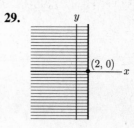

29.

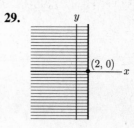

31. $x + 4y \geq 12$

$$y \geq -\frac{1}{4}x + 3$$

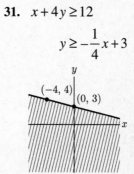

33. $4x - 5y + 25 \geq 0$

$$y \leq \frac{4}{5}x + 5$$

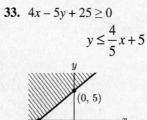

35. $\frac{1}{2}x - \frac{1}{3}y \leq 1$

$$y \geq \frac{3}{2}x - 3$$

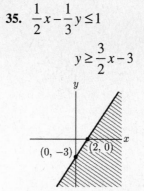

37. $0.5x + 0.4y \leq 2$

$$y \leq 1.25x + 5$$

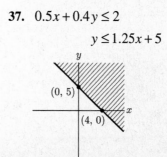

39. $\begin{cases} y \le 2x - 4 \\ y \ge 0 \end{cases}$

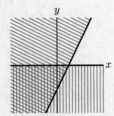

41. $\begin{cases} x + 2y \ge 2 \\ 3x - y \ge 3 \end{cases}$

$\begin{cases} y \ge -\dfrac{1}{2}x + 1 \\ y \le 3x - 3 \end{cases}$

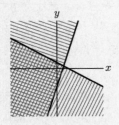

43. $\begin{cases} x + 5y \le 10 \\ x + y \le 3 \\ x \ge 0,\ y \ge 0 \end{cases}$

$\begin{cases} y \le -\dfrac{1}{5}x + 2 \\ y \le -x + 3 \\ x \ge 0,\ y \ge 0 \end{cases}$

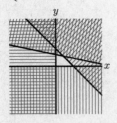

45. $\begin{cases} 6(8) + 3(7) \le 96 \\ 8 + 7 \le 18 \\ 2(8) + 6(7) \le 72 \\ 8 \ge 0,\ 7 \ge 0 \end{cases}$

$\begin{cases} 69 \le 96 \\ 15 \le 18 \\ 58 \le 72 \\ 8 \ge 0,\ 7 \ge 0 \end{cases}$

Yes

47. $\begin{cases} 6(9) + 3(10) \le 96 \\ 9 + 10 \le 18 \\ 2(9) + 6(10) \le 72 \\ 9 \ge 0,\ 10 \ge 0 \end{cases}$

$\begin{cases} 84 \le 96 \\ 19 \le 18 \\ 78 \le 72 \\ 9 \ge 0,\ 10 \ge 0 \end{cases}$

No

49. For $x = 3$, $y = 2(3) + 5 = 11$.
So (3, 9) is below.

51. $7 - 4x + 5y = 0$

$$y = \frac{4}{5}x - \frac{7}{5}$$

For $x = 0$, $y = \dfrac{4}{5}(0) - \dfrac{7}{5} = -\dfrac{7}{5}$.

So (0, 0) is above.

53. $8x - 4y = 4$

$\quad\quad y = 2x - 1$

$\quad 8x - 4y = 0$

$\quad\quad\quad y = 2x$

$\begin{cases} y \ge 2x - 1 \\ y \le 2x \end{cases}$

55. d

57. e

59. $x + 2y = 11$

$$y = -\frac{1}{2}x + \frac{11}{2}$$

a.

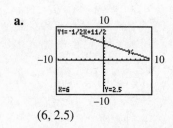

(6, 2.5)

b. Above, because (6, 2.5) is on the line.

61. $\begin{cases} 3x + 6y \geq 24 \\ 3x + y \geq 6 \end{cases}$

$\begin{cases} y \geq -\dfrac{1}{2}x + 4 \\ y \geq -3x + 6 \end{cases}$

Exercises 1.3

1. $4x - 5 = -2x + 7$

$6x = 12$

$x = 2$

$y = 4(2) - 5 = 3$

(2, 3)

3. $x = 4y - 2$

$x = -2y + 4$

$4y - 2 = -2y + 4$

$6y = 6$

$y = 1$

$x = 4(1) - 2 = 2$

(2, 1)

5. $y = \dfrac{1}{3}(12) - 1 = 3$

(12, 3)

7. $\begin{cases} 6 - 3(4) = -6 \\ 3(6) - 2(4) = 10 \end{cases}$

$\begin{cases} -6 = -6 \\ 10 = 10 \end{cases}$

Yes

9. $\begin{cases} y = -2x + 7 \\ y = x - 3 \end{cases}$

$-2x + 7 = x - 3$

$-3x = -10$

$x = \dfrac{10}{3}$

$y = \dfrac{10}{3} - 3 = \dfrac{1}{3}$

$x = \dfrac{10}{3},\ y = \dfrac{1}{3}$

11. $\begin{cases} y = \dfrac{5}{2}x - \dfrac{1}{2} \\ y = -2x - 4 \end{cases}$

$\dfrac{5}{2}x - \dfrac{1}{2} = -2x - 4$

$\dfrac{9}{2}x = -\dfrac{7}{2}$

$x = -\dfrac{7}{9}$

$y = -2\left(-\dfrac{7}{9}\right) - 4 = -\dfrac{22}{9}$

$x = -\dfrac{7}{9},\ y = -\dfrac{22}{9}$

13. $\begin{cases} x = 3 \\ 2x + 3y = 18 \end{cases}$

$y = -\dfrac{2}{3}x + 6 = -\dfrac{2}{3}(3) + 6 = 4$

$A = (3, 4)$

$\begin{cases} y = 2 \\ 2x + 3y = 18 \end{cases}$

$x = -\dfrac{3}{2}y + 9 = -\dfrac{3}{2}(2) + 9 = 6$

$B = (6, 2)$

15. $A = (0, 0)$

$\begin{cases} y = 2x \\ y = \dfrac{1}{2}x + 3 \end{cases}$

$2x = \dfrac{1}{2}x + 3$

$x = 2$

$y = 2(2) = 4$

$B = (2, 4)$

$\begin{cases} y = \dfrac{1}{2}x + 3 \\ x = 5 \end{cases}$

$y = \dfrac{1}{2}(5) + 3 = \dfrac{11}{2}$

$C = \left(5, \dfrac{11}{2}\right)$

$D = (5, 0)$

17. $\begin{cases} 2y - x \le 6 \\ x + 2y \ge 10 \\ x \le 6 \end{cases}$

$\begin{cases} y \le \dfrac{1}{2}x + 3 \\ y \ge -\dfrac{1}{2}x + 5 \\ x \le 6 \end{cases}$

$\begin{cases} y = \dfrac{1}{2}x + 3 \\ y = -\dfrac{1}{2}x + 5 \end{cases} \Rightarrow (2, 4)$

$\begin{cases} y = -\dfrac{1}{2}x + 5 \\ x = 6 \end{cases} \Rightarrow (6, 2)$

$\begin{cases} y = \dfrac{1}{2}x + 3 \\ x = 6 \end{cases} \Rightarrow (6, 6)$

19. $\begin{cases} x + 3y \le 18 \\ 2x + y \le 16 \\ x \ge 0,\ y \ge 0 \end{cases}$

$\begin{cases} y \le -\dfrac{1}{3}x + 6 \\ y \le -2x + 16 \\ x \ge 0,\ y \ge 0 \end{cases}$

$\begin{cases} y = -\dfrac{1}{3}x + 6 \\ y = -2x + 16 \end{cases} \Rightarrow (6, 4)$

$\begin{cases} y = -\dfrac{1}{3}x + 6 \\ x = 0 \end{cases} \Rightarrow (0, 6)$

$$\begin{cases} y = -2x + 16 \\ y = 0 \end{cases} \Rightarrow (8, 0)$$

$$\begin{cases} x = 0 \\ y = 0 \end{cases} \Rightarrow (0, 0)$$

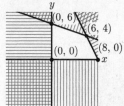

21. $\begin{cases} 4x + y \geq 8 \\ x + y \geq 5 \\ x + 3y \geq 9 \\ x \geq 0, \ y \geq 0 \end{cases}$

$$\begin{cases} y \geq -4x + 8 \\ y \geq -x + 5 \\ y \geq -\dfrac{1}{3}x + 3 \\ x \geq 0, \ y \geq 0 \end{cases}$$

$$\begin{cases} y = -4x + 8 \\ y = -x + 5 \end{cases} \Rightarrow (1, 4)$$

$$\begin{cases} y = -x + 5 \\ y = -\dfrac{1}{3}x + 3 \end{cases} \Rightarrow (3, 2)$$

$$\begin{cases} y = -\dfrac{1}{3}x + 3 \\ y = 0 \end{cases} \Rightarrow (9, 0)$$

$$\begin{cases} y = -4x + 8 \\ x = 0 \end{cases} \Rightarrow (0, 8)$$

23. a. $p = .0001(19,500) + .05$

$= \$2.00$

b. $p = .0001(0) + .05$

$= \$.05$

No units will be supplied for $\$.05$ or less.

25. $\begin{cases} p = .0001q + .05 \\ p = -.001q + 32.5 \end{cases}$

$0.0001q + 0.05 = -0.001q + 32.5$

$.0011q = 32.45$

$q = 29,500$ units

$p = .0001(29,500) + .05$

$p = \$3.00$

27. $p = \dfrac{1}{300}q + 13$

$p = -.03q + 19$

$\dfrac{1}{300}q + 13 = -.03q + 19$

$\dfrac{1}{30}q = 6$

$q = 180$ books

$p = -.03(180) + 19$

$p = \$13.60$

29. Method A: $y = 0.45 + 0.01x$

Method B: $y = 0.035x$

Intersection point:

$0.45 + 0.01x = 0.035x$

$0.45 = 0.025x$

$18 = x$

For a call lasting 18 minutes, the costs for either method will be the same, $y = 0.035(18) = 63$. The cost will be 63cents.

31. Let x = number of 15" TVs sold
y = number of 19" TVs sold

$$\begin{cases} y = x + 5 \\ 280x + 400y = 15600 \end{cases}$$

$$\begin{cases} y = x + 5 \\ y = -\dfrac{7}{10}x + 39 \end{cases}$$

$$x + 5 = -\dfrac{7}{10}x + 39$$

$$\dfrac{17}{10}x = 34$$

$$x = 20 \text{ TV sets}$$
$$y = 20 + 5$$

$$= 25 \text{ TV sets}$$
Total = 20 + 25 = 45 TV sets
Answer (d) is correct.

33.

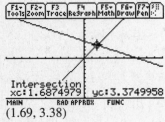

(1.69, 3.38)

35. $\begin{cases} 2x + 3y = 5 \\ -4x + 5y = 1 \end{cases}$

$$\begin{cases} y = -\dfrac{2}{3}x + \dfrac{5}{3} \\ y = \dfrac{4}{5}x + \dfrac{1}{5} \end{cases}$$

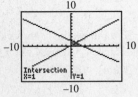

(1, 1)

37. $\begin{cases} 2x + y \geq 5 \\ x - 2y \leq 0 \end{cases}$

$$\begin{cases} y \geq -2x + 5 \\ y \geq \dfrac{1}{2}x \end{cases}$$

a.

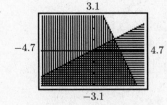

b. (2, 1)

c.

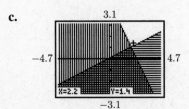

d. Yes

Exercises 1.4

1. $m = \dfrac{2}{3}$

3. $y - 3 = 5(x + 4)$
$y = 5x + 23$
$m = 5$

5.

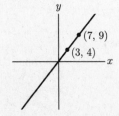

$$m = \dfrac{9 - 4}{7 - 3} = \dfrac{5}{4}$$

7.

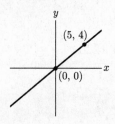

$$m = \frac{4-0}{5-0} = \frac{4}{5}$$

9. The slope of a vertical line is undefined.

11.

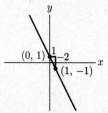

13.

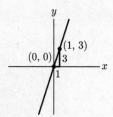

15. $m = \dfrac{-2}{1} = -2$

$$y - 3 = -2(x - 2)$$
$$y = -2x + 7$$

17. $m = \dfrac{0-2}{2-1} = -2$

$$y - 0 = -2(x - 2)$$
$$y = -2x + 4$$

19. $m = -\dfrac{1}{-4} = \dfrac{1}{4}$

$$y - 2 = \frac{1}{4}(x - 2)$$

$$y = \frac{1}{4}x + \frac{3}{2}$$

21. $m = -1$

$$y - 0 = -1(x - 0)$$
$$y = -x$$

23. $m = 0$

$$y - 3 = 0(x - 2)$$
$$y = 3$$

25. $y - 6 = \dfrac{3}{5}(x - 5)$

$$y = \frac{3}{5}x + 3$$

y-intercept: $(0, 3)$

27. Each unit sold yields a commission of $5. In addition, she receives $60 per week base pay.

29. a. p-intercept: $(0, 1200)$; at $1200 no one will buy the item.

    b. $0 = -3q + 1200$
       $q = 400$ units
       q-intercept: $(400, 0)$; even if the item is given away, only 400 will be taken.

    c. -3; to sell an additional item, the price must be reduced by $3.

    d. $p = -3(350) + 1200 = \$150$

    e. $300 = -3q + 1200$

       $q = 300$ items

    f.

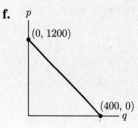

31. a. Let x = quantity and y = cost.

$$m = \frac{9500 - 6800}{50 - 20} = 90$$

$$y - 6800 = 90(x - 20)$$

$$y = 90x + 5000$$

b. $5000

c. $90

d.
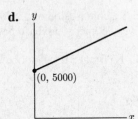
(0, 5000)

33. a. $100(300) = \$30,000$

b. $6000 = 100x$

$x = 60$ coats

c. $y = 100(0) = 0$
(0, 0); if no coats are sold, there is no revenue.

d. 100; each additional coat yields an additional $100 in revenue.

35. a.
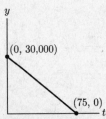
(0, 30,000)

(75, 0)

b. On February 1, 31 days have elapsed since January 1. The amount of oil y = 30,000 – 400(31) = 17,600 gallons.

c. On February 15, 45 days have elapsed since January 1. Therefore, the amount of oil would be y = 30,000- 400(45) = 12,000 gallons.

d. The significance of the y-intercept is that amount of oil present initially on January 1. This amount is 30,000 gallons.

e. The t-intercept is (75,0) and corresponds to the number of days at which the oil will be depleted.

37. a. $y = 0.10x + 220$

b. $y = 0.10(2000) + 220$
$y = 420$

c. $540 = 0.10x + 220$
$x = \$3200$

39. $m = 3, b = -1$
$y = 3x - 1$

41. $m = 1$
$y - 2 = 1(x - 1)$
$y = x + 1$

43. $m = -7$
$y - 0 = -7(x - 5)$
$y = -7x + 35$

45. $m = 0$
$y - 4 = 0(x - 7)$
$y = 4$

47. $m = \dfrac{2-1}{4-2} = \dfrac{1}{2}$

$y - 1 = \dfrac{1}{2}(x - 2)$

$y = \dfrac{1}{2}x$

49. $m = \dfrac{-2-0}{1-0} = -2$
$y = -2x$

51. Changes in x-coordinate: 1, –1, –2
Changes in y-coordinate are m times that or 2, –2, –4: new y values are 5, 1, –1

53. The slope is $\dfrac{-1}{4}$ Changes in x coordinates are 1, 2, –1. Changes in y coordinates are m times the x coordinate changes. New y coordinates are

$\dfrac{-5}{4}, \dfrac{-3}{2}, \dfrac{-3}{4}$

55. a. $x + y = 1$
   $y = -x + 1$
   (C)

   b. $x - y = 1$
   $y = x - 1$
   (B)

   c. $x + y = -1$
   $y = -x - 1$
   (D)

   d. $x - y = -1$
   $y = x + 1$
   (A)

57. One possible equation is $y = x + 1$.

59. One possible equation is $y = 5$.

61. One possible equation is $y = -\dfrac{2}{3}x$.

63. $m = \dfrac{212 - 32}{100 - 0} = \dfrac{9}{5}$

$F - 32 = \dfrac{9}{5}(C - 0)$

$F = \dfrac{9}{5}C + 32$

65. Let 1995 correspond to x = 0. So in 2006, x = 11. When x = 0, tuition is 2848. When x = 11, tuition is 5685. Using (0,2848) and (11,5685) as ordered pairs, find the slope of the line containing these points:
$\dfrac{5685 - 2848}{11 - 0} = 257.91$. Since the y-intercept is 2848, the equation becomes y = 257.91x + 2848. Therefore, in 2000 when x = 5, the tuition should approximately be
$y = 257.91(5) + 2848 = 4137.55$.

67. Let x = number of pounds tires are under inflated. When x = 0, the miles per gallon (y) is 25. When x = 1, mpg decreases to 24.5. The equation is $y = -\dfrac{1}{2}x + 25$. Thus, when x = 8 pounds the miles per gallon will be
$y = -\dfrac{1}{2}(8) + 25 = 21$ mpg.

69. Let 1991 correspond to x = 0 and 2006 correspond to x = 15. Then, the two ordered pairs are on the line: (0, 249,165) and (15,318,042). The slope of the line is
$\dfrac{318{,}042 - 249{,}165}{15 - 0} = 4591.8$ The equation of the line is therefore $y = 4591.8x + 249{,}165$. In the year 2011, x = 20, so the number of Bachelor's degrees awarded can be estimated as $y = 4591.8(20) + 249{,}165 = 341{,}001$.

71. Let 2005 correspond to x = 5 and 2008 correspond to x = 8. Then, the two ordered pairs are on the line: (5, 2.4) and (8,2.7). The slope of the line is $\dfrac{2.7 - 2.4}{8 - 5} = 0.1$. The equation of the line is therefore $y = 0.1x + 1.9$. In the year 2007, x = 7, so the cost of a 30-second advertising slot (in millions) can be estimated as $y = 0.1(7) + 1.9 = \$2.6$ million.

73. $y \geq 4x + 3$

75. $m_1 = \dfrac{3 - 4}{2 - 0} = -\dfrac{1}{2}$

$y = -\dfrac{1}{2}x + 4$

$m_2 = \dfrac{1 - 3}{4 - 2} = -1$

$y - 1 = -(x - 4)$

$y = -x + 5$

$m_3 = \dfrac{1 - 0}{4 - 3} = 1$

$y = x - 3$

$\begin{cases} y \leq -\dfrac{1}{2}x + 4 \\ y \leq -x + 5 \\ y \geq x - 3 \\ x \geq 0,\ y \geq 0 \end{cases}$

77. Set two slopes equal:
$\dfrac{7 - 5}{2 - 1} = \dfrac{k - 7}{3 - 2}$

$2 = k - 7$

$k = 9$

79. Make slopes negative inverses of each other:

$$\frac{-3.1-1}{2-a} = -\frac{1}{\frac{2.4-0}{3.8-(-1)}}$$

$$\frac{-4.1}{2-a} = -2$$

$$4.1 = 4 - 2a$$

$$a = -0.05$$

81. $l_1 : y = m_1 x$

$l_2 : y = m_2 x$

So the vertical segment lies on $x = 1$.
Then

$$1^2 + m_1^2 = a^2$$

$$1^2 + m_2^2 = b^2$$

Add equations and rearrange:

$$a^2 + b^2 - (m_1^2 + m_2^2) = 2$$

l_1 and l_2 are perpendicular if and only if

$$a^2 + b^2 = (m_1 - m_2)^2 = m_1^2 + m_2^2 - 2m_1 m_2$$

or $a^2 + b^2 - (m_1^2 + m_2^2) = -2m_1 m_2$

Substitute: $2 = -2m_1 m_2$

Therefore, the product of the slopes are -1.

83. Let x = weight

y = cost

$$m = \frac{38-5}{60-0} = \frac{11}{20}$$

$$y = \frac{11}{20}x + 5$$

$$y = \frac{11}{20}(20) + 5 = \$16$$

The answer is (c).

85. Let x = number of units

profit = revenue – cost

$$2,000,000 = 130x - (100x + 1,000,000)$$

$$3,000,000 = 30x$$

$$x = 100,000 \text{ units}$$

Answer (e) is correct.

87. $n = 2200 - 25(8)$

$= 2000$ cameras

revenue = $8(2000) = \$16,000$

Answer (c) is correct.

89. Let x = variable costs

For 2008: profit = revenue – costs

$$300,000 = 100(50,000) - (50,000x + 800,000)$$

$$50,000x = 3,900,000$$

$$x = \$78 \text{ per unit}$$

For 2009:

Let y = 2009 price

profit = revenue – cost

$$300,000 = 50,000y -$$

$$[78(50,000) + 800,000 + 200,000]$$

$$5,200,000 = 50,000y$$

$$y = \$104$$

Answer (d) is correct.

91.

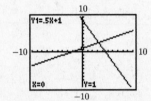

No, do not appear perpendicular

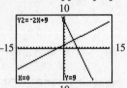

Do appear perpendicular

93.

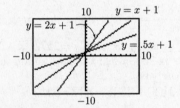

The steeper the line, the greater the slope m in $y = mx + b$ form.

Exercises 1.5

1.

Data Point	Point on Line	Vertical Distance
(1, 3)	(1, 4)	1
(2, 6)	(2, 7)	1
(3, 11)	(3, 10)	1
(4, 12)	(4, 13)	1

$1^2 + 1^2 + 1^2 + 1^2 = 4$

5.

x	y	xy	x^2
1	7	7	1
2	6	12	4
3	4	12	9
4	3	12	16
$\sum x = 10$	$\sum y = 20$	$\sum xy = 43$	$\sum x^2 = 30$

$m = \dfrac{4 \cdot 43 - 10 \cdot 20}{4 \cdot 30 - 10^2} = -1.4$

$b = \dfrac{20 - (-1.4)(10)}{4} = 8.5$

7. $\sum x = 6, \sum y = 18, \sum xy = 45, \sum x^2 = 14$

$m = \dfrac{3 \cdot 45 - 6 \cdot 18}{3 \cdot 14 - 6^2} = 4.5$

$b = \dfrac{18 - (4.5)(6)}{3} = -3$

$y = 4.5x - 3$

9. $\sum x = 10, \sum y = 26, \sum xy = 55,$
$\sum x^2 = 30$

$m = \dfrac{4 \cdot 55 - 10 \cdot 26}{4 \cdot 30 - 10^2} = -2$

$b = \dfrac{26 - (-2)(10)}{4} = 11.5$

$y = -2x + 11.5$

3. $E_1^2 = [1.1(1) + 3 - 3]^2 = 1.21$

$E_2^2 = [1.1(2) + 3 - 6]^2 = 0.64$

$E_3^2 = [1.1(3) + 3 - 8]^2 = 2.89$

$E_4^2 = [1.1(4) + 3 - 6]^2 = 1.96$

$E = 1.21 + 0.64 + 2.89 + 1.96 = 6.70$

11. a.
```
LinReg
y=ax+b
a=.3383317713
b=21.62136832
```
$y = .338x + 21.6$

b. $.338(1100) + 21.6 = 393.4$
About 393 deaths per million males

13. a. Let x be the number of years after 1980, then
$y = .419x + 17.1$

b. $.419(23) + 17.1 = 26.73$
About 26.7%

c. $30 = .419x + 17.1$

$x \approx 30.78$
The year 2011 or late 2010

15. a. $y = 0.153x + 73.5$

    b. $0.153(30) + 73.5 = 78.09$
       About 78.09 years

    c. $0.153(50) + 73.5 = 81.15$
       About 81.15 years

    d. $0.153(90) + 73.5 = 87.27$
       About 87.27 years.(This is an example of a
       fit that is not capable of extrapolating
       beyond the given data)

17. a. Let x be the number of years after 1993, then
       $y = 0.048x + 2.89$

    b. $0.048(6) + 2.89 \approx 3.178$
       About $3.18

    c. $3.85 = 0.048x + 2.89$
       $x = 20$
       The year 2013

19. $\sum x = 12, \sum y = 7, \sum xy = 41, \sum x^2 = 74$

$$m = \frac{2 \cdot 41 - 12 \cdot 7}{2 \cdot 74 - 12^2} = -0.5$$

$$b = \frac{7 - (-0.5)(12)}{2} = 6.5$$

$y = -0.5x + 6.5$
$4 = -0.5(5) + 6.5$
$3 = -0.5(7) + 6.5$

Chapter 1 Supplementary Exercises

1. $x = 0$

2.

$(2, -1)$

3. $\begin{cases} x - 5y = 6 \\ \quad\ 3x = 6 \end{cases}$

$\begin{cases} x = 5y + 6 \\ x = 2 \end{cases}$

$5y + 6 = 2$

$$y = -\frac{4}{5}$$

$$\left(2, -\frac{4}{5}\right)$$

4. $3x - 4y = 8$

$$y = \frac{3}{4}x - 2$$

$$m = \frac{3}{4}$$

5. $m = \frac{0-5}{10-0} = -\frac{1}{2}, \ b = 5$

$$y = -\frac{1}{2}x + 5$$

6. $x - 3y \geq 12$

$$y \leq \frac{1}{3}x - 4$$

$(-3, -5)$ $(0, -4)$

7. $3(1) + 4(2) \geq 11$
     $3 + 8 \geq 11$
       $11 \geq 11$
  Yes

8. $\begin{cases} 2x - y = 1 \\ x + 2y = 13 \end{cases}$

$\begin{cases} y = 2x - 1 \\ y = -\dfrac{1}{2}x + \dfrac{13}{2} \end{cases}$

$$2x - 1 = -\frac{1}{2}x + \frac{13}{2}$$

$$\frac{5}{2}x = \frac{15}{2}$$

$$x = 3$$

$y = 2(3) - 1 = 5$
$(3, 5)$

9. $2x - 10y = 7$

$$y = \frac{1}{5}x - \frac{7}{10}$$

$$m = \frac{1}{5}$$

$$y - 16 = \frac{1}{5}(x - 15)$$

$$y = \frac{1}{5}x + 13$$

10. $y = 3(1) + 7 = 10$

11. $(5, 0)$

12.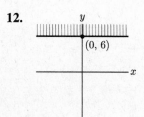

13. $\begin{cases} 3x - 2y = 1 \\ 2x + y = 24 \end{cases}$

$$\begin{cases} y = \frac{3}{2}x - \frac{1}{2} \\ y = -2x + 24 \end{cases}$$

$$\frac{3}{2}x - \frac{1}{2} = -2x + 24$$

$$\frac{7}{2}x = \frac{49}{2}$$

$$x = 7$$

$y = -2(7) + 24 = 10$
$(7, 10)$

14. $\begin{cases} 2y + 7x \geq 28 \\ 2y - x \geq 0 \\ y \leq 8 \end{cases}$

$$\begin{cases} y \geq -\frac{7}{2}x + 14 \\ y \geq \frac{1}{2}x \\ y \leq 8 \end{cases}$$

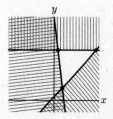

15. $y - 9 = \frac{1}{2}(x - 4)$

$$y = \frac{1}{2}x + 7$$

$b = 7$
$(0, 7)$

16. The rate is \$35 per hour plus a flat fee of \$20.

17. $m_1 = \frac{0 - 2}{2 - 1} = -2$

$$m_2 = \frac{1 - 0}{3 - 2} = 1$$

$m_1 \neq m_2$
No

18. $m = \frac{-2 - 0}{0 - 3} = \frac{2}{3}, b = -2$

$$y = \frac{2}{3}x - 2$$

19. $x + 7y = 30$
$$-2y + 7y = 30$$
$$5y = 30$$
$$y = 6$$
Answer (d) is correct.

20. $y \leq \frac{2}{3}x + \frac{3}{2}$

21. $m = \dfrac{8.6-(-1)}{6-2} = 2.4$

$y + 1 \geq 2.4(x-2)$

$y \geq 2.4x - 5.8$

22. $\begin{cases} 1.2x + 2.4y = .6 \\ 4.8y - 1.6x = 2.4 \end{cases}$

$\begin{cases} y = -.5x + .25 \\ y = \dfrac{1}{3}x + .5 \end{cases}$

$-.5x + .25 = \dfrac{1}{3}x + .5$

$-\dfrac{5}{6}x = 0.25$

$x = -0.3$

$y = \dfrac{1}{3}(-.3) + .5 = 0.4$

23. $\begin{cases} y = -x + 1 \\ y = 2x + 3 \end{cases}$

$-x + 1 = 2x + 3$

$-3x = 2$

$x = -\dfrac{2}{3}$

$y = -\left(-\dfrac{2}{3}\right) + 1 = \dfrac{5}{3}$

$\left(-\dfrac{2}{3}, \dfrac{5}{3}\right)$

$m = \dfrac{\frac{5}{3}-1}{-\frac{2}{3}-1} = -\dfrac{2}{5}$

$y - 1 = -\dfrac{2}{5}(x-1)$

$y = -\dfrac{2}{5}x + \dfrac{7}{5}$

24. $2x + 3(x-2) \geq 0$

$5x \geq 6$

$x \geq \dfrac{6}{5}$

25. $x + \dfrac{1}{2}y = 4$

$y = -2x + 8$

$m = -2$

y-intercept: $(0, 8)$

$0 = -2x + 8$

$x = 4$

x-intercept: $(4, 0)$

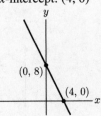

26. $\begin{cases} 5x + 2y = 0 \\ x + y = 1 \end{cases}$

$\begin{cases} y = -\dfrac{5}{2}x \\ y = -x + 1 \end{cases}$

$-\dfrac{5}{2}x = -x + 1$

$-\dfrac{3}{2}x = 1$

$x = -\dfrac{2}{3}$

$y = -\left(-\dfrac{2}{3}\right) + 1 = \dfrac{5}{3}$

Substitute $x = -\dfrac{2}{3}$ and $y = \dfrac{5}{3}$ in

$2x - 3y = 1$

$2\left(-\dfrac{2}{3}\right) - 3\left(\dfrac{5}{3}\right) = 1$

$-\dfrac{19}{3} = 1$

No

27. $\begin{cases} 2x - 3y = 1 \\ 3x + 2y = 4 \end{cases}$

$\begin{cases} y = \dfrac{2}{3}x - \dfrac{1}{3} \\ y = -\dfrac{3}{2}x + 2 \end{cases}$

$m_1 = -\dfrac{1}{m_2}$

28. a. $x + y \geq 1$
$\qquad y \geq -x + 1$
$\qquad$ (C)

b. $x + y \leq 1$
$\qquad y \leq -x + 1$
$\qquad$ (A)

c. $x - y \leq 1$
$\qquad y \geq x - 1$
$\qquad$ (B)

d. $y - x \leq -1$
$\qquad y \leq x - 1$
$\qquad$ (D)

29. a. $4x + y = 17$
$\qquad\qquad y = -4x + 17$
$\qquad L_3$

b. $y = x + 2$
$\qquad L_1$

c. $2x + 3y = 11$
$\qquad\qquad y = -\dfrac{2}{3}x + \dfrac{11}{3}$
$\qquad L_2$

30. $m_1 = \dfrac{\frac{3}{2} - 5}{4 - 0} = -\dfrac{7}{8}, \; b_1 = 5$

$y = -\dfrac{7}{8}x + 5$

$m_2 = -\dfrac{1}{m_1} = \dfrac{8}{7}$

$y - \dfrac{3}{2} = \dfrac{8}{7}(x - 4)$

$y = \dfrac{8}{7}x - \dfrac{43}{14}$

$\begin{cases} y \leq -\dfrac{7}{8}x + 5 \\ y \geq \dfrac{8}{7}x - \dfrac{43}{14} \\ x \geq 0, \; y \geq 0 \end{cases}$

$0 = \dfrac{8}{7}x - \dfrac{43}{14}$

$x = \dfrac{43}{16}$

$\left(\dfrac{43}{16}, 0 \right)$

31. Supply curve is $p = .005q + .5$
Demand curve is $p = -.01q + 5$

$\begin{cases} p = .005q + .5 \\ p = -.01q + 5 \end{cases}$

$.005q + .5 = -.01q + 5$

$.015q = 4.5$

$q = 300 \text{ units}$

$p = .005(300) + .5 = \$2$

32. $\begin{cases} x \geq 0 \\ y \geq 0 \end{cases}$

$(0, 0)$

$\begin{cases} y \geq 0 \\ 5x + y \leq 50 \end{cases}$

$\begin{cases} y \geq 0 \\ y \leq -5x + 50 \end{cases}$

$0 = -5x + 50$

$x = 10$

$(10, 0)$

$\begin{cases} 5x + y \leq 50 \\ 2x + 3y \leq 33 \end{cases}$

$\begin{cases} y \leq -5x + 50 \\ y \leq -\dfrac{2}{3}x + 11 \end{cases}$

$-5x + 50 = -\dfrac{2}{3}x + 11$

$-\dfrac{13}{3}x = -39$

$x = 9$

$y = -5(9) + 50 = 5$

$(9, 5)$

$$\begin{cases} 2x + 3y \leq 33 \\ x - 2y \geq -8 \end{cases}$$

$$\begin{cases} y \leq -\dfrac{2}{3}x + 11 \\ y \leq \dfrac{1}{2}x + 4 \end{cases}$$

$$-\dfrac{2}{3}x + 11 = \dfrac{1}{2}x + 4$$

$$-\dfrac{7}{6}x = -7$$

$$x = 6$$

$$x = \dfrac{1}{2}(6) + 4 = 7$$

$(6, 7)$

$$\begin{cases} x - 2y \geq -8 \\ x \geq 0 \end{cases}$$

$$\begin{cases} x \geq 2y - 8 \\ x \geq 0 \end{cases}$$

$$2y - 8 = 0$$

$$y = 4$$

$(0, 4)$

33. a. In 2000, 8.8% of college freshmen intended to obtain a medical degree.

b. $2008 - 2000 = 8$

$y = 0.1(8) + 8.8$

$y = 9.6$

9.6% of college freshmen in 2008 intended to obtain a medical degree

c. $9.2 = 0.1x + 8.8$

$x = 4$

$2000 + 4 = 2004$

In 2004, the percent of college freshmen that intended to obtain a medical degree was 9.2.

34. a. $m = 10$

$y - 4000 = 10(x - 1000)$

$y = 10x - 6000$

b. $0 = 10x - 6000$

$x = 600$

x-intercept: $(600, 0)$

y-intercept: $(0, -6000)$

c.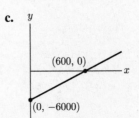

35. a. A: $y = .1x + 50$

B: $y = .2x + 40$

b. A: $.1(80) + 50 = 58$

B: $.2(80) + 40 = 56$

Company B

c. A: $.1(160) + 50 = 66$

B: $.2(160) + 40 = 72$

Company A

d. $.1x + 50 = .2x + 40$

$-.1x = -10$

$x = 100$ miles

36. a. $m = \dfrac{1.21 - 0.69}{17 - 0} = 0.031$

$y - 0.69 = 0.031(x - 0)$

$y = 0.031x + 0.69$

b. $1.00 = 0.031x + 0.69$

$x = 10$

The year $1990 + 10 = 2000$

37. $x \leq 3y + 2$

$y \geq \dfrac{1}{3}x - \dfrac{2}{3}$

38. $0.03x + 200 = 0.05x + 100$

$$-0.02x = -100$$
$$x = \$5000$$

39. $m_1 = \dfrac{5-0}{0-(-4)} = \dfrac{5}{4}, \; b_1 = 5$

$$y = \dfrac{5}{4}x + 5$$

$m_2 = \dfrac{0-2}{5-0} = -\dfrac{2}{5}, \; b_2 = 2$

$$y = -\dfrac{2}{5}x + 2$$

$m_3 = \dfrac{0-(-3)}{5-0} = \dfrac{3}{5}, \; b_3 = -3$

$$y = \dfrac{3}{5}x - 3$$

$m_4 = \dfrac{-5-0}{0-(-2)} = -\dfrac{5}{2}, \; b_4 = -5$

$$y = -\dfrac{5}{2}x - 5$$

$$\begin{cases} y \le \dfrac{5}{4}x + 5 \\[2mm] y \le -\dfrac{2}{5}x + 2 \\[2mm] y \ge \dfrac{3}{5}x - 3 \\[2mm] y \ge -\dfrac{5}{2} - 5 \end{cases}$$

40. $m_1 = \dfrac{2-0}{0-3} = -\dfrac{2}{3}, \; b_1 = 2$

$$y = -\dfrac{2}{3}x + 2$$

The other lines are $x = -2$, $x = 4$, and $y = -3$.

$$\begin{cases} y \le -\dfrac{2}{3}x + 2 \\[2mm] x \ge -2 \\[2mm] x \le 4 \\[2mm] y \ge -3 \end{cases}$$

41. $(0, 417{,}000)$; in 2016: $(10, 565{,}000)$

$$\mathrm{m} = m = \dfrac{565{,}000 - 417{,}000}{10 - 0} = 14800$$
$$y - 417{,}000 = 14800(x - 0)$$
$$y = 14800x + 417{,}000$$

For the year 2012, x=6:
$$y = 14800(6) + 417{,}000 = 505{,}800.$$

42. Slope of line is –237.93. Equation of line is: $y = -237.93x + 110{,}807$. In 2010, x = 19 so $y = 106{,}286$.

43. Let x = 0 correspond to year 2000. Then y = 20.4. When x = 7, y = 17.2. The rate of change (slope) = (17.2 – 20.4)/(7 – 0) = –0.46. The equation of the line that predicts the percentage of market is y = –0.46x + 20.4. When x = 5, y = 18.1%.

44. a. $y = 1.06x + 1.71$

    b. $1.06(77.2) + 1.71 = 83.54$
About 83.5 years

    c. $84.2 = 1.06x + 1.71$

        $x \approx 77.82$
About 77.8 years

45. a. $y = 0.18x + 3.06$

    b. $0.18(9) + 3.06 = 4.68$
About 4.68%

    c. $5.4 = 0.18x + 3.06$

        $x = 13$
13 years after 2000 or 2013

46. a.

```
LinReg
y=ax+b
a=.1517702501
b=-3.063197325
```

      $y = .152x - 3.063$

    b. $.152(160) - 3.063 = 21.257$
About 21 deaths per 100,000

    c. $22 = .152x - 3.063$
        $x \approx 164.888$
About 165 grams

47. Up; the value of b is the y-intercept

48. Counter - Clockwise

49. When the line passes through the origin.

50. A line with undefined sloe is a vertical line and a line with zero slope is a horizontal line.

51. a. No; A line that is parallel to the x axis will not have an x intercept.

   b. No; A line that is parallel to the y axis will not have a y intercept

Chapter 1 Chapter Test

1.

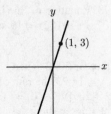

2. $y = -2\left(\dfrac{1}{2}\right) + 6$

$y = 5$

3. $m = -2$

$y - 3 = -2(x + 1)$

$y = -2x + 1$

4. $\begin{cases} 2x - 3y = 9 \\ -3x + 7y = -11 \end{cases}$

$\begin{cases} y = \dfrac{2}{3}x - 3 \\ y = \dfrac{3}{7}x - \dfrac{11}{7} \end{cases}$

$\dfrac{2}{3}x - 3 = \dfrac{3}{7}x - \dfrac{11}{7}$

$\dfrac{5}{21}x = \dfrac{10}{7}$

$x = 6$

$y = \dfrac{2}{3}(6) - 3 = 1$

$(6, 1)$

5. $3x - y = 1$

$y = 3x - 1$

$m = 3$

$-\dfrac{1}{3}x - 4 = y$

$m = -\dfrac{1}{3}$

The lines are perpendicular.

6.

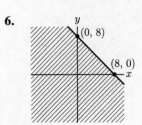

7. $\begin{cases} 4x + 5y = 11 \\ 2x - 3y = 7 \end{cases}$

$\begin{cases} y = -\dfrac{4}{5}x + \dfrac{11}{5} \\ y = \dfrac{2}{3}x - \dfrac{7}{3} \end{cases}$

$-\dfrac{4}{5}x + \dfrac{11}{5} = \dfrac{2}{3}x - \dfrac{7}{3}$

$\dfrac{68}{15} = \dfrac{22}{15}x$

$x = \dfrac{68}{22}$

$= \dfrac{34}{11}$

$y = -\dfrac{4}{5}\left(\dfrac{34}{11}\right) + \dfrac{11}{5}$

$y = -\dfrac{3}{11}$

$\left(\dfrac{34}{11}, -\dfrac{3}{11}\right)$

$y - \left(-\dfrac{3}{11}\right) = 2\left(x - \dfrac{34}{11}\right)$

$y = 2x - \dfrac{71}{11}$

8.

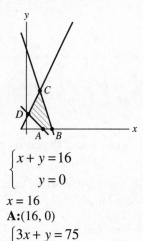

$$\begin{cases} x + y = 16 \\ y = 0 \end{cases}$$

$x = 16$

A:$(16, 0)$

$$\begin{cases} 3x + y = 75 \\ y = 0 \end{cases}$$

$3x = 75$

$x = 25$

B:$(25, 0)$

$$\begin{cases} 3x + y = 75 \\ -2x + y = 10 \end{cases}$$

$$\begin{cases} y = 75 - 3x \\ y = 2x + 10 \end{cases}$$

$75 - 3x = 2x + 10$

$65 = 5x$

$x = 13$

$y = 2(13) + 10 = 36$

C:$(13, 36)$

$$\begin{cases} x + y = 16 \\ -2x + y = 10 \end{cases}$$

$$\begin{cases} y = 16 - x \\ y = 2x + 10 \end{cases}$$

$16 - x = 2x + 10$

$6 = 3x$

$x = 2$

$y = 2(2) + 10 = 14$

D:$(2, 14)$

9. Let x = volume of sales

$250 + .03x > 200 + .05x$

$50 > .02x$

$2500 > x$

Fred's sister is correct for sales less than \$2500.
She is incorrect for sales greater than \$2500.

10. a. Let x be the number of years after 1999, then
$y = 0.033x + 0.819$

b. $0.033(5) + 0.819 \approx 0.984$
About \$0.98

c. $1.25 = 0.033x + 0.819$
$x \approx 13.06$
In the year 2012

Chapter 2

1. $\begin{cases} \dfrac{1}{2}x - 3y = 2 \\ 5x + 4y = 1 \end{cases}$

$\xrightarrow{\;2R_1\;} \begin{cases} x - 6y = 4 \\ 5x + 4y = 1 \end{cases}$

3. $\quad 5(\text{first}) \quad 5x + 10y = 15$

$+ (\text{second}) \quad \dfrac{-5x + 4y = 1}{14y = 16}$

$\begin{cases} x + 2y = 3 \\ -5x + 4y = 1 \end{cases}$

$\xrightarrow{\;R_2 + 5R_1\;} \begin{cases} x + 2y = 3 \\ \quad\;\; 14y = 16 \end{cases}$

5. $\quad -4(\text{first}) \quad -4x + 8y - 4z = 0$

$+ (\text{third}) \quad \dfrac{4x + y + 3z = 5}{9y - z = 5}$

$\begin{cases} x - 2y + z = 0 \\ \quad\;\; y - 2z = 4 \\ 4x + y + 3z = 5 \end{cases}$

$\xrightarrow{\;R_3 + (-4)R_1\;} \begin{cases} x - 2y + z = 0 \\ \quad\;\; y - 2z = 4 \\ \quad\;\; 9y - z = 5 \end{cases}$

7. $\begin{bmatrix} 1 & -\frac{1}{2} & 3 \\ 0 & 1 & 4 \end{bmatrix} \xrightarrow{\;R_1 + \frac{1}{2}R_2\;} \begin{bmatrix} 1 & 0 & 5 \\ 0 & 1 & 4 \end{bmatrix}$

9. $\begin{bmatrix} -3 & 4 & -2 \\ 1 & -7 & 8 \end{bmatrix}$

11. $\begin{bmatrix} 1 & 13 & -2 & 0 \\ 2 & 0 & -1 & 3 \\ 0 & 1 & 0 & 5 \end{bmatrix}$

13. $\begin{cases} -2y = 3 \\ x + 7y = -4 \end{cases}$

15. $\begin{cases} 3x + 2y = -3 \\ \quad\;\; y - 6z = 4 \\ -5x - y + 7z = 0 \end{cases}$

17. Multiply the second row of the matrix by $\dfrac{1}{3}$.

19. Change the first row of the matrix by adding to it 3 times the second row.

21. Interchange rows 2 and 3.

23. $\begin{bmatrix} 1 & 2 & 0 \\ 0 & 10 & 5 \end{bmatrix}$

25. $\begin{bmatrix} 1 & 2 & 3 \\ 3 & -2 & 0 \end{bmatrix}$

27. $\begin{bmatrix} 1 & 3 & -5 \\ 0 & 1 & 7 \end{bmatrix}$

29. Use $R_2 + 2R_1$ to change the -2 to a 0.

31. Use R_1 to $(-2)\,R_2$ to change the 2 to a 0.

33. Interchange rows 1 and 2 or rows 1 and 3 to make the first entry in row 1 nonzero.

35. Use $R_1 + (-3)\,R_3$ to change the 3 to a 0.

37. $\begin{bmatrix} 1 & 9 & 8 \\ 2 & 8 & 6 \end{bmatrix}$

$\xrightarrow{\;R_2 + (-2)R_1\;} \begin{bmatrix} 1 & 9 & 8 \\ 0 & -10 & -10 \end{bmatrix}$

$\xrightarrow{\;-\frac{1}{10}R_2\;} \begin{bmatrix} 1 & 9 & 8 \\ 0 & 1 & 1 \end{bmatrix}$

$\xrightarrow{\;R_1 + (-9)R_2\;} \begin{bmatrix} 1 & 0 & -1 \\ 0 & 1 & 1 \end{bmatrix}$

$x = -1,\; y = 1$

39. $\begin{bmatrix} 1 & -3 & 4 & | & 1 \\ 4 & -10 & 10 & | & 4 \\ -3 & 9 & -5 & | & -6 \end{bmatrix}$

$\xrightarrow{R_2+(-4)R_1} \begin{bmatrix} 1 & -3 & 4 & | & 1 \\ 0 & 2 & -6 & | & 0 \\ -3 & 9 & -5 & | & -6 \end{bmatrix}$

$\xrightarrow{R_3+3R_1} \begin{bmatrix} 1 & -3 & 4 & | & 1 \\ 0 & 2 & -6 & | & 0 \\ 0 & 0 & 7 & | & -3 \end{bmatrix}$

$\xrightarrow{\frac{1}{2}R_2} \begin{bmatrix} 1 & -3 & 4 & | & 1 \\ 0 & 1 & -3 & | & 0 \\ 0 & 0 & 7 & | & -3 \end{bmatrix}$

$\xrightarrow{R_1+3R_2} \begin{bmatrix} 1 & 0 & -5 & | & 1 \\ 0 & 1 & -3 & | & 0 \\ 0 & 0 & 7 & | & -3 \end{bmatrix}$

$\xrightarrow{\frac{1}{7}R_3} \begin{bmatrix} 1 & 0 & -5 & | & 1 \\ 0 & 1 & -3 & | & 0 \\ 0 & 0 & 1 & | & -\frac{3}{7} \end{bmatrix}$

$\xrightarrow{R_1+5R_3} \begin{bmatrix} 1 & 0 & 0 & | & -\frac{8}{7} \\ 0 & 1 & -3 & | & 0 \\ 0 & 0 & 1 & | & -\frac{3}{7} \end{bmatrix}$

$\xrightarrow{R_2+3R_3} \begin{bmatrix} 1 & 0 & 0 & | & -\frac{8}{7} \\ 0 & 1 & 0 & | & -\frac{9}{7} \\ 0 & 0 & 1 & | & -\frac{3}{7} \end{bmatrix}$

$x = -\dfrac{8}{7},\ y = -\dfrac{9}{7},\ z = -\dfrac{3}{7}$

41. $\begin{bmatrix} 2 & -2 & | & -4 \\ 3 & 4 & | & 1 \end{bmatrix} \xrightarrow{\frac{1}{2}R_1} \begin{bmatrix} 1 & -1 & | & -2 \\ 3 & 4 & | & 1 \end{bmatrix}$

$\xrightarrow{R_2+(-3)R_1} \begin{bmatrix} 1 & -1 & | & -2 \\ 0 & 7 & | & 7 \end{bmatrix}$

$\xrightarrow{\frac{1}{7}R_2} \begin{bmatrix} 1 & -1 & | & -2 \\ 0 & 1 & | & 1 \end{bmatrix}$

$\xrightarrow{R_1+1R_2} \begin{bmatrix} 1 & 0 & | & -1 \\ 0 & 1 & | & 1 \end{bmatrix}$

$x = -1,\ y = 1$

43. $\begin{bmatrix} 4 & -4 & 4 & | & -8 \\ 1 & -2 & -2 & | & -1 \\ 2 & 1 & 3 & | & 1 \end{bmatrix}$

$\xrightarrow{\frac{1}{4}R_1} \begin{bmatrix} 1 & -1 & 1 & | & -2 \\ 1 & -2 & -2 & | & -1 \\ 2 & 1 & 3 & | & 1 \end{bmatrix}$

$\xrightarrow{R_2+(-1)R_1} \begin{bmatrix} 1 & -1 & 1 & | & -2 \\ 0 & -1 & -3 & | & 1 \\ 2 & 1 & 3 & | & 1 \end{bmatrix}$

$\xrightarrow{R_3+(-2)R_1} \begin{bmatrix} 1 & -1 & 1 & | & -2 \\ 0 & -1 & -3 & | & 1 \\ 0 & 3 & 1 & | & 5 \end{bmatrix}$

$\xrightarrow{(-1)R_2} \begin{bmatrix} 1 & -1 & 1 & | & -2 \\ 0 & 1 & 3 & | & -1 \\ 0 & 3 & 1 & | & 5 \end{bmatrix}$

$\xrightarrow{R_1+R_2} \begin{bmatrix} 1 & 0 & 4 & | & -3 \\ 0 & 1 & 3 & | & -1 \\ 0 & 3 & 1 & | & 5 \end{bmatrix}$

$\xrightarrow{R_3+(-3)R_2} \begin{bmatrix} 1 & 0 & 4 & | & -3 \\ 0 & 1 & 3 & | & -1 \\ 0 & 0 & -8 & | & 8 \end{bmatrix}$

$\xrightarrow{(-\frac{1}{8})R_3} \begin{bmatrix} 1 & 0 & 4 & | & -3 \\ 0 & 1 & 3 & | & -1 \\ 0 & 0 & 1 & | & -1 \end{bmatrix}$

$\xrightarrow{R_1+(-4)R_3} \begin{bmatrix} 1 & 0 & 0 & | & 1 \\ 0 & 1 & 3 & | & -1 \\ 0 & 0 & 1 & | & -1 \end{bmatrix}$

$\xrightarrow{R_2+(-3)R_3} \begin{bmatrix} 1 & 0 & 0 & | & 1 \\ 0 & 1 & 0 & | & 2 \\ 0 & 0 & 1 & | & -1 \end{bmatrix}$

$x = 1,\ y = 2,\ z = -1$

45. $\begin{bmatrix} .2 & .3 & | & 4 \\ .6 & 1.1 & | & 15 \end{bmatrix}$

$\xrightarrow{5R_1} \begin{bmatrix} 1 & 1.5 & | & 20 \\ .6 & 1.1 & | & 15 \end{bmatrix}$

$\xrightarrow{R_2+(-.6)R_1} \begin{bmatrix} 1 & 1.5 & | & 20 \\ 0 & .2 & | & 3 \end{bmatrix}$

$\xrightarrow{5R_2} \begin{bmatrix} 1 & 1.5 & | & 20 \\ 0 & 1 & | & 15 \end{bmatrix}$

$\xrightarrow{R_1+(-1.5)R_2} \begin{bmatrix} 1 & 0 & | & -2.5 \\ 0 & 1 & | & 15 \end{bmatrix}$

$x = -2.5,\ y = 15$

47. $\begin{bmatrix} 1 & 1 & 4 & | & 3 \\ 4 & 1 & -2 & | & -6 \\ -3 & 0 & 2 & | & 1 \end{bmatrix}$

$\xrightarrow{R_2+(-4)R_1} \begin{bmatrix} 1 & 1 & 4 & | & 3 \\ 0 & -3 & -18 & | & -18 \\ -3 & 0 & 2 & | & 1 \end{bmatrix}$

$\xrightarrow{R_3+3R_1} \begin{bmatrix} 1 & 1 & 4 & | & 3 \\ 0 & -3 & -18 & | & -18 \\ 0 & 3 & 14 & | & 10 \end{bmatrix}$

$\xrightarrow{(-\frac{1}{3})R_2} \begin{bmatrix} 1 & 1 & 4 & | & 3 \\ 0 & 1 & 6 & | & 6 \\ 0 & 3 & 14 & | & 10 \end{bmatrix}$

$\xrightarrow{R_1+(-1)R_2} \begin{bmatrix} 1 & 0 & -2 & | & -3 \\ 0 & 1 & 6 & | & 6 \\ 0 & 3 & 14 & | & 10 \end{bmatrix}$

$\xrightarrow{R_3+(-3)R_2} \begin{bmatrix} 1 & 0 & -2 & | & -3 \\ 0 & 1 & 6 & | & 6 \\ 0 & 0 & -4 & | & -8 \end{bmatrix}$

$\xrightarrow{(-\frac{1}{4})R_3} \begin{bmatrix} 1 & 0 & -2 & | & -3 \\ 0 & 1 & 6 & | & 6 \\ 0 & 0 & 1 & | & 2 \end{bmatrix}$

$\xrightarrow{R_1+2R_3} \begin{bmatrix} 1 & 0 & 0 & | & 1 \\ 0 & 1 & 6 & | & 6 \\ 0 & 0 & 1 & | & 2 \end{bmatrix}$

$\xrightarrow{R_2+(-6)R_3} \begin{bmatrix} 1 & 0 & 0 & | & 1 \\ 0 & 1 & 0 & | & -6 \\ 0 & 0 & 1 & | & 2 \end{bmatrix}$

$x = 1,\ y = -6,\ z = 2$

49. $\begin{bmatrix} -1 & 1 & 0 & | & -1 \\ 1 & 0 & 1 & | & 4 \\ 6 & -3 & 2 & | & 10 \end{bmatrix}$

$\xrightarrow{(-1)R_1} \begin{bmatrix} 1 & -1 & 0 & | & 1 \\ 1 & 0 & 1 & | & 4 \\ 6 & -3 & 2 & | & 10 \end{bmatrix}$

$\xrightarrow{R_2+(-1)R_1} \begin{bmatrix} 1 & -1 & 0 & | & 1 \\ 0 & 1 & 1 & | & 3 \\ 6 & -3 & 2 & | & 10 \end{bmatrix}$

$\xrightarrow{R_3+(-6)R_1} \begin{bmatrix} 1 & -1 & 0 & | & 1 \\ 0 & 1 & 1 & | & 3 \\ 0 & 3 & 2 & | & 4 \end{bmatrix}$

$\xrightarrow{R_1+1R_2} \begin{bmatrix} 1 & 0 & 1 & | & 4 \\ 0 & 1 & 1 & | & 3 \\ 0 & 3 & 2 & | & 4 \end{bmatrix}$

$\xrightarrow{R_3+(-3)R_2} \begin{bmatrix} 1 & 0 & 1 & | & 4 \\ 0 & 1 & 1 & | & 3 \\ 0 & 0 & -1 & | & -5 \end{bmatrix}$

$\xrightarrow{(-1)R_3} \begin{bmatrix} 1 & 0 & 1 & | & 4 \\ 0 & 1 & 1 & | & 3 \\ 0 & 0 & 1 & | & 5 \end{bmatrix}$

$\xrightarrow{R_1+(-1)R_3} \begin{bmatrix} 1 & 0 & 0 & | & -1 \\ 0 & 1 & 1 & | & 3 \\ 0 & 0 & 1 & | & 5 \end{bmatrix}$

$\xrightarrow{R_2+(-1)R_3} \begin{bmatrix} 1 & 0 & 0 & | & -1 \\ 0 & 1 & 0 & | & -2 \\ 0 & 0 & 1 & | & 5 \end{bmatrix}$

$x = -1,\ y = -2,\ z = 5$

51. Let x = grams of cheddar cheese
     y = grams of potato

$$\begin{cases} x + y = 180 \\ .25x + .02y = 10.5 \end{cases}$$

$$\begin{bmatrix} 1 & 1 & | & 180 \\ .25 & .02 & | & 10.5 \end{bmatrix}$$

$$\xrightarrow{R_2 + (-.25)R_1} \begin{bmatrix} 1 & 1 & | & 180 \\ 0 & -.23 & | & -34.5 \end{bmatrix}$$

$$\xrightarrow{-\frac{1}{23}R_2} \begin{bmatrix} 1 & 1 & | & 180 \\ 0 & 1 & | & 150 \end{bmatrix}$$

$$\xrightarrow{R_1 + (-1)R_2} \begin{bmatrix} 1 & 0 & | & 30 \\ 0 & 1 & | & 150 \end{bmatrix}$$

30 grams of cheddar cheese
Answer (b) is correct.

53. Let x = cost of golf balls
and y = cost of golf glove.
Then $x + y = 20$.
Using **Statement I**: $y = 3x$
$$x + 3x = 20$$
$$4x = 20$$
$$x = 5.$$
Using **Statement II**: $y = 15$
$$x + 15 = 20$$
$$x = 5.$$
The box of balls costs $5. Either statement is sufficient, so the answer is (d).

55. Let x = number of short sleeve shirts
     y = number of long sleeve shirts
$$x + y = 350$$
$$10x + 14y = 4300$$

$$\begin{bmatrix} 1 & 1 & | & 350 \\ 10 & 14 & | & 4300 \end{bmatrix}$$

$$\xrightarrow{R_2 + (-10)R_1} \begin{bmatrix} 1 & 1 & | & 350 \\ 0 & 4 & | & 800 \end{bmatrix}$$

$$\xrightarrow{\frac{1}{4}R_2} \begin{bmatrix} 1 & 1 & | & 350 \\ 0 & 1 & | & 200 \end{bmatrix}$$

$$\xrightarrow{R_1 + (-1)R_2} \begin{bmatrix} 1 & 0 & | & 150 \\ 0 & 1 & | & 200 \end{bmatrix}$$

150 short sleeve, 200 long sleeve

57. Let x = adults, y = children

$$\begin{cases} x + y = 350 \\ 9.25x + 6.25y = 2973.50 \end{cases}$$

$$\begin{bmatrix} 1 & 1 & | & 350 \\ 9.25 & 6.25 & | & 2973.50 \end{bmatrix}$$

$$\xrightarrow{R_2 + (-9.25)R_1} \begin{bmatrix} 1 & 1 & | & 350 \\ 0 & -3 & | & -264 \end{bmatrix}$$

$$\xrightarrow{-\frac{1}{3}R_2} \begin{bmatrix} 1 & 1 & | & 350 \\ 0 & 1 & | & 88 \end{bmatrix}$$

$$\xrightarrow{R_1 + (-1)R_2} \begin{bmatrix} 1 & 0 & | & 262 \\ 0 & 1 & | & 88 \end{bmatrix}$$

262 adults, 88 children

59.
$$\begin{cases} x + y + z = 100{,}000 \\ .08x + .07y + .1z = 8000 \\ x + y - 3z = 0 \end{cases}$$

$$\begin{bmatrix} 1 & 1 & 1 & | & 100{,}000 \\ .08 & .07 & .1 & | & 8000 \\ 1 & 1 & -3 & | & 0 \end{bmatrix}$$

$$\xrightarrow{R_2 + (-.08)R_1} \begin{bmatrix} 1 & 1 & 1 & | & 100{,}000 \\ 0 & -.01 & .02 & | & 0 \\ 1 & 1 & -3 & | & 0 \end{bmatrix}$$

$$\xrightarrow{R_3 + (-1)R_1} \begin{bmatrix} 1 & 1 & 1 & | & 100{,}000 \\ 0 & -.01 & .02 & | & 0 \\ 0 & 0 & -4 & | & -100{,}000 \end{bmatrix}$$

$$\xrightarrow{(-100)R_2} \begin{bmatrix} 1 & 1 & 1 & | & 100{,}000 \\ 0 & 1 & -2 & | & 0 \\ 0 & 0 & -4 & | & -100{,}000 \end{bmatrix}$$

$$\xrightarrow{R_1 + (-1)R_2} \begin{bmatrix} 1 & 0 & 3 & | & 100{,}000 \\ 0 & 1 & -2 & | & 0 \\ 0 & 0 & -4 & | & -100{,}000 \end{bmatrix}$$

$$\xrightarrow{\left(-\frac{1}{4}\right)R_3} \begin{bmatrix} 1 & 0 & 3 & | & 100{,}000 \\ 0 & 1 & -2 & | & 0 \\ 0 & 0 & 1 & | & 25{,}000 \end{bmatrix}$$

$$\xrightarrow{R_1 + (-3)R_3} \begin{bmatrix} 1 & 0 & 0 & | & 25{,}000 \\ 0 & 1 & -2 & | & 0 \\ 0 & 0 & 1 & | & 25{,}000 \end{bmatrix}$$

$$\xrightarrow{R_2+2R_3}\begin{bmatrix}1&0&0&|&25{,}000\\0&1&0&|&50{,}000\\0&0&1&|&25{,}000\end{bmatrix}$$

$x = \$25{,}000,\ y = \$50{,}000,\ z = \$25{,}000$

61. Let x = pounds of first type

y = pounds of second type

z = pounds of third type.

$.4x +\ \ \ \ \ \ .4z = 90$

$.6x + .3y + .3z = 100$

$.7y + .3z = 120$

$$\begin{bmatrix}.4&0&.4&|&90\\.6&.3&.3&|&100\\0&.7&.3&|&120\end{bmatrix}$$

$$\xrightarrow{\frac{1}{.4}R_1}\begin{bmatrix}1&0&1&|&225\\.6&.3&.3&|&100\\0&.7&.3&|&120\end{bmatrix}$$

$$\xrightarrow{R_2+(-.6)R_1}\begin{bmatrix}1&0&1&|&225\\0&.3&-.3&|&-35\\0&.7&.3&|&120\end{bmatrix}$$

$$\xrightarrow{\frac{1}{.3}R_2}\begin{bmatrix}1&0&1&|&225\\0&1&-1&|&-\frac{350}{3}\\0&.7&.3&|&120\end{bmatrix}$$

$$\xrightarrow{R_3+(-.7)R_2}\begin{bmatrix}1&0&1&|&225\\0&1&-1&|&-\frac{350}{3}\\0&0&1&|&\frac{605}{3}\end{bmatrix}$$

$$\xrightarrow{R_1+(-1)R_3}\begin{bmatrix}1&0&0&|&\frac{70}{3}\\0&1&-1&|&\frac{-350}{3}\\0&0&1&|&\frac{605}{3}\end{bmatrix}$$

$$\xrightarrow{R_2+R_3}\begin{bmatrix}1&0&0&|&\frac{70}{3}\\0&1&0&|&85\\0&0&1&|&\frac{605}{3}\end{bmatrix}$$

$\dfrac{70}{3}$ pounds of the first type, 85 pounds of the

second type, and $\dfrac{605}{3}$ pounds of the third type

63.
```
[A]
     [[1  2  3]
      [-5 4  11]]
*row+(5,[A],1,2)
     [[1  2  3 ]
      [0  14 16]]
```

65.
```
[A]
     [[1 -.5 3]
      [0  1  4]]
rowSwap([A],1,2)
     [[0  1  4]
      [1 -.5 3]]
```

67.

	A	B
1	1	3
2	-6	-6
3	2	1
4		
5	Cell	Content
6	B1	=x+y+4*z
7	B2	=4*x+y-2*z
8	B3	=-3*x+2*z

Exercises 2.2

1. $\begin{bmatrix}2&-4&6\\3&7&1\end{bmatrix}\xrightarrow[R_2+(-3)R_1]{\frac{1}{2}R_1}\begin{bmatrix}1&-2&3\\0&13&-8\end{bmatrix}$

3. $\begin{bmatrix}7&1&4&5\\-1&1&2&6\\4&0&2&3\end{bmatrix}$

$$\xrightarrow[R_3+(-2)R_2]{\substack{\frac{1}{2}R_2\\R_1+(-4)R_2}}\begin{bmatrix}9&-1&0&-7\\-\frac{1}{2}&\frac{1}{2}&1&3\\5&-1&0&-3\end{bmatrix}$$

5. $\begin{bmatrix}2&3\\6&0\\1&5\end{bmatrix}\xrightarrow[R_3+(-1)R_1]{\substack{\frac{1}{2}R_1\\R_2+(-6)R_1}}\begin{bmatrix}1&\frac{3}{2}\\0&-9\\0&\frac{7}{2}\end{bmatrix}$

7. $\begin{bmatrix}4&3&0\\\frac{2}{3}&0&-2\\1&3&6\end{bmatrix}\xrightarrow[R_2+2R_3]{\frac{1}{6}R_3}\begin{bmatrix}4&3&0\\1&1&0\\\frac{1}{6}&\frac{1}{2}&1\end{bmatrix}$

9. $\begin{bmatrix} 2 & -4 & | & 6 \\ -1 & 2 & | & -3 \end{bmatrix}$

$\begin{bmatrix} 1 & -2 & | & 3 \\ 0 & 0 & | & 0 \end{bmatrix}$

$\begin{cases} x - 2y = 3 \\ \quad\quad 0 = 0 \end{cases}$

y = any value, $x = 2y + 3$

11. $\begin{bmatrix} 1 & 2 & | & 5 \\ 3 & -1 & | & 1 \\ -1 & 3 & | & 5 \end{bmatrix}$

$\begin{bmatrix} 1 & 2 & | & 5 \\ 0 & -7 & | & -14 \\ 0 & 5 & | & 10 \end{bmatrix}$

$\begin{bmatrix} 1 & 0 & | & 1 \\ 0 & 1 & | & 2 \\ 0 & 0 & | & 0 \end{bmatrix}$

$x = 1, y = 2$

13. $\begin{bmatrix} 1 & -1 & 3 & | & 3 \\ -2 & 3 & -11 & | & -4 \\ 1 & -2 & 8 & | & 6 \end{bmatrix}$

$\begin{bmatrix} 1 & -1 & 3 & | & 3 \\ 0 & 1 & -5 & | & 2 \\ 0 & -1 & 5 & | & 3 \end{bmatrix}$

$\begin{bmatrix} 1 & 0 & -2 & | & 5 \\ 0 & 1 & -5 & | & 2 \\ 0 & 0 & 0 & | & 5 \end{bmatrix}$

$\begin{cases} x - 2z = 5 \\ y - 5z = 2 \\ \quad\quad 0 = 5 \end{cases}$

No solution

15. $\begin{bmatrix} 1 & 1 & 1 & | & -1 \\ 2 & 3 & 2 & | & 3 \\ 2 & 1 & 2 & | & -7 \end{bmatrix}$

$\begin{bmatrix} 1 & 1 & 1 & | & -1 \\ 0 & 1 & 0 & | & 5 \\ 0 & -1 & 0 & | & -5 \end{bmatrix}$

$\begin{bmatrix} 1 & 0 & 1 & | & -6 \\ 0 & 1 & 0 & | & 5 \\ 0 & 0 & 0 & | & 0 \end{bmatrix}$

$\begin{cases} x + z = -6 \\ \quad\quad y = 5 \\ \quad\quad 0 = 0 \end{cases}$

z = any value, $x = -z - 6, y = 5$

17. $\begin{bmatrix} 1 & 2 & 3 & | & 4 \\ 5 & 6 & 7 & | & 8 \\ 1 & 2 & 3 & | & 5 \end{bmatrix}$

$\begin{bmatrix} 1 & 2 & 3 & | & 4 \\ 0 & -4 & -8 & | & -12 \\ 0 & 0 & 0 & | & 1 \end{bmatrix}$

$\begin{bmatrix} 1 & 0 & -1 & | & -2 \\ 0 & 1 & 2 & | & 3 \\ 0 & 0 & 0 & | & 1 \end{bmatrix}$

$\begin{cases} x - z = -2 \\ y + 2z = 3 \\ \quad\quad 0 = 1 \end{cases}$

No solution

19. $\begin{bmatrix} \underline{1} & 1 & -2 & 2 & | & 5 \\ 2 & 1 & -4 & 1 & | & 5 \\ 3 & 4 & -6 & 9 & | & 20 \\ 4 & 4 & -8 & 8 & | & 20 \end{bmatrix}$

$\begin{bmatrix} 1 & 1 & -2 & 2 & | & 5 \\ 0 & \underline{-1} & 0 & -3 & | & -5 \\ 0 & 1 & 0 & 3 & | & 5 \\ 0 & 0 & 0 & 0 & | & 0 \end{bmatrix}$

$\begin{bmatrix} 1 & 0 & -2 & -1 & | & 0 \\ 0 & 1 & 0 & 3 & | & 5 \\ 0 & 0 & 0 & 0 & | & 0 \\ 0 & 0 & 0 & 0 & | & 0 \end{bmatrix}$

$\begin{cases} x - 2z - w = 0 \\ \quad\quad y + 3w = 5 \\ \quad\quad\quad\quad 0 = 0 \\ \quad\quad\quad\quad 0 = 0 \end{cases}$

z = any value, w = any value, $x = 2z + w$, $y = -3w + 5$

21. $\begin{bmatrix} \underline{6} & -4 & | & 2 \\ -3 & 3 & | & 6 \\ 5 & 2 & | & 39 \end{bmatrix}$

$\begin{bmatrix} 1 & -\frac{2}{3} & | & \frac{1}{3} \\ 0 & \underline{1} & | & 7 \\ 0 & \frac{16}{3} & | & \frac{112}{3} \end{bmatrix}$

$\begin{bmatrix} 1 & 0 & | & 5 \\ 0 & 1 & | & 7 \\ 0 & 0 & | & 0 \end{bmatrix}$

$x = 5, y = 7$

23. $\begin{bmatrix} \underline{1} & 2 & 1 & | & 5 \\ 0 & 1 & 3 & | & 9 \end{bmatrix}$

$\begin{bmatrix} 1 & 2 & 1 & | & 5 \\ 0 & \underline{1} & 3 & | & 9 \end{bmatrix}$

$\begin{bmatrix} 1 & 0 & -5 & | & -13 \\ 0 & 1 & 3 & | & 9 \end{bmatrix}$

$\begin{cases} x - 5z = -13 \\ y + 3z = 9 \end{cases}$

z = any value, $x = 5z - 13$, $y = -3z + 9$
Possible answers: $z = 0$, $x = -13$, $y = 9$;
$z = 1$, $x = -8$, $y = 6$; $z = 2$, $x = -3$, $y = 3$

25. $\begin{bmatrix} \underline{1} & 7 & -3 & | & 8 \\ 0 & 0 & 1 & | & 5 \end{bmatrix}$

$\begin{bmatrix} 1 & 7 & -3 & | & 8 \\ 0 & 0 & \underline{1} & | & 5 \end{bmatrix}$

$\begin{bmatrix} 1 & 7 & 0 & | & 23 \\ 0 & 0 & 1 & | & 5 \end{bmatrix}$

$\begin{cases} x + 7y = 23 \\ \quad\quad z = 5 \end{cases}$

y = any value, $x = -7y + 23$, $z = 5$
Possible answers: $y = 0$, $x = 23$, $z = 5$;
$y = 1$, $x = 16$, $z = 5$; $y = 2$, $x = 9$, $z = 5$

27. Let x = food 1, y = food 2, and z = food 3.

$$\begin{cases} 2x + 4y + 6z = 1000 \\ 3x + 7y + 10z = 1600 \\ 5x + 9y + 14z = 2400 \end{cases}$$

$$\begin{bmatrix} \underline{2} & 4 & 6 & | & 1000 \\ 3 & 7 & 10 & | & 1600 \\ 5 & 9 & 14 & | & 2400 \end{bmatrix}$$

$$\begin{bmatrix} 1 & 2 & 3 & | & 500 \\ 0 & \underline{1} & 1 & | & 100 \\ 0 & -1 & -1 & | & -100 \end{bmatrix}$$

$$\begin{bmatrix} 1 & 0 & 1 & | & 300 \\ 0 & 1 & 1 & | & 100 \\ 0 & 0 & 0 & | & 0 \end{bmatrix}$$

$$\begin{cases} x + z = 300 \\ y + z = 100 \\ 0 = 0 \end{cases}$$

z = any amount, $x = 300 - z$, $y = 100 - z$
$(0 \le z \le 100)$

29. $\begin{cases} g + b + f = 8 \times 12 \\ g + b - 15f = 0 \\ 3g + 3b + 5f = 300 \end{cases}$

$$\begin{bmatrix} \underline{1} & 1 & 1 & | & 96 \\ 1 & 1 & -15 & | & 0 \\ 3 & 3 & 5 & | & 300 \end{bmatrix}$$

$$\begin{bmatrix} 1 & 1 & 1 & | & 96 \\ 0 & 0 & \underline{-16} & | & -96 \\ 0 & 0 & 2 & | & 12 \end{bmatrix}$$

$$\begin{bmatrix} 1 & 1 & 0 & | & 90 \\ 0 & 0 & 1 & | & 6 \\ 0 & 0 & 0 & | & 0 \end{bmatrix}$$

$$\begin{cases} g + b = 90 \\ f = 6 \\ 0 = 0 \end{cases}$$

6 floral squares, the other 90 any mix of solid green and blue

31. $\begin{bmatrix} \underline{1} & 1 & 1 & | & 14 \\ 1 & -1 & 2 & | & 15 \\ 1 & 2 & 3 & | & 36 \end{bmatrix}$

$$\begin{bmatrix} 1 & 1 & 1 & | & 14 \\ 0 & \underline{-2} & 1 & | & 1 \\ 0 & 1 & 2 & | & 22 \end{bmatrix}$$

$$\begin{bmatrix} 1 & 0 & \frac{3}{2} & | & \frac{29}{2} \\ 0 & 1 & -\frac{1}{2} & | & -\frac{1}{2} \\ 0 & 0 & \frac{5}{2} & | & \frac{45}{2} \end{bmatrix}$$

$$\begin{bmatrix} 1 & 0 & 0 & | & 1 \\ 0 & 1 & 0 & | & 4 \\ 0 & 0 & 1 & | & 9 \end{bmatrix}$$

$$\begin{cases} x^2 = 1 \\ y^2 = 4 \\ z^2 = 9 \end{cases}$$

$x = \pm 1$, $y = \pm 2$, $z = \pm 3$

33. $\begin{bmatrix} 2 & -3 & | & 4 \\ -6 & 9 & | & k \end{bmatrix}$

$$\begin{bmatrix} 1 & -\frac{3}{2} & | & 2 \\ 0 & 0 & | & 12 + k \end{bmatrix}$$

No solution if $0 \ne 12 + k$, which happens when $k \ne -12$.
Infinitely many if $0 = 12 + k$, which happens when $k = -12$.

35. (b); There is no point that satisfies all three equations at the same time.

37.

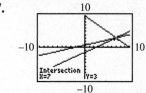

One solution when $x = 7$ and $y = 3$

39.

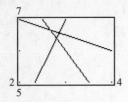

No solution

Exercises 2.3

1. 2 by 3

3. 1 by 3, row matrix

5. 2 by 2, square matrix

7. $a_{12} = -4$; $a_{21} = 0$

9. $i = 1$; $j = 3$

11. $\begin{bmatrix} 4+5 & -2+5 \\ 3+4 & 0+(-1) \end{bmatrix} = \begin{bmatrix} 9 & 3 \\ 7 & -1 \end{bmatrix}$

13. $\begin{bmatrix} 2-1 & 8-5 \\ \frac{4}{3}-\frac{1}{3} & 4-2 \\ 1-(-3) & -2-0 \end{bmatrix} = \begin{bmatrix} 1 & 3 \\ 1 & 2 \\ 4 & -2 \end{bmatrix}$

15. $\begin{bmatrix} 5 \cdot 1 + 3 \cdot 2 \end{bmatrix} = \begin{bmatrix} 11 \end{bmatrix}$

17. $\begin{bmatrix} 6 \cdot \frac{1}{2} + 1(-3) + 5 \cdot 2 \end{bmatrix} = [10]$

19. Yes, columns of A = rows of B; 3, therefore the product will be size 3×5

21. No, columns of $A \neq$ rows of B

23. Yes, columns of A = rows of B; 3, therefore the product will be size 3×1

25. $\begin{bmatrix} 3 \cdot 1 + 1 \cdot 3 & 3 \cdot 4 + 1 \cdot 5 \\ 0 \cdot 1 + 2 \cdot 3 & 0 \cdot 4 + 2 \cdot 5 \end{bmatrix} = \begin{bmatrix} 6 & 17 \\ 6 & 10 \end{bmatrix}$

27. $\begin{bmatrix} 4 \cdot 5 + 1 \cdot 1 + 0 \cdot 2 \\ -2 \cdot 5 + 0 \cdot 1 + 3 \cdot 2 \\ 1 \cdot 5 + 5 \cdot 1 + (-1)2 \end{bmatrix} = \begin{bmatrix} 21 \\ -4 \\ 8 \end{bmatrix}$

29. Multiplication by identity matrix: $\begin{bmatrix} 5 & 6 \\ 7 & 8 \end{bmatrix}$

31.

$\begin{bmatrix} (.6)(.6)+(.3)(.4) & (.6)(.3)+(.3)(.7) \\ (.4)(.6)+(.7)(.4) & (.4)(.3)+(.7)(.7) \end{bmatrix}$

$= \begin{bmatrix} .48 & .39 \\ .52 & .61 \end{bmatrix}$

33. $\begin{bmatrix} 2 \cdot 4 + (-1)3 + 4 \cdot 5 & 2 \cdot 8 + (-1)(-1) + 4 \cdot 0 & 2 \cdot 0 + (-1)2 + 4 \cdot 1 \\ 0 \cdot 4 + 1 \cdot 3 + 0 \cdot 5 & 0 \cdot 8 + 1(-1) + 0 \cdot 0 & 0 \cdot 0 + 1 \cdot 2 + 0 \cdot 1 \\ \frac{1}{2} \cdot 4 + 3 \cdot 3 + (-2)5 & \frac{1}{2} \cdot 8 + 3(-1) + (-2)0 & \frac{1}{2} \cdot 0 + 3 \cdot 2 + (-2) \cdot 1 \end{bmatrix} = \begin{bmatrix} 25 & 17 & 2 \\ 3 & -1 & 2 \\ 1 & 1 & 4 \end{bmatrix}$

35. $\begin{cases} 2x + 3y = 6 \\ 4x + 5y = 7 \end{cases}$

37. $\begin{cases} x + 2y + 3z = 10 \\ 4x + 5y + 6z = 11 \\ 7x + 8y + 9z = 12 \end{cases}$

39. $\begin{bmatrix} 3 & 2 \\ 7 & -1 \end{bmatrix} \begin{bmatrix} x \\ y \end{bmatrix} = \begin{bmatrix} -1 \\ 2 \end{bmatrix}$

41. $\begin{bmatrix} 1 & -2 & 3 \\ 0 & 1 & 1 \\ 0 & 0 & 1 \end{bmatrix}\begin{bmatrix} x \\ y \\ z \end{bmatrix} = \begin{bmatrix} 5 \\ 6 \\ 2 \end{bmatrix}$

43. $\left(\begin{bmatrix} 1 & 2 \\ 0 & 3 \end{bmatrix} + \begin{bmatrix} 3 & -2 \\ 4 & 5 \end{bmatrix}\right)\begin{bmatrix} 1 & 6 \\ 2 & 0 \end{bmatrix} = \begin{bmatrix} 4 & 0 \\ 4 & 8 \end{bmatrix}\begin{bmatrix} 1 & 6 \\ 2 & 0 \end{bmatrix} = \begin{bmatrix} 4 & 24 \\ 20 & 24 \end{bmatrix}$

$\begin{bmatrix} 1 & 2 \\ 0 & 3 \end{bmatrix}\begin{bmatrix} 1 & 6 \\ 2 & 0 \end{bmatrix} + \begin{bmatrix} 3 & -2 \\ 4 & 5 \end{bmatrix}\begin{bmatrix} 1 & 6 \\ 2 & 0 \end{bmatrix} = \begin{bmatrix} 5 & 6 \\ 6 & 0 \end{bmatrix} + \begin{bmatrix} -1 & 18 \\ 14 & 24 \end{bmatrix} = \begin{bmatrix} 4 & 24 \\ 20 & 24 \end{bmatrix}$

45. $\begin{bmatrix} 3\cdot 1 + (-1)2 & 3\cdot 2 + (-1)6 \\ -1\cdot 1 + \frac{1}{2}\cdot 2 & -1\cdot 2 + \frac{1}{2}\cdot 6 \end{bmatrix} = \begin{bmatrix} 1 & 0 \\ 0 & 1 \end{bmatrix}$

47. a. $\begin{bmatrix} 6 & 8 & 2 \\ 2 & 5 & 3 \end{bmatrix}\begin{bmatrix} 20 \\ 15 \\ 50 \end{bmatrix} = \begin{bmatrix} 340 \\ 265 \end{bmatrix}$

b. Mike's clothes are worth $340; Don's clothes are worth $265.

49. a. $\begin{bmatrix} 210 & 175 & 135 \end{bmatrix}\begin{bmatrix} 3 & 3 & 5.8 \\ 2.5 & 3.5 & 6 \\ 9 & 8 & 9.5 \end{bmatrix}$

$= \begin{bmatrix} 2282.50 & 2322.50 & 3550.50 \end{bmatrix}$

The total value of the store's plain items was $2282.50, of the milk chocolate-covered items was $2322.50, and of the dark chocolate covered items was $3550.50.

b. $\begin{bmatrix} 3 & 3 & 5.8 \\ 2.5 & 3.5 & 6 \\ 9 & 8 & 9.5 \end{bmatrix}\begin{bmatrix} 105 \\ 390 \\ 285 \end{bmatrix} = \begin{bmatrix} 3138.00 \\ 3337.50 \\ 6772.50 \end{bmatrix}$

The store's weekly sales of peanuts was $3138.00, of raisins was $3337.50, and of espresso beans was $6772.50.

51. a. $\begin{bmatrix} .25 & .35 & .30 & .10 & 0 \\ .10 & .20 & .40 & .20 & .10 \\ .05 & .10 & .20 & .40 & .25 \end{bmatrix}\begin{bmatrix} 4 \\ 3 \\ 2 \\ 1 \\ 0 \end{bmatrix} = \begin{bmatrix} 2.75 \\ 2.00 \\ 1.30 \end{bmatrix}$

I: 2.75; II: 2, III: 1.3

b. $\begin{bmatrix} 240 & 120 & 40 \end{bmatrix}\begin{bmatrix} .25 & .35 & .30 & .10 & 0 \\ .10 & .20 & .40 & .20 & .10 \\ .05 & .10 & .20 & .40 & .25 \end{bmatrix}$

$= \begin{bmatrix} 74 & 112 & 128 & 64 & 22 \end{bmatrix}$

A: 74, B: 112, C: 128, D: 64, F: 22

53.

$\begin{bmatrix} 6000 & 8000 & 4000 \end{bmatrix}\begin{bmatrix} .65 & .35 \\ .55 & .45 \\ .45 & .55 \end{bmatrix} = \begin{bmatrix} 10{,}100 & 7900 \end{bmatrix}$

10,100 voting Democratic, 7900 voting Republican

55. $\begin{bmatrix} 50 & 20 & 10 \\ 30 & 30 & 15 \\ 20 & 20 & 5 \end{bmatrix}\begin{bmatrix} 10 \\ 15 \\ 20 \end{bmatrix} = \begin{bmatrix} 1000 \\ 1050 \\ 600 \end{bmatrix}$

Carpenters: $1000, bricklayers: $1050, plumbers: $600

57. a. $BN = \begin{bmatrix} 162 & 150 & 143 \end{bmatrix}$, number of units of each nutrient consumed at breakfast

b. $LN = \begin{bmatrix} 186 & 200 & 239 \end{bmatrix}$, number of units of each nutrient consumed at lunch

c. $DN = \begin{bmatrix} 288 & 300 & 344 \end{bmatrix}$, number of units of each nutrient consumed at dinner

d. $B + L + D = \begin{bmatrix} 5 & 8 \end{bmatrix}$, total number of ounces of each food that Mikey eats during a day

e. $(B + L + D)N = \begin{bmatrix} 636 & 650 & 726 \end{bmatrix}$, number of units of each nutrient consumed per day

59. a. $AP = \begin{bmatrix} 720 \\ 646 \end{bmatrix}$, the average amount taken in daily by the pool and the weight room

b. $720

61. a.

Boston Cream Pie   Carrot Cake

$$T = \begin{bmatrix} 30 & 45 \\ 30 & 50 \\ 15 & 10 \end{bmatrix} \begin{matrix} \text{Preparation} \\ \text{Baking} \\ \text{Finishing} \end{matrix}$$

b. $S = \begin{bmatrix} 20 \\ 8 \end{bmatrix} \begin{matrix} \text{Boston Cream Pie} \\ \text{Carrot Cake} \end{matrix}$

$$TS = \begin{bmatrix} 960 \\ 1000 \\ 380 \end{bmatrix} \begin{matrix} \text{Preparation} \\ \text{Baking} \\ \text{Finishing} \end{matrix}$$

c. Total baking time is 1000 minutes or $16\frac{2}{3}$ hours. Total finishing time is 380 minutes or $6\frac{1}{3}$ hours.

63. a.

Cutting   Sewing   Finishing

$$T = \begin{bmatrix} 2 & 3 & 2 \\ 1.5 & 2 & 1 \end{bmatrix} \begin{matrix} \text{Huge One} \\ \text{Regular Joe} \end{matrix}$$

b. $S = \begin{bmatrix} 32 \\ 24 \end{bmatrix} \begin{matrix} \text{Huge One} \\ \text{Regular Joe} \end{matrix}$

c.

Huge One   Regular Joe

$$A = \begin{bmatrix} 27 & 56 \end{bmatrix}$$

Cutting   Sewing   Finishing

$$AT = \begin{bmatrix} 138 & 193 & 110 \end{bmatrix}$$

$$AS = \begin{bmatrix} 2208 \end{bmatrix}$$

d. 193 hours are needed for sewing.

e. The total revenue would be $2208.

65. answers will vary.

67. $(A+B) - A = \begin{bmatrix} 3 & -2 & 1 \\ -5 & 6 & 7 \end{bmatrix}$

69. 4×4

71. The matrix product:

$$\begin{bmatrix} 7998 & 56,685 & 96,158 \end{bmatrix} \begin{bmatrix} 15.1 \\ 15.2 \\ 12.6 \end{bmatrix} \text{ gives the}$$

total number of pupils in the three states.

73. $AB = \begin{bmatrix} 27.9 & 130.6 & -69.88 \\ 106.75 & -149.44 & 26.1 \\ -47.5 & 336.2 & -18.7 \end{bmatrix}$

75. $A(B+C) = \begin{bmatrix} -69.14 & 147.9 & -43.26 \\ 158.05 & -3.69 & 33.46 \\ -176.1 & 259.5 & 59.3 \end{bmatrix}$

77. $A^2 = AA = \begin{bmatrix} 160.16 & -26.7 & 4 \\ 2.7 & 150.85 & -53 \\ 187.4 & -35.5 & 48.6 \end{bmatrix}$

79. They match.

Exercises 2.4

1. $\begin{bmatrix} 1 & -2 \\ -\frac{1}{2} & 2 \end{bmatrix} \begin{bmatrix} 4 \\ 1 \end{bmatrix} = \begin{bmatrix} 2 \\ 0 \end{bmatrix}$

$x = 2, y = 0$

3. $D = 7 \cdot 1 - 3 \cdot 2 = 1$

$\begin{bmatrix} \frac{1}{1} & -\frac{2}{1} \\ -\frac{3}{1} & \frac{7}{1} \end{bmatrix} = \begin{bmatrix} 1 & -2 \\ -3 & 7 \end{bmatrix}$

5. $D = 6 \cdot 2 - 5 \cdot 2 = 2$

$\begin{bmatrix} \frac{2}{2} & -\frac{2}{2} \\ -\frac{5}{2} & \frac{6}{2} \end{bmatrix} = \begin{bmatrix} 1 & -1 \\ -\frac{5}{2} & 3 \end{bmatrix}$

7. $D = 0.7 \cdot 0.8 - 0.3 \cdot 0.2 = 0.5$

$\begin{bmatrix} \frac{.8}{.5} & -\frac{.2}{.5} \\ -\frac{.3}{.5} & \frac{.7}{.5} \end{bmatrix} = \begin{bmatrix} 1.6 & -.4 \\ -.6 & 1.4 \end{bmatrix}$

9. For a 1×1 matrix $[a]$ $(a \neq 0)$, $[a]^{-1} = \begin{bmatrix} \dfrac{1}{a} \end{bmatrix}$.

$\begin{bmatrix} \dfrac{1}{3} \end{bmatrix}$

11. $\begin{bmatrix} 1 & 2 \\ 2 & 6 \end{bmatrix}^{-1} \begin{bmatrix} 3 \\ 5 \end{bmatrix} = \begin{bmatrix} 3 & -1 \\ -1 & \frac{1}{2} \end{bmatrix} \begin{bmatrix} 3 \\ 5 \end{bmatrix} = \begin{bmatrix} 4 \\ -\frac{1}{2} \end{bmatrix}$

$x = 4, \ y = -\dfrac{1}{2}$

13. $\begin{bmatrix} \frac{1}{2} & 2 \\ 3 & 16 \end{bmatrix}^{-1} \begin{bmatrix} 4 \\ 0 \end{bmatrix} = \begin{bmatrix} 8 & -1 \\ -\frac{3}{2} & \frac{1}{4} \end{bmatrix} \begin{bmatrix} 4 \\ 0 \end{bmatrix} = \begin{bmatrix} 32 \\ -6 \end{bmatrix}$

$x = 32, y = -6$

15. a. $\begin{bmatrix} .8 & .3 \\ .2 & .7 \end{bmatrix} \begin{bmatrix} x \\ y \end{bmatrix} = \begin{bmatrix} m \\ s \end{bmatrix}$

b. $\begin{bmatrix} x \\ y \end{bmatrix} = \begin{bmatrix} .8 & .3 \\ .2 & .7 \end{bmatrix}^{-1} \begin{bmatrix} m \\ s \end{bmatrix}$

$= \begin{bmatrix} 1.4 & -.6 \\ -.4 & 1.6 \end{bmatrix} \begin{bmatrix} m \\ s \end{bmatrix}$

c. $\begin{bmatrix} 1.4 & -.6 \\ -.4 & 1.6 \end{bmatrix} \begin{bmatrix} 100,000 \\ 50,000 \end{bmatrix} = \begin{bmatrix} 110,000 \\ 40,000 \end{bmatrix}$

110,000 married; 40,000 single

d. $\begin{bmatrix} 1.4 & -.6 \\ -.4 & 1.6 \end{bmatrix} \begin{bmatrix} 110,000 \\ 40,000 \end{bmatrix} = \begin{bmatrix} 130,000 \\ 20,000 \end{bmatrix}$

130,000 married; 20,000 single

17. a.
$$\begin{bmatrix} .7 & .1 \\ .3 & .9 \end{bmatrix}\begin{bmatrix} x \\ y \end{bmatrix} = \begin{bmatrix} u \\ v \end{bmatrix}$$

b.
$$\begin{bmatrix} x \\ y \end{bmatrix} = \begin{bmatrix} .7 & .1 \\ .3 & .9 \end{bmatrix}^{-1}\begin{bmatrix} u \\ v \end{bmatrix} = \begin{bmatrix} \frac{3}{2} & -\frac{1}{6} \\ -\frac{1}{2} & \frac{7}{6} \end{bmatrix}\begin{bmatrix} u \\ v \end{bmatrix}$$

c.
$$\begin{bmatrix} \frac{3}{2} & -\frac{1}{6} \\ -\frac{1}{2} & \frac{7}{6} \end{bmatrix}\begin{bmatrix} 6000 \\ 3000 \end{bmatrix} = \begin{bmatrix} 8500 \\ 500 \end{bmatrix}$$

$$\begin{bmatrix} .7 & .1 \\ .3 & .9 \end{bmatrix}\begin{bmatrix} 6000 \\ 3000 \end{bmatrix} = \begin{bmatrix} 4500 \\ 4500 \end{bmatrix}$$
8500; 4500

19.
$$\begin{bmatrix} 5 & -2 & -2 \\ -1 & 1 & 0 \\ -1 & 0 & 1 \end{bmatrix}\begin{bmatrix} 1 \\ -1 \\ -1 \end{bmatrix} = \begin{bmatrix} 9 \\ -2 \\ -2 \end{bmatrix}$$
$x = 9, y = -2, z = -2$

21.
$$\begin{bmatrix} 1 & 0 & -2 & 0 \\ 0 & 1 & 0 & -5 \\ -4 & 0 & 9 & 0 \\ 0 & 2 & 1 & -9 \end{bmatrix}\begin{bmatrix} 1 \\ 0 \\ 0 \\ -1 \end{bmatrix} = \begin{bmatrix} 1 \\ 5 \\ -4 \\ 9 \end{bmatrix}$$
$x = 1, y = 5, z = -4, w = 9$

23. Suppose $\begin{bmatrix} 6 & 3 \\ 2 & 1 \end{bmatrix}^{-1} = \begin{bmatrix} s & t \\ u & v \end{bmatrix}.$

Then $\begin{bmatrix} 6s + 3u & 6t + 3v \\ 2s + u & 2t + v \end{bmatrix} = \begin{bmatrix} 1 & 0 \\ 0 & 1 \end{bmatrix}.$

Then $\dfrac{6s + 3u}{3} = 2s + u = \dfrac{1}{3}$, which contradicts
$2s + u = 0.$

25. a. $\begin{cases} x + 2y = a \\ \quad .9x = b \end{cases}$

$$\begin{bmatrix} 1 & 2 \\ .9 & 0 \end{bmatrix}\begin{bmatrix} x \\ y \end{bmatrix} = \begin{bmatrix} a \\ b \end{bmatrix}$$

b.
$$\begin{bmatrix} 1 & 2 \\ .9 & 0 \end{bmatrix}\begin{bmatrix} 450,000 \\ 360,000 \end{bmatrix} = \begin{bmatrix} 1,170,000 \\ 405,000 \end{bmatrix}$$
After 1 year:
1,170,000 in group I,
405,000 in group II

$$\begin{bmatrix} 1 & 2 \\ .9 & 0 \end{bmatrix}\begin{bmatrix} 1,170,000 \\ 405,000 \end{bmatrix} = \begin{bmatrix} 1,980,000 \\ 1,053,000 \end{bmatrix}$$
After 2 years:
1,980,000 in group I,
1,053,000 in group II

c.
$$\begin{bmatrix} 1 & 2 \\ .9 & 0 \end{bmatrix}^{-1}\begin{bmatrix} 810,000 \\ 630,000 \end{bmatrix} = \begin{bmatrix} 700,000 \\ 55,000 \end{bmatrix}$$
700,000 in group I, 55,000 in group II

27. If $AB = 0$ (zero matrix) and A has an inverse the
B is a matrix of all zeros:
Proof : Assume $AB = 0$ and A has an inverse.
$$A^{-1}(AB) = A^{-1}(0) \rightarrow$$
$$(A^{-1}A)B = A^{-1}(0) \rightarrow$$
$$(I_n)B = 0 \rightarrow$$
$$B = 0$$

29. One example:
$$AX = B$$
Let $A = \begin{bmatrix} 3 & 4 \\ 6 & 8 \end{bmatrix}$ $X = \begin{bmatrix} x \\ y \end{bmatrix}$ $B = \begin{bmatrix} 2 \\ 3 \end{bmatrix}$

The system $\begin{cases} 3x + 4y = 2 \\ 6x + 8y = 3 \end{cases}$ has no solution.

31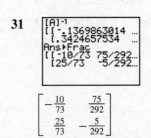

$$\begin{bmatrix} -\dfrac{10}{73} & \dfrac{75}{292} \\ \dfrac{25}{73} & -\dfrac{5}{292} \end{bmatrix}$$

33.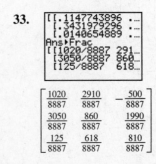

$$\begin{bmatrix} \dfrac{1020}{8887} & \dfrac{2910}{8887} & -\dfrac{500}{8887} \\ \dfrac{3050}{8887} & \dfrac{860}{8887} & \dfrac{1990}{8887} \\ \dfrac{125}{8887} & \dfrac{618}{8887} & \dfrac{810}{8887} \end{bmatrix}$$

35.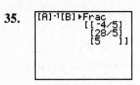

$$x = -\frac{4}{5},\ y = \frac{28}{5},\ z = 5$$

37.
```
[A]⁻¹[B]▸Frac
         [[0]
          [2]
          [0]
          [2]]
```

$x = 0,\ y = 2,\ z = 0,\ w = 2$

39. With the message ERR:INVALID DIM

Exercises 2.5

1. $\begin{bmatrix} 7 & 3 & | & 1 & 0 \\ 5 & 2 & | & 0 & 1 \end{bmatrix}$

$\begin{bmatrix} 1 & \frac{3}{7} & | & \frac{1}{7} & 0 \\ 0 & -\frac{1}{7} & | & -\frac{5}{7} & 1 \end{bmatrix}$

$\begin{bmatrix} 1 & 0 & | & -2 & 3 \\ 0 & 1 & | & 5 & -7 \end{bmatrix}$

$\begin{bmatrix} -2 & 3 \\ 5 & -7 \end{bmatrix}$

3. $\begin{bmatrix} 2 & 3 & | & 1 & 0 \\ -4 & -7 & | & 0 & 1 \end{bmatrix}$

$\begin{bmatrix} 1 & \frac{3}{2} & | & \frac{1}{2} & 0 \\ 0 & -1 & | & 2 & 1 \end{bmatrix}$

$\begin{bmatrix} 1 & 0 & | & \frac{7}{2} & \frac{3}{2} \\ 0 & 1 & | & -2 & -1 \end{bmatrix}$

$\begin{bmatrix} \frac{7}{2} & \frac{3}{2} \\ -2 & -1 \end{bmatrix}$

5. $\begin{bmatrix} 2 & -4 & | & 1 & 0 \\ -1 & 2 & | & 0 & 1 \end{bmatrix}$

$\begin{bmatrix} 1 & -2 & | & \frac{1}{2} & 0 \\ 0 & 0 & | & \frac{1}{2} & 1 \end{bmatrix}$

No inverse

7. $\begin{bmatrix} 1 & 2 & -2 & | & 1 & 0 & 0 \\ 1 & 1 & 1 & | & 0 & 1 & 0 \\ 0 & 0 & 1 & | & 0 & 0 & 1 \end{bmatrix}$

$\begin{bmatrix} 1 & 2 & -2 & | & 1 & 0 & 0 \\ 0 & -1 & 3 & | & -1 & 1 & 0 \\ 0 & 0 & 1 & | & 0 & 0 & 1 \end{bmatrix}$

$\begin{bmatrix} 1 & 0 & 4 & | & -1 & 2 & 0 \\ 0 & 1 & -3 & | & 1 & -1 & 0 \\ 0 & 0 & 1 & | & 0 & 0 & 1 \end{bmatrix}$

$\begin{bmatrix} 1 & 0 & 0 & | & -1 & 2 & -4 \\ 0 & 1 & 0 & | & 1 & -1 & 3 \\ 0 & 0 & 1 & | & 0 & 0 & 1 \end{bmatrix}$

$\begin{bmatrix} -1 & 2 & -4 \\ 1 & -1 & 3 \\ 0 & 0 & 1 \end{bmatrix}$

9. $\begin{bmatrix} -2 & 5 & 2 & | & 1 & 0 & 0 \\ 1 & -3 & -1 & | & 0 & 1 & 0 \\ -1 & 2 & 1 & | & 0 & 0 & 1 \end{bmatrix}$

$\begin{bmatrix} 1 & -\frac{5}{2} & -1 & | & -\frac{1}{2} & 0 & 0 \\ 0 & -\frac{1}{2} & 0 & | & \frac{1}{2} & 1 & 0 \\ 0 & -\frac{1}{2} & 0 & | & -\frac{1}{2} & 0 & 1 \end{bmatrix}$

$\begin{bmatrix} 1 & 0 & -1 & | & -3 & -5 & 0 \\ 0 & 1 & 0 & | & -1 & -2 & 0 \\ 0 & 0 & 0 & | & -1 & -1 & 1 \end{bmatrix}$

No inverse

11.
$$\left[\begin{array}{cccc|cccc} \underline{1} & 6 & 0 & 0 & 1 & 0 & 0 & 0 \\ 1 & 5 & 0 & 0 & 0 & 1 & 0 & 0 \\ 0 & 0 & 4 & 2 & 0 & 0 & 1 & 0 \\ 0 & 0 & 50 & 2 & 0 & 0 & 0 & 1 \end{array}\right]$$

$$\left[\begin{array}{cccc|cccc} 1 & 6 & 0 & 0 & 1 & 0 & 0 & 0 \\ 0 & \underline{-1} & 0 & 0 & -1 & 1 & 0 & 0 \\ 0 & 0 & 4 & 2 & 0 & 0 & 1 & 0 \\ 0 & 0 & 50 & 2 & 0 & 0 & 0 & 1 \end{array}\right]$$

$$\left[\begin{array}{cccc|cccc} 1 & 0 & 0 & 0 & -5 & 6 & 0 & 0 \\ 0 & 1 & 0 & 0 & 1 & -1 & 0 & 0 \\ 0 & 0 & \underline{4} & 2 & 0 & 0 & 1 & 0 \\ 0 & 0 & 50 & 2 & 0 & 0 & 0 & 1 \end{array}\right]$$

$$\left[\begin{array}{cccc|cccc} 1 & 0 & 0 & 0 & -5 & 6 & 0 & 0 \\ 0 & 1 & 0 & 0 & 1 & -1 & 0 & 0 \\ 0 & 0 & 1 & \frac{1}{2} & 0 & 0 & \frac{1}{4} & 0 \\ 0 & 0 & 0 & \underline{-23} & 0 & 0 & -\frac{25}{2} & 1 \end{array}\right]$$

$$\left[\begin{array}{cccc|cccc} 1 & 0 & 0 & 0 & -5 & 6 & 0 & 0 \\ 0 & 1 & 0 & 0 & 1 & -1 & 0 & 0 \\ 0 & 0 & 1 & 0 & 0 & 0 & -\frac{1}{46} & \frac{1}{46} \\ 0 & 0 & 0 & 1 & 0 & 0 & \frac{25}{46} & -\frac{1}{23} \end{array}\right]$$

$$\left[\begin{array}{cccc} -5 & 6 & 0 & 0 \\ 1 & -1 & 0 & 0 \\ 0 & 0 & -\frac{1}{46} & \frac{1}{46} \\ 0 & 0 & \frac{25}{46} & -\frac{1}{23} \end{array}\right]$$

13. Find the inverse of $\begin{bmatrix} 1 & 1 & 2 \\ 3 & 2 & 2 \\ 1 & 1 & 3 \end{bmatrix}$.

$$\left[\begin{array}{ccc|ccc} 1 & 1 & 2 & 1 & 0 & 0 \\ 3 & 2 & 2 & 0 & 1 & 0 \\ 1 & 1 & 3 & 0 & 0 & 1 \end{array}\right]$$

$$\left[\begin{array}{ccc|ccc} 1 & 1 & 2 & 1 & 0 & 0 \\ 0 & -1 & -4 & -3 & 1 & 0 \\ 0 & 0 & 1 & -1 & 0 & 1 \end{array}\right]$$

$$\left[\begin{array}{ccc|ccc} 1 & 0 & -2 & -2 & 1 & 0 \\ 0 & 1 & 4 & 3 & -1 & 0 \\ 0 & 0 & 1 & -1 & 0 & 1 \end{array}\right]$$

$$\left[\begin{array}{ccc|ccc} 1 & 0 & 0 & -4 & 1 & 2 \\ 0 & 1 & 0 & 7 & -1 & -4 \\ 0 & 0 & 1 & -1 & 0 & 1 \end{array}\right]$$

$$\begin{bmatrix} 1 & 1 & 2 \\ 3 & 2 & 2 \\ 1 & 1 & 3 \end{bmatrix}^{-1}\begin{bmatrix} 3 \\ 4 \\ 5 \end{bmatrix} = \begin{bmatrix} -4 & 1 & 2 \\ 7 & -1 & -4 \\ -1 & 0 & 1 \end{bmatrix}\begin{bmatrix} 3 \\ 4 \\ 5 \end{bmatrix} = \begin{bmatrix} 2 \\ -3 \\ 2 \end{bmatrix}$$

$x = 2,\ y = -3,\ z = 2$

15. Find the inverse of $\begin{bmatrix} 1 & 0 & -2 & -2 \\ 0 & 1 & 0 & -5 \\ -4 & 0 & 9 & 9 \\ 0 & 2 & 1 & -8 \end{bmatrix}$.

$$\left[\begin{array}{cccc|cccc} 1 & 0 & -2 & -2 & 1 & 0 & 0 & 0 \\ 0 & 1 & 0 & -5 & 0 & 1 & 0 & 0 \\ -4 & 0 & 9 & 9 & 0 & 0 & 1 & 0 \\ 0 & 2 & 1 & -8 & 0 & 0 & 0 & 1 \end{array}\right]$$

$$\left[\begin{array}{cccc|cccc} 1 & 0 & -2 & -2 & 1 & 0 & 0 & 0 \\ 0 & 1 & 0 & -5 & 0 & 1 & 0 & 0 \\ 0 & 0 & 1 & 1 & 4 & 0 & 1 & 0 \\ 0 & 2 & 1 & -8 & 0 & 0 & 0 & 1 \end{array}\right]$$

$$\left[\begin{array}{cccc|cccc} 1 & 0 & -2 & -2 & 1 & 0 & 0 & 0 \\ 0 & 1 & 0 & -5 & 0 & 1 & 0 & 0 \\ 0 & 0 & 1 & 1 & 4 & 0 & 1 & 0 \\ 0 & 0 & 1 & 2 & 0 & -2 & 0 & 1 \end{array}\right]$$

$$\left[\begin{array}{cccc|cccc} 1 & 0 & 0 & 0 & 9 & 0 & 2 & 0 \\ 0 & 1 & 0 & -5 & 0 & 1 & 0 & 0 \\ 0 & 0 & 1 & 1 & 4 & 0 & 1 & 0 \\ 0 & 0 & 0 & 1 & -4 & -2 & -1 & 1 \end{array}\right]$$

$$\left[\begin{array}{cccc|cccc} 1 & 0 & 0 & 0 & 9 & 0 & 2 & 0 \\ 0 & 1 & 0 & 0 & -20 & -9 & -5 & 5 \\ 0 & 0 & 1 & 0 & 8 & 2 & 2 & -1 \\ 0 & 0 & 0 & 1 & -4 & -2 & -1 & 1 \end{array}\right]$$

$$\begin{bmatrix} 1 & 0 & -2 & -2 \\ 0 & 1 & 0 & -5 \\ -4 & 0 & 9 & 9 \\ 0 & 2 & 1 & -8 \end{bmatrix}^{-1}\begin{bmatrix} 0 \\ 1 \\ 2 \\ 3 \end{bmatrix}$$

$$= \begin{bmatrix} 9 & 0 & 2 & 0 \\ -20 & -9 & -5 & 5 \\ 8 & 2 & 2 & -1 \\ -4 & -2 & -1 & 1 \end{bmatrix} \begin{bmatrix} 0 \\ 1 \\ 2 \\ 3 \end{bmatrix}$$

$$= \begin{bmatrix} 4 \\ -4 \\ 3 \\ -1 \end{bmatrix}$$

$x = 4, y = -4, z = 3, w = -1$

17. If $A^{-1} = \begin{bmatrix} 2 & 7 \\ 1 & 3 \end{bmatrix}$,

Fact: $A^{-1}A = I = A \cdot A^{-1}$

Let $A = \begin{bmatrix} a & b \\ c & d \end{bmatrix}$.

Then $\begin{bmatrix} a & b \\ c & d \end{bmatrix} \cdot \begin{bmatrix} 2 & 7 \\ 1 & 3 \end{bmatrix} = \begin{bmatrix} 1 & 0 \\ 0 & 1 \end{bmatrix}$

$2a + b = 1$ and $7a + 3b = 0$

Also: $2c + d = 0$ and $7c + 3d = 1$

Solving for a, b, c, d, we have $a = -3$, $b = 7$, $c = 1$ and $d = -2$

$$A = \begin{bmatrix} -3 & 7 \\ 1 & -2 \end{bmatrix}$$

19. $A \cdot \begin{bmatrix} 2 & 5 \\ 1 & 3 \end{bmatrix} = \begin{bmatrix} -1 & 0 \\ 4 & 2 \end{bmatrix}$

$A = A \cdot \begin{bmatrix} 2 & 5 \\ 1 & 3 \end{bmatrix} \begin{bmatrix} 2 & 5 \\ 1 & 3 \end{bmatrix}^{-1}$

$= \begin{bmatrix} -1 & 0 \\ 4 & 2 \end{bmatrix} \begin{bmatrix} 2 & 5 \\ 1 & 3 \end{bmatrix}^{-1}$

$= \begin{bmatrix} -1 & 0 \\ 4 & 2 \end{bmatrix} \begin{bmatrix} 3 & -5 \\ -1 & 2 \end{bmatrix}$

$= \begin{bmatrix} -3 & 5 \\ 10 & -16 \end{bmatrix}$

Exercises 2.6

1. $A_{21} = .2$, 20 cents of energy are required to produce \$1 worth of manufactured goods.

3. $A_{31} = .1$, $A_{32} = .2$, $A_{33} = .15$

$A_{32} > A_{33} > A_{31}$

The energy sector uses the greatest amount of services in order to produce \$1 worth of output.

5. $D_{new} = \begin{bmatrix} 4 \\ 1.5 \\ 9 \end{bmatrix}$

$(I - A)^{-1} D_{new} = \begin{bmatrix} 1.01 & .20 & .50 \\ .02 & 1.05 & .23 \\ .01 & .09 & 1.08 \end{bmatrix} \begin{bmatrix} 4 \\ 1.5 \\ 9 \end{bmatrix}$

$= \begin{bmatrix} 8.84 \\ 3.725 \\ 9.895 \end{bmatrix}$

Coal: \$8.84 billion, steel: \$3.725 billion, electricity: \$9.895 billion

7. $D_{\text{new}} = \begin{bmatrix} 3 \\ 1 \\ 4 \end{bmatrix}$

$(I - A)^{-1} D_{\text{new}} = \begin{bmatrix} 1.04 & .02 & .10 \\ .21 & 1.01 & .02 \\ .11 & .02 & 1.02 \end{bmatrix} \begin{bmatrix} 3 \\ 1 \\ 4 \end{bmatrix} = \begin{bmatrix} 3.54 \\ 1.72 \\ 4.43 \end{bmatrix}$

Computers: $354 million,
semiconductors: $172 million

9.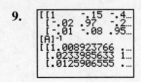

11. a. $\begin{array}{cc} T & E \end{array}$

$A = \begin{bmatrix} .25 & .30 \\ .20 & .15 \end{bmatrix} \begin{array}{c} T \\ E \end{array}$

b. $(I - A)^{-1} = \left(\begin{bmatrix} 1 & 0 \\ 0 & 1 \end{bmatrix} - \begin{bmatrix} .25 & .30 \\ .20 & .15 \end{bmatrix} \right)^{-1} \approx \begin{bmatrix} 1.47 & .52 \\ .35 & 1.30 \end{bmatrix}$

c. $(I - A)^{-1} D \approx \begin{bmatrix} 1.47 & .52 \\ .35 & 1.30 \end{bmatrix} \begin{bmatrix} 5 \\ 3 \end{bmatrix} = \begin{bmatrix} 8.91 \\ 5.65 \end{bmatrix}$

Transportation should produce $8.91 billion worth of output and Energy should produce $5.65 billion.

13. $\begin{array}{cc} P & I \end{array}$

$A = \begin{bmatrix} .02 & .01 \\ .10 & .05 \end{bmatrix} \begin{array}{c} P \\ I \end{array}$

$D = \begin{bmatrix} 930 \\ 465 \end{bmatrix}$

$(I - A)^{-1} D = \begin{bmatrix} 955 \\ 590 \end{bmatrix}$

Plastics: $955,000,
industrial equipment: $590,000

15. $(I - A)^{-1} \begin{bmatrix} 100 \\ 80 \\ 200 \end{bmatrix} \approx \begin{bmatrix} 398 \\ 313 \\ 452 \end{bmatrix}$

manufacturing: $398 million,
transportation: $313 million,
agriculture: $452 million

17. $(I-A)^{-1}\begin{bmatrix} 3 \\ 1 \\ 3 \end{bmatrix} = (I-A)^{-1}\left(\begin{bmatrix} 2 \\ 1 \\ 3 \end{bmatrix} + \begin{bmatrix} 1 \\ 0 \\ 0 \end{bmatrix}\right)$ When

the matrix $(I-A)^{-1}$ is multiplied by the column

matrix $\begin{bmatrix} 1 \\ 0 \\ 0 \end{bmatrix}$, the entries give the first column of

the matrix $(I-A)^{-1}$, which represents the additional amounts that must be produced by the three industries.

$= (I-A)^{-1} D_{new} - (I-A)^{-1} D_{old}$

$= (I-A)^{-1}\left(\begin{bmatrix} 3 \\ 1 \\ 3 \end{bmatrix} - \begin{bmatrix} 2 \\ 1 \\ 3 \end{bmatrix}\right)$

$= (I-A)^{-1}\left(\begin{bmatrix} 1 \\ 0 \\ 0 \end{bmatrix}\right)$

$= \begin{bmatrix} 1.01 & .20 & .50 \\ .02 & 1.05 & .23 \\ .01 & .09 & 1.08 \end{bmatrix}\begin{bmatrix} 1 \\ 0 \\ 0 \end{bmatrix} = \begin{bmatrix} 1.01 \\ .02 \\ .01 \end{bmatrix}$

19.
```
round((identity(
3)-[A])⁻¹[D],2)
        [[10.25]
         [13.82]
         [8.65 ]]
```

$\begin{bmatrix} 10.25 \\ 13.82 \\ 8.65 \end{bmatrix}$

Chapter 2 Supplementary Exercises

1. $\begin{bmatrix} 3 & -6 & 1 \\ 2 & 4 & 6 \end{bmatrix} \xrightarrow{\frac{1}{3}R_1} \begin{bmatrix} 1 & -2 & \frac{1}{3} \\ 2 & 4 & 6 \end{bmatrix}$

$\xrightarrow{R_2+(-2)R_1} \begin{bmatrix} 1 & -2 & \frac{1}{3} \\ 0 & 8 & \frac{16}{3} \end{bmatrix}$

2. $\begin{bmatrix} -5 & -3 & 1 \\ 4 & 2 & 0 \\ 0 & 6 & 7 \end{bmatrix} \xrightarrow{\frac{1}{2}R_2} \begin{bmatrix} -5 & -3 & 1 \\ 2 & 1 & 0 \\ 0 & 6 & 7 \end{bmatrix}$

$\xrightarrow{R_1+3R_2} \begin{bmatrix} 1 & 0 & 1 \\ 2 & 1 & 0 \\ 0 & 6 & 7 \end{bmatrix}$

$\xrightarrow{R_3+(-6)R_2} \begin{bmatrix} 1 & 0 & 1 \\ 2 & 1 & 0 \\ -12 & 0 & 7 \end{bmatrix}$

3. $\begin{bmatrix} \frac{1}{2} & -1 & | & -3 \\ 4 & -5 & | & -9 \end{bmatrix}$

$\begin{bmatrix} 1 & -2 & | & -6 \\ 0 & 3 & | & 15 \end{bmatrix}$

$\begin{bmatrix} 1 & 0 & | & 4 \\ 0 & 1 & | & 5 \end{bmatrix}$

$x = 4,\ y = 5$

4. $\begin{bmatrix} 3 & 0 & 9 & | & 42 \\ 2 & 1 & 6 & | & 30 \\ -1 & 3 & -2 & | & -20 \end{bmatrix}$

$\begin{bmatrix} 1 & 0 & 3 & | & 14 \\ 0 & 1 & 0 & | & 2 \\ 0 & 3 & 1 & | & -6 \end{bmatrix}$

$\begin{bmatrix} 1 & 0 & 3 & | & 14 \\ 0 & 1 & 0 & | & 2 \\ 0 & 0 & 1 & | & -12 \end{bmatrix}$

$\begin{bmatrix} 1 & 0 & 0 & | & 50 \\ 0 & 1 & 0 & | & 2 \\ 0 & 0 & 1 & | & -12 \end{bmatrix}$

$x = 50,\ y = 2,\ z = -12$

5. $\begin{bmatrix} 3 & -6 & 6 & | & -5 \\ -2 & 3 & -5 & | & \frac{7}{3} \\ 1 & 1 & 10 & | & 3 \end{bmatrix}$

$\begin{bmatrix} 1 & -2 & 2 & | & -\frac{5}{3} \\ 0 & -1 & -1 & | & -1 \\ 0 & 3 & 8 & | & \frac{14}{3} \end{bmatrix}$

$\begin{bmatrix} 1 & 0 & 4 & | & \frac{1}{3} \\ 0 & 1 & 1 & | & 1 \\ 0 & 0 & 5 & | & \frac{5}{3} \end{bmatrix}$

$\begin{bmatrix} 1 & 0 & 0 & | & -1 \\ 0 & 1 & 0 & | & \frac{2}{3} \\ 0 & 0 & 1 & | & \frac{1}{3} \end{bmatrix}$

$x = -1, \ y = \dfrac{2}{3}, \ z = \dfrac{1}{3}$

6. $\begin{bmatrix} \underline{3} & 6 & -9 & | & 1 \\ 2 & 4 & -6 & | & 1 \\ 3 & 4 & 5 & | & 0 \end{bmatrix}$

$\begin{bmatrix} 1 & 2 & -3 & | & \frac{1}{3} \\ 0 & 0 & 0 & | & \frac{1}{3} \\ 0 & -2 & 14 & | & -1 \end{bmatrix}$

$\begin{bmatrix} 1 & 2 & -3 & | & \frac{1}{3} \\ 0 & \underline{-2} & 14 & | & -1 \\ 0 & 0 & 0 & | & \frac{1}{3} \end{bmatrix}$

$\begin{bmatrix} 1 & 0 & 11 & | & -\frac{2}{3} \\ 0 & 1 & -7 & | & \frac{1}{2} \\ 0 & 0 & 0 & | & \frac{1}{3} \end{bmatrix}$

No solution

7. $\begin{bmatrix} \underline{1} & 2 & -5 & 3 & | & 16 \\ -5 & -7 & 13 & -9 & | & -50 \\ -1 & 1 & -7 & 2 & | & 9 \\ 3 & 4 & -7 & 6 & | & 33 \end{bmatrix}$

$\begin{bmatrix} 1 & 2 & -5 & 3 & | & 16 \\ 0 & \underline{3} & -12 & 6 & | & 30 \\ 0 & 3 & -12 & 5 & | & 25 \\ 0 & -2 & 8 & -3 & | & -15 \end{bmatrix}$

$\begin{bmatrix} 1 & 0 & 3 & -1 & | & -4 \\ 0 & 1 & -4 & 2 & | & 10 \\ 0 & 0 & 0 & \underline{-1} & | & -5 \\ 0 & 0 & 0 & 1 & | & 5 \end{bmatrix}$

$\begin{bmatrix} 1 & 0 & 3 & 0 & | & 1 \\ 0 & 1 & -4 & 0 & | & 0 \\ 0 & 0 & 0 & 1 & | & 5 \\ 0 & 0 & 0 & 0 & | & 0 \end{bmatrix}$

$\begin{cases} x + 3z = 1 \\ y - 4z = 0 \\ \quad\quad w = 5 \\ \quad\quad 0 = 0 \end{cases}$

$z = $ any value, $x = 1 - 3z$, $y = 4z$, $w = 5$

8. $\begin{bmatrix} 5 & -10 & | & 5 \\ 3 & -8 & | & -3 \\ -3 & 7 & | & 0 \end{bmatrix}$

$\begin{bmatrix} 1 & -2 & | & 1 \\ 0 & -2 & | & -6 \\ 0 & 1 & | & 3 \end{bmatrix}$

$\begin{bmatrix} 1 & 0 & | & 7 \\ 0 & 1 & | & 3 \\ 0 & 0 & | & 0 \end{bmatrix}$

$x = 7, y = 3$

9. $\begin{bmatrix} 2 + 3 \\ -1 + 4 \\ 0 + 7 \end{bmatrix} = \begin{bmatrix} 5 \\ 3 \\ 7 \end{bmatrix}$

10. $\begin{bmatrix} 1\cdot3+3\cdot1+(-2)0 & 1\cdot5+3\cdot0+(-2)(-6) \\ 4\cdot3+0\cdot1+(-1)0 & 4\cdot5+0\cdot0+(-1)(-6) \end{bmatrix}$

$= \begin{bmatrix} 6 & 17 \\ 12 & 26 \end{bmatrix}$

11. $\begin{bmatrix} 3 & 2 \\ 5 & 4 \end{bmatrix}^{-1} = \begin{bmatrix} \frac{4}{2} & -\frac{2}{2} \\ -\frac{5}{2} & \frac{3}{2} \end{bmatrix} = \begin{bmatrix} 2 & -1 \\ -\frac{5}{2} & \frac{3}{2} \end{bmatrix}$

$\begin{bmatrix} 2 & -1 \\ -\frac{5}{2} & \frac{3}{2} \end{bmatrix}\begin{bmatrix} 0 \\ 2 \end{bmatrix} = \begin{bmatrix} -2 \\ 3 \end{bmatrix}$

$x = -2, y = 3$

12. a. $\begin{bmatrix} 4 & -2 & 3 \\ 8 & -3 & 5 \\ 7 & -2 & 4 \end{bmatrix}\begin{bmatrix} 1 \\ 0 \\ 3 \end{bmatrix} = \begin{bmatrix} 13 \\ 23 \\ 19 \end{bmatrix}$

$x = 13, y = 23, z = 19$

b. $\begin{bmatrix} -2 & 2 & -1 \\ 3 & -5 & 4 \\ 5 & -6 & 4 \end{bmatrix}\begin{bmatrix} 0 \\ -1 \\ 2 \end{bmatrix} = \begin{bmatrix} -4 \\ 13 \\ 14 \end{bmatrix}$

$x = -4, y = 13, z = 14$

13. $\begin{bmatrix} \underline{2} & 6 & | & 1 & 0 \\ 1 & 2 & | & 0 & 1 \end{bmatrix}$

$\begin{bmatrix} 1 & 3 & | & \frac{1}{2} & 0 \\ 0 & \underline{-1} & | & -\frac{1}{2} & 1 \end{bmatrix}$

$\begin{bmatrix} 1 & 0 & | & -1 & 3 \\ 0 & 1 & | & \frac{1}{2} & -1 \end{bmatrix}$

$\begin{bmatrix} -1 & 3 \\ \frac{1}{2} & -1 \end{bmatrix}$

14. $\begin{bmatrix} \underline{1} & 1 & 1 & | & 1 & 0 & 0 \\ 3 & 4 & 3 & | & 0 & 1 & 0 \\ 1 & 1 & 2 & | & 0 & 0 & 1 \end{bmatrix}$

$\begin{bmatrix} 1 & 1 & 1 & | & 1 & 0 & 0 \\ 0 & \underline{1} & 0 & | & -3 & 1 & 0 \\ 0 & 0 & 1 & | & -1 & 0 & 1 \end{bmatrix}$

$\begin{bmatrix} 1 & 0 & 1 & | & 4 & -1 & 0 \\ 0 & 1 & 0 & | & -3 & 1 & 0 \\ 0 & 0 & \underline{1} & | & -1 & 0 & 1 \end{bmatrix}$

$\begin{bmatrix} 1 & 0 & 0 & | & 5 & -1 & -1 \\ 0 & 1 & 0 & | & -3 & 1 & 0 \\ 0 & 0 & 1 & | & -1 & 0 & 1 \end{bmatrix}$

$\begin{bmatrix} 5 & -1 & -1 \\ -3 & 1 & 0 \\ -1 & 0 & 1 \end{bmatrix}$

15. $\begin{cases} c + w + s = 1000 \\ 206c + 85w + 97s = 151{,}500 \\ c - w - s = 0 \end{cases}$

$\begin{bmatrix} 1 & 1 & 1 \\ 206 & 85 & 97 \\ 1 & -1 & -1 \end{bmatrix}^{-1}\begin{bmatrix} 1000 \\ 151{,}500 \\ 0 \end{bmatrix} = \begin{bmatrix} 500 \\ 0 \\ 500 \end{bmatrix}$

Corn: 500 acres, wheat: 0 acres, soybeans: 500 acres

16. a. $AC = \begin{bmatrix} 15 & 20 & 8 \\ 10 & 17 & 12 \end{bmatrix}\begin{bmatrix} 165 \\ 65 \\ 210 \end{bmatrix} = \begin{bmatrix} 5455 \\ 5275 \end{bmatrix}$

The total cost to make the equipment in the first store was $5455 and in the second store $5275.

b. $AS = \begin{bmatrix} 15 & 20 & 8 \\ 10 & 17 & 12 \end{bmatrix}\begin{bmatrix} 200 \\ 80 \\ 250 \end{bmatrix} = \begin{bmatrix} 6600 \\ 6360 \end{bmatrix}$

The total revenue of the equipment in the first store was $6600 and in the second store $6360.

c. $S - C = \begin{bmatrix} 200 \\ 80 \\ 250 \end{bmatrix} - \begin{bmatrix} 165 \\ 65 \\ 210 \end{bmatrix} = \begin{bmatrix} 35 \\ 15 \\ 40 \end{bmatrix}$

The entries represent the profit per unit of each item.

d. $A(S - C) = \begin{bmatrix} 15 & 20 & 8 \\ 10 & 17 & 12 \end{bmatrix}\begin{bmatrix} 35 \\ 15 \\ 40 \end{bmatrix} = \begin{bmatrix} 1145 \\ 1085 \end{bmatrix}$

The total profit for the first store was $1145 and for the second store $1085.

17. a. $BA = \begin{bmatrix} 5000 & 8000 & 10{,}000 \end{bmatrix} \begin{bmatrix} 0.50 & 0.43 & 0.07 \\ 0.45 & 0.26 & 0.29 \\ 0.40 & 0.40 & 0.20 \end{bmatrix} = \begin{bmatrix} 10{,}100 & 8230 & 4670 \end{bmatrix}$

   Total amount invested in bonds, stocks , and the conservative fixed income fund, respectively.

   b. $BC = \begin{bmatrix} 5000 & 8000 & 10{,}000 \end{bmatrix} \begin{bmatrix} 0.0032 & 0.1119 \\ 0.0233 & 0.0976 \\ 0.0320 & 0.0467 \end{bmatrix} = \begin{bmatrix} 522.40 & 1807.30 \end{bmatrix}$

   total return on the investments for one year and five years, respectively.

   c. $8230 is the total amount invested in stocks.

   d. $522.40 is the total return after one year.

18. a. $AB = \begin{bmatrix} 11 & 7 & 12 \\ 9 & 5 & 16 \\ 13 & 8 & 9 \\ 13 & 7 & 10 \end{bmatrix} \begin{bmatrix} 8 \\ 6 \\ 9 \end{bmatrix} = \begin{bmatrix} 238 \\ 246 \\ 233 \\ 236 \end{bmatrix} \begin{matrix} Sara \\ Quinn \\ Tamia \\ Zack \end{matrix}$

   total amount earned by each person for the week.

   b. Quinn earned the most and Tamia earned the least.

   c. $AB = \begin{bmatrix} 11 & 7 & 12 \\ 9 & 5 & 16 \\ 13 & 8 & 9 \\ 13 & 7 & 10 \end{bmatrix} \begin{bmatrix} 9 \\ 6 \\ 8 \end{bmatrix} = \begin{bmatrix} 237 \\ 239 \\ 237 \\ 239 \end{bmatrix} \begin{matrix} Sara \\ Quinn \\ Tamia \\ Zack \end{matrix}$

   Quinn and Zack both earn $239.

   d. Sara worked 11 hours in concessions, 7 hours at the front desk, and 12 hours cleaning for a total of 30 hours.

19. Let x = number of apples
    y = number of bananas
    z = number of oranges

$$x + y + z = 18$$
$$0.85x + 0.20y + 0.76z = 9$$
$$-x + y - z = 0$$

$\begin{bmatrix} 1 & 1 & 1 & | & 18 \\ 0.85 & 0.20 & 0.76 & | & 9 \\ -1 & 1 & -1 & | & 0 \end{bmatrix}$

$\xrightarrow{R_3 + R_1} \begin{bmatrix} 1 & 1 & 1 & | & 18 \\ 0.85 & 0.20 & 0.76 & | & 9 \\ 0 & 2 & 0 & | & 18 \end{bmatrix}$

$\xrightarrow{R_2 + (-0.85)R_1} \begin{bmatrix} 1 & 1 & 1 & | & 18 \\ 0 & -0.65 & -0.09 & | & -6.3 \\ 0 & 2 & 0 & | & 18 \end{bmatrix}$

$\xrightarrow{R_2 \leftrightarrow R_3} \begin{bmatrix} 1 & 1 & 1 & | & 18 \\ 0 & 2 & 0 & | & 18 \\ 0 & -0.65 & -0.09 & | & -6.3 \end{bmatrix}$

$\xrightarrow{\frac{1}{2}R_2} \begin{bmatrix} 1 & 1 & 1 & | & 18 \\ 0 & 1 & 0 & | & 9 \\ 0 & -0.65 & -0.09 & | & -6.3 \end{bmatrix}$

$\xrightarrow{R_1 + (-1)R_2} \begin{bmatrix} 1 & 0 & 1 & | & 9 \\ 0 & 1 & 0 & | & 9 \\ 0 & -0.65 & -0.09 & | & -6.3 \end{bmatrix}$

$$R_3 + (0.65)R_2 \longrightarrow \begin{bmatrix} 1 & 0 & 1 & | & 9 \\ 0 & 1 & 0 & | & 9 \\ 0 & 0 & -0.09 & | & -0.45 \end{bmatrix}$$

$$-\frac{1}{0.09}R_3 \longrightarrow \begin{bmatrix} 1 & 0 & 1 & | & 9 \\ 0 & 1 & 0 & | & 9 \\ 0 & 0 & 1 & | & 5 \end{bmatrix}$$

$$R_1 + (-1)R_3 \longrightarrow \begin{bmatrix} 1 & 0 & 0 & | & 4 \\ 0 & 1 & 0 & | & 9 \\ 0 & 0 & 1 & | & 5 \end{bmatrix}$$

4 apples, 9 bananas, 5 oranges

20. $x = A$'s current stockpile,
$y = B$'s current stockpile,
$a = A$'s next-year's stockpile,
$b = B$'s next-year's stockpile
$.8x + .2y = a$
$.1x + .9y = b$

a. Next year: $\begin{bmatrix} .8 & .2 \\ .1 & .9 \end{bmatrix} \begin{bmatrix} 10{,}000 \\ 7000 \end{bmatrix} = \begin{bmatrix} 9400 \\ 7300 \end{bmatrix}$

Two years: $\begin{bmatrix} .8 & .2 \\ .1 & .9 \end{bmatrix}^2 \begin{bmatrix} 10{,}000 \\ 7000 \end{bmatrix} = \begin{bmatrix} 8980 \\ 7510 \end{bmatrix}$

A: 9400, 8980; B: 7300, 7510

b. Previous year:
$\begin{bmatrix} .8 & .2 \\ .1 & .9 \end{bmatrix}^{-1} \begin{bmatrix} 10{,}000 \\ 7000 \end{bmatrix} \approx \begin{bmatrix} 10{,}857 \\ 6571 \end{bmatrix}$

Two years ago:
$\left(\begin{bmatrix} .8 & .2 \\ .1 & .9 \end{bmatrix}^{-1} \right)^2 \begin{bmatrix} 10{,}000 \\ 7000 \end{bmatrix} \approx \begin{bmatrix} 12{,}082 \\ 5959 \end{bmatrix}$

A: 10,857, 12,082; B: 6571, 5959

c. $a - b = (.8x + .2y) - (.1x + .9y)$
$\quad = .7x - .7y = .7(x - y)$
Thus, the "missile gap" of the following year is 70% of the previous, so the "missile gap" decreases 30% each year.
$a + b = .9x + 1.1y < x + y$
when $.1y < .1x$ or $y < x$.
$a + b = .9x + 1.1y > x + y$ when $y > x$.

21. $\left(\begin{bmatrix} 1 & 0 \\ 0 & 1 \end{bmatrix} - \begin{bmatrix} .4 & .2 \\ .1 & .3 \end{bmatrix} \right)^{-1} \begin{bmatrix} 8 \\ 12 \end{bmatrix} = \begin{bmatrix} 20 \\ 20 \end{bmatrix}$

Industry I: 20; industry II: 20

22. Let x = number of nickels
$\quad y$ = number of dimes
$\quad z$ = number of quarters

$\begin{cases} x + y + z = 30 \\ .05x + .10y + .25z = 3.30 \\ y = 5z \end{cases}$

$$\begin{bmatrix} \underline{1} & 1 & 1 & | & 30 \\ .05 & .10 & .25 & | & 3.30 \\ 0 & 1 & -5 & | & 0 \end{bmatrix}$$

$$\begin{bmatrix} 1 & 1 & 1 & | & 30 \\ 0 & \underline{.05} & .20 & | & 1.8 \\ 0 & 1 & -5 & | & 0 \end{bmatrix}$$

$$\begin{bmatrix} 1 & 1 & 1 & | & 30 \\ 0 & \underline{1} & 4 & | & 36 \\ 0 & 1 & -5 & | & 0 \end{bmatrix}$$

$$\begin{bmatrix} 1 & 0 & -3 & | & -6 \\ 0 & 1 & 4 & | & 36 \\ 0 & 0 & \underline{-9} & | & -36 \end{bmatrix}$$

$$\begin{bmatrix} 1 & 0 & -3 & | & -6 \\ 0 & 1 & 4 & | & 36 \\ 0 & 0 & 1 & | & 4 \end{bmatrix}$$

$$\begin{bmatrix} 1 & 0 & 0 & | & 6 \\ 0 & 1 & 0 & | & 20 \\ 0 & 0 & 1 & | & 4 \end{bmatrix}$$

Joe has 4 quarters.
Answer (a) is correct.

23. a. True; a system of equations has no solution, exactly one solution, or infinitely many solutions.

b. False; a system of equations could have two or more equations that are multiples of each other.

c. True; At least one variable must be dependent on another when there are less equations than variables.

24. One possible system with infinitely many

solutions is: $\begin{cases} 2x + 3y = 4 \\ 4x + 6y = 8 \end{cases}$.

One possible system with no solution is:

$\begin{cases} 2x + 3y = 4 \\ 2x + 3y = 5 \end{cases}$.

25. No; Since matrix multiplication involves adding as well as multiplying, it is possible that the matrices have the correct positive and negative numbers to have the zero matrix.

26. If the matrix has no inverse, there will be a row of zeros in the resulting Gauss-Jordan matrix.

27. The column values should add to less than one in an input-output matrix because they reflect the amount of input required from each industry and all the industries are dependent on each other.

Chapter 2 Chapter Test

1. $\begin{bmatrix} 1 & 2 & 1 & | & 5 \\ 2 & -1 & 1 & | & -5 \\ -3 & 1 & -2 & | & 8 \end{bmatrix}$

$\begin{bmatrix} 1 & 2 & 1 & | & 5 \\ 0 & -5 & -1 & | & -15 \\ 0 & 7 & 1 & | & 23 \end{bmatrix}$

$\begin{bmatrix} 1 & 2 & 1 & | & 5 \\ 0 & 1 & \frac{1}{5} & | & 3 \\ 0 & 7 & 1 & | & 23 \end{bmatrix}$

$\begin{bmatrix} 1 & 0 & \frac{3}{5} & | & -1 \\ 0 & 1 & \frac{1}{5} & | & 3 \\ 0 & 0 & -\frac{2}{5} & | & 2 \end{bmatrix}$

$\begin{bmatrix} 1 & 0 & \frac{3}{5} & | & -1 \\ 0 & 1 & \frac{1}{5} & | & 3 \\ 0 & 0 & 1 & | & -5 \end{bmatrix}$

$\begin{bmatrix} 1 & 0 & 0 & | & 2 \\ 0 & 1 & 0 & | & 4 \\ 0 & 0 & 1 & | & -5 \end{bmatrix}$

$x = 2, y = 4, z = -5$

2. a. $x = 4, y = -3, z = 6$

b. $x = 2, y = 3, z = 5$

c. No solution

d. $z = $ any value
$x = z + 2, y = -2z + 2$

e. $y = $ any value
$x = -2y, z = 0$

3. Answers may vary.
Sample answer:
$w = 1, z = 2, x = 3, y = 5$

4. Answers may vary.
Sample answer:
(1) $z = 0, y = 6, x = 9$
(2) $z = 1, y = 2, x = 15$
(3) $z = -1, y = 10, x = 3$

5. $A + B$ is not defined.

$A + C = \begin{bmatrix} 2 & 1 & 0 \\ 3 & 2 & 1 \end{bmatrix} + \begin{bmatrix} 0 & 1 & 2 \\ -1 & 1 & 1 \end{bmatrix} = \begin{bmatrix} 2 & 2 & 2 \\ 2 & 3 & 2 \end{bmatrix}$

$AB = \begin{bmatrix} 2 & 1 & 0 \\ 3 & 2 & 1 \end{bmatrix} \begin{bmatrix} -1 & 1 \\ 0 & 1 \\ 1 & -1 \end{bmatrix} = \begin{bmatrix} -2 & 3 \\ -2 & 4 \end{bmatrix}$

AC is not defined.

$BC = \begin{bmatrix} -1 & 1 \\ 0 & 1 \\ 1 & -1 \end{bmatrix} \begin{bmatrix} 0 & 1 & 2 \\ -1 & 1 & 1 \end{bmatrix} = \begin{bmatrix} -1 & 0 & -1 \\ -1 & 1 & 1 \\ 1 & 0 & 1 \end{bmatrix}$

6. a. $.5x + y + z = m$
$3x + 2.5y + 2z = v$
$4x + 3y + 2z = p$

b. $\begin{bmatrix} .5 & 1 & 1 \\ 3 & 2.5 & 2 \\ 4 & 3 & 2 \end{bmatrix} \cdot \begin{bmatrix} x \\ y \\ z \end{bmatrix} = \begin{bmatrix} m \\ v \\ p \end{bmatrix}$

c. $\begin{bmatrix} m \\ v \\ p \end{bmatrix} = \begin{bmatrix} .5 & 1 & 1 \\ 3 & 2.5 & 2 \\ 4 & 3 & 2 \end{bmatrix} \begin{bmatrix} 100 \\ 300 \\ 200 \end{bmatrix} = \begin{bmatrix} 550 \\ 1450 \\ 1700 \end{bmatrix}$

550 hours molding time, 1450 hours oven time, 1700 hours painting time

7. $\begin{bmatrix} \underline{1} & 2 & 1 & | & 1 & 0 & 0 \\ 0 & 1 & 1 & | & 0 & 1 & 0 \\ 1 & -1 & 0 & | & 0 & 0 & 1 \end{bmatrix}$

$\begin{bmatrix} 1 & 2 & 1 & | & 1 & 0 & 0 \\ 0 & \underline{1} & 1 & | & 0 & 1 & 0 \\ 0 & -3 & -1 & | & -1 & 0 & 1 \end{bmatrix}$

$\begin{bmatrix} 1 & 0 & -1 & | & 1 & -2 & 0 \\ 0 & 1 & 1 & | & 0 & 1 & 0 \\ 0 & 0 & \underline{2} & | & -1 & 3 & 1 \end{bmatrix}$

$\begin{bmatrix} 1 & 0 & -1 & | & 1 & -2 & 0 \\ 0 & 1 & 1 & | & 0 & 1 & 0 \\ 0 & 0 & \underline{1} & | & -\frac{1}{2} & \frac{3}{2} & \frac{1}{2} \end{bmatrix}$

$\begin{bmatrix} 1 & 0 & 0 & | & \frac{1}{2} & -\frac{1}{2} & \frac{1}{2} \\ 0 & 1 & 0 & | & \frac{1}{2} & -\frac{1}{2} & -\frac{1}{2} \\ 0 & 0 & 1 & | & -\frac{1}{2} & \frac{3}{2} & \frac{1}{2} \end{bmatrix}$

$A^{-1} = \begin{bmatrix} \frac{1}{2} & -\frac{1}{2} & \frac{1}{2} \\ \frac{1}{2} & -\frac{1}{2} & -\frac{1}{2} \\ -\frac{1}{2} & \frac{3}{2} & \frac{1}{2} \end{bmatrix}$

8. Let x = number of out-of-state students
y = number of in-state students

$\begin{cases} x + y = 1500 \\ 10,000x + 4500y = 12,800,000 \end{cases}$

$\begin{bmatrix} \underline{1} & 1 & | & 1500 \\ 10,000 & 4500 & | & 12,800,000 \end{bmatrix}$

$\begin{bmatrix} 1 & 1 & | & 1500 \\ 0 & \underline{-5500} & | & -2,200,000 \end{bmatrix}$

$\begin{bmatrix} 1 & 1 & | & 1500 \\ 0 & \underline{1} & | & 400 \end{bmatrix}$

$\begin{bmatrix} 1 & 0 & | & 1100 \\ 0 & 1 & | & 400 \end{bmatrix}$

400 students

9.

	Wood	Steel	Coal
Wood	.30	0	.10
Steel	.20	.30	.20
Coal	.10	.20	.05

$= A$

$D = \begin{bmatrix} 1 \\ 4 \\ 2 \end{bmatrix}$

$X = (I - A)^{-1}D$

$= \begin{bmatrix} .70 & 0 & -.10 \\ -.20 & .70 & -.20 \\ -.10 & -.20 & .95 \end{bmatrix} \begin{bmatrix} 1 \\ 4 \\ 2 \end{bmatrix}$

$= \begin{bmatrix} 1.47 & .05 & .16 \\ .49 & 1.54 & .38 \\ .26 & .33 & 1.15 \end{bmatrix} \begin{bmatrix} 1 \\ 4 \\ 2 \end{bmatrix}$

$= \begin{bmatrix} 1.98 \\ 7.39 \\ 3.87 \end{bmatrix}$

wood: $1.98, steel: $7.39, coal: $3.87

Chapter 3

Exercises 3.1

1. (8, 7)

$$\begin{cases} 6(8) + 3(7) \le 96 \\ 8 + 7 \le 18 \\ 2(8) + 6(7) \le 72 \\ 8 \ge 0, \ 7 \ge 0 \end{cases}$$

$$\begin{cases} 69 \le 96 & \text{true} \\ 15 \le 18 & \text{true} \\ 58 \le 72 & \text{true} \\ 8 \ge 0, \ 7 \ge 0 & \text{true} \end{cases}$$

Yes

3. (9, 10)

$$\begin{cases} 6(9) + 3(10) \le 96 \\ 9 + 10 \le 18 \\ 2(9) + 6(10) \le 72 \\ 9 \ge 0, \ 3 \ge 0 \end{cases}$$

$$\begin{cases} 84 \le 96 & \text{true} \\ 19 \le 18 & \text{false} \\ 78 \le 72 & \text{false} \\ 9 \ge 0, \ 3 \ge 0 & \text{true} \end{cases}$$

No

5. a.

	A	B	*Truck capacity*
Volume	4 cubic feet	3 cubic feet	300 cubic feet
Weight	100 pounds	200 pounds	10,000 pounds
Earnings	$13	$9	

b. Volume: $4x + 3y \le 300$
Weight: $100x + 200y \le 10{,}000$

c. $y \le 2x$, $x \ge 0$, $y \ge 0$

d. $13x + 9y$

e. In standard form, the inequalities from (b) and (c) are:

$$\begin{cases} y \le -\dfrac{4}{3}x + 100 \\ y \le -\dfrac{1}{2}x + 50 \\ y \le 2x \\ x \ge 0, \ y \ge 0 \end{cases}$$

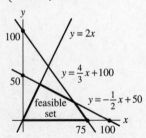

3-1

7. a.

	Essay questions	*Short-answer Questions*	*Available*
Time to answer	10 minutes	2 minutes	90 minutes
Quantity	10	50	
Required	3	10	
Worth	20 points	5 points	

b. $10x + 2y \leq 90$

c. $3 \leq x \leq 10, \ 10 \leq y \leq 50$

d. $20x + 5y$

e.   In standard form, the inequality from (b) is $y \leq -5x + 45$. Graph the system:
$$\begin{cases} y \leq -5x + 45 \\ 3 \leq x \leq 10 \\ 10 \leq y \leq 50 \end{cases}$$

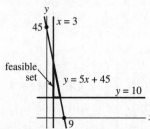

Note that the conditions $x \leq 10$ and $y \leq 50$ are superfluous because they are automatically assured if the other inequalities hold.

9. a.

	Alfalfa	*Corn*	*Requirements*
Protein	.13 pound	.065 pound	4550 pounds
TDN	.48 pound	.96 pound	26,880 pounds
Vitamin A	2.16 IUs	0 IUs	43,200 IUs
Cost/lb	$.01	$.016	

b.   Protein: $.13x + .065y \geq 4550$
TDN: $.48x + .96y \geq 26{,}880$
Vitamin A: $2.16x \geq 43{,}200$
Other: $y \geq 0$
(The condition $x \geq 0$ is unnecessary because it is automatically assured if the inequality for Vitamin A holds.)

c. In standard form, the inequalities are:

$$\begin{cases} y \geq -2x + 70{,}000 \\ y \geq -.5x + 28{,}000 \\ x \geq 20{,}000 \\ y \geq 0 \end{cases}$$

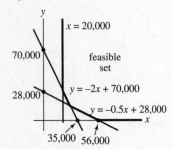

d. $.01x + .016y$

Exercises 3.2

1.

Vertex	$4x + 3y$
(0, 0)	$4(0) + 3(0) = 0$
(0, 20)	$4(0) + 3(20) = 60$
(20, 0)	$4(20) + 3(0) = 80$

The objective function is maximized at (20, 0).

3. Find the vertices.

Lower left corner: (0, 0)

y-intercept of $y = -\dfrac{1}{2}x + 4$: (0, 4)

Intersection of $y = -\dfrac{1}{2}x + 4$ and $y = -x + 6$:

$$-\frac{1}{2}x + 4 = -x + 6$$

$$\frac{1}{2}x = 2$$

$$x = 4$$

$$y = -\frac{1}{2}x + 4 = -\frac{1}{2}(4) + 4 = 2$$

(4, 2)

x-intercept of $y = -x + 6$: (6, 0)

Vertex	$4x + 3y$
(0, 0)	$4(0) + 3(0) = 0$
(0, 4)	$4(0) + 3(4) = 12$
(4, 2)	$4(4) + 3(2) = 22$
(6, 0)	$4(6) + 3(0) = 24$

The objective function is maximized at (6, 0).

5.

Vertex	$x + 2y$
(0, 0)	$0 + 2(0) = 0$
(0, 5)	$0 + 2(5) = 10$
(3, 3)	$3 + 2(3) = 9$
(4, 0)	$4 + 2(0) = 4$

The objective function is maximized at (0, 5).

7.

Vertex	$2x + y$
(0, 0)	$2(0) + 0 = 0$
(0, 5)	$2(0) + 5 = 5$
(3, 3)	$2(3) + 3 = 9$
(4, 0)	$2(4) + 0 = 8$

The objective function is maximized at (3, 3).

9.

Vertex	$8x + y$
$(0, 7)$	$8(0) + 7 = 7$
$(1, 2)$	$8(1) + 2 = 10$
$(2, 1)$	$8(2) + 1 = 17$
$(6, 0)$	$8(6) + 0 = 48$

The objective function is minimized at $(0, 7)$.

11.

Vertex	$2x + 3y$
$(0, 7)$	$2(0) + 3(7) = 21$
$(1, 2)$	$2(1) + 3(2) = 8$
$(2, 1)$	$2(2) + 3(1) = 7$
$(6, 0)$	$2(6) + 3(0) = 12$

The objective function is minimized at $(2, 1)$.

13. Find the vertices.

Lower left corner: $(0, 0)$

Intersection of $y = 2x$ and $y = -\dfrac{1}{2}x + 50$:

$$2x = -\frac{1}{2}x + 50$$

$$\frac{5}{2}x = 50$$

$$x = 20$$

$y = 2x = 2(20) = 40$

$(20, 40)$

Intersection of $y = -\dfrac{1}{2}x + 50$ and

$y = -\dfrac{4}{3}x + 100$:

$$-\frac{1}{2}x + 50 = -\frac{4}{3}x + 100$$

$$\frac{5}{6}x = 50$$

$$x = 60$$

$y = -\dfrac{1}{2}x + 50 = -\dfrac{1}{2}(60) + 50 = 20$

$(60, 20)$

x-intercept of $y = -\dfrac{4}{3}x + 100$: $(75, 0)$

Vertex	*Earnings* $= 13x + 9y$
$(0, 0)$	$13(0) + 9(0) = 0$
$(20, 40)$	$13(20) + 9(40) = 620$
$(60, 20)$	$13(60) + 9(20) = 960$
$(75, 0)$	$13(75) + 9(0) = 975$

The earnings are maximized at $(75, 0)$. Ship 75 crates of cargo A and no crates of cargo B.

15.

Vertex	*Score* $= 20x + 5y$
$(3, 10)$	$20(3) + 5(10) = 110$
$(3, 30)$	$20(3) + 5(30) = 210$
$(7, 10)$	$20(7) + 5(10) = 190$

The score is maximized at $(3, 30)$. Answer 3 essay questions and 30 short-answer questions.

17. Find the vertices.

x-intercept of $y = -.5x + 28{,}000$: $(56{,}000, 0)$

Intersection of $y = -.5x + 28{,}000$ and $y = -2x + 70{,}000$:

$$-.5x + 28{,}000 = -2x + 70{,}000$$

$$1.5x = 42{,}000$$

$$x = 28{,}000$$

$y = -2x + 70{,}000 = -2(28{,}000) + 70{,}000 = 14{,}000$

$(28{,}000, 14{,}000)$

Intersection of $y = -2x + 70{,}000$ and $x = 20{,}000$:

$y = -2x + 70{,}000 = -2(20{,}000) + 70{,}000 = 30{,}000$

$(20{,}000, 30{,}000)$

Vertex	*cost* $= .01x + .016y$
$(56{,}000, 0)$	$.01(56{,}000) + .016(0) = 560$
$(28{,}000, 14{,}000)$	$.01(28{,}000) + .016(14{,}000) = 504$
$(20{,}000, 30{,}000)$	$.01(20{,}000) + .016(30{,}000) = 680$

The cost is minimized at $(28{,}000, 14{,}000)$. Buy 28,000 pounds of alfalfa and 14,000 pounds of corn.

19.

Vertex	Profit $= 150x + 70y$
$(0, 0)$	$150(0) + 70(0) = 0$
$(0, 12)$	$150(0) + 70(12) = 840$
$(9, 9)$	$150(9) + 70(9) = 1980$
$(14, 4)$	$150(14) + 70(4) = 2380$
$(16, 0)$	$150(16) + 70(0) = 2400$

The profit is maximized at $(16, 0)$.
Make 16 chairs and no sofas.

21. In standard form the inequalities are:

$$\begin{cases} y \geq -2x + 10 \\ y \geq -\dfrac{1}{2}x + 7 \\ x \geq 0, \ y \geq 0 \end{cases}$$

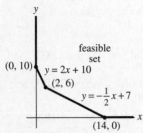

Find the vertices.
y-intercept of $y = -2x + 10$: $(0, 10)$

Intersection of $y = -2x + 10$ and $y = -\dfrac{1}{2}x + 7$:

$$-\frac{1}{2}x + 7 = -2x + 10$$

$$\frac{3}{2}x = 3$$

$$x = 2$$

$y = -2x + 10 = -2(2) + 10 = 6$
$(2, 6)$

x-intercept of $y = -\dfrac{1}{2}x + 7$: $(14, 0)$

Vertex	$3x + 4y$
$(0, 10)$	$3(0) + 4(10) = 40$
$(2, 6)$	$3(2) + 4(6) = 30$
$(14, 0)$	$3(14) + 4(0) = 42$

The minimum value is 30 and occurs at $(2, 6)$.

23. In standard form the inequalities are:

$$\begin{cases} y \leq -\dfrac{1}{2}x + 10 \\ y \geq -\dfrac{3}{2}x + 12 \\ x \leq 6 \\ x \geq 0, \ y \geq 0 \end{cases}$$

Find the vertices.

Intersection of $y = -\dfrac{1}{2}x + 10$ and

$y = -\dfrac{3}{2}x + 12$:

$$-\frac{1}{2}x + 10 = -\frac{3}{2}x + 12$$

$$x = 2$$

$$y = -\frac{1}{2}x + 10 = -\frac{1}{2}(2) + 10 = 9$$

$(2, 9)$

Intersection of $y = -\dfrac{1}{2}x + 10$ and $x = 6$:

$x = 6$

$$y = -\frac{1}{2}x + 10 = -\frac{1}{2}(6) + 10 = 7$$

$(6, 7)$

Intersection of $y = -\dfrac{3}{2}x + 12$ and $x = 6$:

$x = 6$

$$y = -\frac{3}{2}x + 12 = -\frac{3}{2}(6) + 12 = 3$$

$(6, 3)$

Vertex	$2x + 5y$
$(2, 9)$	$2(2) + 5(9) = 49$
$(6, 7)$	$2(6) + 5(7) = 47$
$(6, 3)$	$2(6) + 5(3) = 27$

The maximum value is 49 and occurs at $(2, 9)$.

25. In standard form the inequalities are:

$$\begin{cases} y \le -\dfrac{1}{3}x + 40 \\[2mm] y \le -\dfrac{7}{2}x + 78 \\[2mm] x \le 20 \\[1mm] x \ge 0, \ y \ge 0 \end{cases}$$

Find the vertices.

Lower left corner: (0, 0)

y-intercept of $y = -\dfrac{1}{3}x + 40$: (0, 40)

Intersection of $y = -\dfrac{1}{3}x + 40$ and

$y = -\dfrac{7}{2}x + 78$:

$$-\frac{1}{3}x + 40 = -\frac{7}{2}x + 78$$

$$\frac{19}{6}x = 38$$

$$x = 12$$

$$y = -\frac{1}{3}x + 40 = -\frac{1}{3}(12) + 40 = 36$$

(12, 36)

Intersection of $y = -\dfrac{7}{2}x + 78$ and $x = 20$:

$x = 20$

$$y = -\frac{7}{2}x + 78 = -\frac{7}{2}(20) + 78 = 8$$

(20, 8)

Intersection of $x = 20$ and $y = 0$: (20, 0)

Vertex	$100x + 150y$
(0, 0)	$100(0) + 150(0) = 0$
(0, 40)	$100(0) + 150(40) = 6000$
(12, 36)	$100(12) + 150(36) = 6600$
(20, 8)	$100(20) + 150(8) = 3200$
(20, 0)	$100(20) + 150(0) = 2000$

The maximum value is 6600 and occurs at (12, 36).

27. The inequalities are given in standard form.

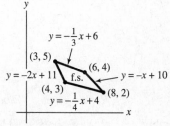

Find the vertices.

Intersection of $y = -2x + 11$ and $y = -\dfrac{1}{3}x + 6$:

$$-\frac{1}{3}x + 6 = -2x + 11$$

$$\frac{5}{3}x = 5$$

$$x = 3$$

$$y = -\frac{1}{3}x + 6 = -\frac{1}{3}(3) + 6 = 5$$

(3, 5)

Intersection of $y = -\dfrac{1}{3}x + 6$ and

$y = -x + 10$:

$$-\frac{1}{3}x + 6 = -x + 10$$

$$\frac{2}{3}x = 4$$

$$x = 6$$

$$y = -x + 10 = -6 + 10 = 4$$

(6, 4)

Intersection of $y = -x + 10$ and $y = -\dfrac{1}{4}x + 4$:

$$-\frac{1}{4}x + 4 = -x + 10$$

$$\frac{3}{4}x = 6$$

$$x = 8$$

$$y = -x + 10 = -8 + 10 = 2$$

(8, 2)

Intersection of $y = -\frac{1}{4}x + 4$ and

$y = -2x + 11$:

$$-\frac{1}{4}x + 4 = -2x + 11$$

$$\frac{7}{4}x = 7$$

$$x = 4$$

$$y = -2x + 11 = -2(4) + 11 = 3$$
(4, 3)

Vertex	$7x + 4y$
(3, 5)	$7(3) + 4(5) = 41$
(6, 4)	$7(6) + 4(4) = 58$
(8, 2)	$7(8) + 4(2) = 64$
(4, 3)	$7(4) + 4(3) = 40$

The minimum value is 40 and occurs at (4, 3).

29.

	Hockey games	*Soccer games*	*Available*
Assembly	2 labor-hours	3 labor-hours	42 labor-hours
Testing	2 labor-hours	1 labor-hour	26 labor-hours

Let x be the number of hockey games produced each day, and let y be the number of soccer games produced each day.
Assembly: $2x + 3y \le 42$
Testing: $2x + y \le 26$
In standard form the equations are:

$$\begin{cases} y \le -\dfrac{2}{3}x + 14 \\ y \le -2x + 26 \\ x \ge 0,\ y \ge 0 \end{cases}$$

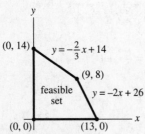

Find the vertices.
Lower left corner: (0, 0)

y-intercept of $y = -\dfrac{2}{3}x + 14$: (0, 14)

Intersection of $y = -\dfrac{2}{3}x + 14$ and $y = -2x + 26$:

$$-\frac{2}{3}x + 14 = -2x + 26$$

$$\frac{4}{3}x = 12$$

$$x = 9$$

$y = -2x + 26 = -2(9) + 26 = 8$

$(9, 8)$

x-intercept of $y = -2x + 26$: $(13, 0)$

The total daily output is simply the total number of games produced, or $x + y$.

Vertex	Output = x + y
(0, 0)	0 + 0 = 0
(0, 14)	0 + 14 = 14
(9, 8)	9 + 8 = 17
(13, 0)	13 + 0 = 13

The maximum output occurs at (9, 8). Produce 9 hockey games and 8 soccer games each day.

31.

	Food A	Food B	Requirement
Protein	4 units	3 units	42 units
Carbohydrates	2 units	6 units	30 units
Fat	2 units	1 unit	18 units
Weight	3 pounds	2 pounds	

Let x be the number of tubes of food A, and let y be the number of tubes of food B.

Protein: $4x + 3y \geq 42$

Carbohydrates: $2x + 6y \geq 30$

Fat: $2x + y \geq 18$

In standard form the inequalities are:

$$\begin{cases} y \geq -\dfrac{4}{3}x + 14 \\ y \geq -\dfrac{1}{3}x + 5 \\ y \geq -2x + 18 \\ x \geq 0, \ y \geq 0 \end{cases}$$

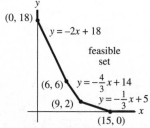

Find the vertices.

y-intercept of $y = -2x + 18$: $(0, 18)$

Intersection of $y = -2x + 18$ and $y = -\dfrac{4}{3}x + 14$:

$$-\dfrac{4}{3}x + 14 = -2x + 18$$

$$\dfrac{2}{3}x = 4$$

$$x = 6$$

$y = -2x + 18 = -2(6) + 18 = 6$

$(6, 6)$

Intersection of $y = -\dfrac{4}{3}x + 14$ and $y = -\dfrac{1}{3}x + 5$:

$$-\dfrac{1}{3}x + 5 = -\dfrac{4}{3}x + 14$$

$$x = 9$$

$$y = -\dfrac{1}{3}x + 5 = -\dfrac{1}{3}(9) + 5 = 2$$

$(9, 2)$

x-intercept of $y = -\dfrac{1}{3}x + 5$: $(15, 0)$

Vertex	Weight $= 3x + 2y$
$(0, 18)$	$3(0) + 2(18) = 36$
$(6, 6)$	$3(6) + 2(6) = 30$
$(9, 2)$	$3(9) + 2(2) = 31$
$(15, 0)$	$3(15) + 2(0) = 45$

The minimum weight is achieved at $(6, 6)$.
Send 6 tubes of food A and 6 tubes of food B.

33.

	Fruit Delight	*Heavenly Punch*	*Available*
Pineapple juice	10 ounces	10 ounces	9000 ounces
Orange juice	3 ounces	2 ounces	2400 ounces
Apricot juice	1 ounce	2 ounces	1400 ounces
Profit	$0.40	$0.60	

Let x be the number of cans of Fruit Delight, and let y be the number of cans of Heavenly Punch produced each week.

Pineapple: $10x + 10y \leq 9000$

Orange: $3x + 2y \leq 2400$

Apricot: $x + 2y \leq 1400$

In standard form the inequalities are:

$$\begin{cases} y \leq -x + 900 \\ y \leq -\dfrac{3}{2}x + 1200 \\ y \leq -\dfrac{1}{2}x + 700 \\ x \geq 0, \ y \geq 0 \end{cases}$$

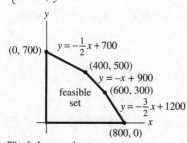

Find the vertices.

Lower left corner: $(0, 0)$

y-intercept of $y = -\dfrac{1}{2}x + 700$: $(0, 700)$

Intersection of $y = -\dfrac{1}{2}x + 700$ and $y = -x + 900$:

$$-\frac{1}{2}x + 700 = -x + 900$$

$$\frac{1}{2}x = 200$$

$$x = 400$$

$y = -x + 900 = -400 + 900 = 500$

$(400, 500)$

Intersection of $y = -x + 900$ and $y = -\dfrac{3}{2}x + 1200$:

$$-x + 900 = -\frac{3}{2}x + 1200$$

$$\frac{1}{2}x = 300$$

$$x = 600$$

$y = -x + 900 = -600 + 900 = 300$

(600, 300)

x-intercept of $y = -\frac{3}{2}x + 1200$: (800, 0)

Vertex	Profit = .4x + .6y
(0, 0)	.4(0) + .6(0) = 0
(0, 700)	.4(0) + .6(700) = 420
(400, 500)	.4(400) + .6(500) = 460
(600, 300)	.4(600) + .6(300) = 420
(800, 0)	.4(800) + .6(0) = 320

The maximum profit is achieved at (400, 500).
Make 400 cans of Fruit Delight and 500 cans of Heavenly Punch.

35. Since the farmer can spend $2400 for labor at $8 per hour, the available labor is $2400 \div 8 = 300$ hours.

	Oats	Corn	Available
Capital	$18	$36	$2100
Labor	2 hours	6 hours	300 hours
Land	1 acre	1 acre	100 acres
Revenue	$55	$125	

Let x be the number of acres of oats, and let y be the number of acres of corn.
Capital: $18x + 36y \le 2100$
Labor: $2x + 6y \le 300$
Land: $x + y \le 100$
In standard form the inequalities are:

$$\begin{cases} y \le -\frac{1}{2}x + \frac{175}{3} \\ y \le -\frac{1}{3}x + 50 \\ y \le -x + 100 \\ x \ge 0, \ y \ge 0 \end{cases}$$

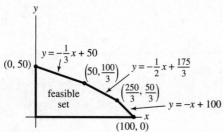

Find the vertices.

Lower left corner: $(0, 0)$

y-intercept of $y = -\dfrac{1}{3}x + 50$: $(0, 50)$

Intersection of $y = -\dfrac{1}{3}x + 50$ and $y = -\dfrac{1}{2}x + \dfrac{175}{3}$:

$$-\frac{1}{3}x + 50 = -\frac{1}{2}x + \frac{175}{3}$$

$$\frac{1}{6}x = \frac{25}{3}$$

$$x = 50$$

$$y = -\frac{1}{3}x + 50 = -\frac{1}{3}(50) + 50 = \frac{100}{3}$$

$$\left(50, \frac{100}{3}\right)$$

Intersection of $y = -\dfrac{1}{2}x + \dfrac{175}{3}$ and $y = -x + 100$:

$$-\frac{1}{2}x + \frac{175}{3} = -x + 100$$

$$\frac{1}{2}x = \frac{125}{3}$$

$$x = \frac{250}{3}$$

$$y = -x + 100 = -\frac{250}{3} + 100 = \frac{50}{3}$$

$$\left(\frac{250}{3}, \frac{50}{3}\right)$$

x-intercept of $y = -x + 100$: $(100, 0)$

Find the objective function for the profit.

Revenue: $55x + 125y$

Leftover capital: $2100 - 18x - 36y$

Leftover labor cash reserve: $2400 - 8(2x + 6y) = 2400 - 16x - 48y$

The profit is the sum of the above: $21x + 41y + 4500$

Vertex	Profit $= 21x + 41y + 4500$
$(0, 0)$	$21(0) + 41(0) + 4500 = 4500$
$(0, 50)$	$21(0) + 41(50) + 4500 = 6550$
$\left(50, \frac{100}{3}\right)$	$21(50) + 41\left(\frac{100}{3}\right) + 4500 \approx 6916.67$
$\left(\frac{250}{3}, \frac{50}{3}\right)$	$21\left(\frac{250}{3}\right) + 41\left(\frac{50}{3}\right) + 4500 \approx 6933.33$
$(100, 0)$	$21(100) + 41(0) + 4500 = 6600$

The maximum profit is achieved at $\left(\dfrac{250}{3}, \dfrac{50}{3}\right)$, or $\left(83\dfrac{1}{3}, 16\dfrac{2}{3}\right)$. The farmer should plant $83\dfrac{1}{3}$ acres of oats

and $16\dfrac{2}{3}$ acres of corn to make a profit of $6933.33.

37. a. Since the conditions are less restrictive than they were in Exercise 35, the optimal solution found in Exercise 35 will still be in the feasible set.

    b. The total cost to plant an acre of oats is $18 capital plus $16 for labor, or $34.

       The total cost to plant an acre of corn is $36 capital plus $48 for labor, or $84.

	Oats	Corn	Available
Cap. And Lab.	$34	$84	$4500
Land	1 acre	1 acre	100 acres
Revenue	$55	$125	

Let x be the number of acres of oats, and let y be the number of acres of corn.

Capital and Labor: $34x + 84y \leq 4500$

Land: $x + y \leq 100$

In standard form the inequalities are:

$$\begin{cases} y \leq -\dfrac{17}{42}x + \dfrac{375}{7} \\ y \leq -x + 100 \\ x \geq, y \geq 0 \end{cases}$$

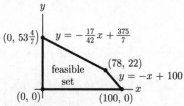

Find the vertices.

Lower left corner: $(0, 0)$

y-intercept of $y = -\dfrac{17}{42}x + \dfrac{375}{7}: \left(0, 53\dfrac{4}{7}\right)$

Intersection of $y = -\dfrac{17}{42}x + \dfrac{375}{7}$ and $y = -x + 100$:

$$-\frac{17}{42}x + \frac{375}{7} = -x + 100$$

$$\frac{25}{42}x = \frac{325}{7}$$

$$x = 78$$

$$y = -x + 100 = -(78) + 100 = 22$$

$$\left(78, 22\right)$$

x-intercept of $y = -x + 100$: $(100, 0)$

Note that the objective function for the profit is the same as in Exercise 35.

Find the objective function for the profit.

Revenue: $55x + 125y$

Leftover capital: $2100 - 18x - 36y$

Leftover labor cash reserve: $2400 - 8(2x + 6y) = 2400 - 16x - 48y$

The profit is the sum of the above: $21x + 41y + 4500$

Vertex	Profit $= 21x + 41y + 4500$
$(0, 0)$	$21(0) + 41(0) + 4500 = 4500$
$(78, 22)$	$21(78) + 41(22) + 4500 = 7040$
$\left(0, \frac{375}{7}\right)$	$21(0) + 41\left(\frac{375}{7}\right) + 4500 \approx 6696.43$
$(100, 0)$	$21(100) + 41(0) + 4500 = 6600$

The maximum profit is achieved at $\left(78, 22\right)$. The farmer should plant 78 acres of oats and 22 acres of corn to make a profit of \$7040. Yes, it provides more profit.

39.

	Regular	Deluxe	*Available*
Capital	$32	$38	$2100
Labor	4 hours	6 hours	280 hours
Price	$46	$55	

Let x be the number of regular bags made each day, and let y be the number of deluxe

$capital: 32x + 38y \le 2100$

$Labor: 4x + 6y \le 280$

$x \ge 0, y \ge 0$

In standard form the inequalities are:

$$\begin{cases} y \le -\dfrac{16}{19}x + \dfrac{1050}{19} \\[2mm] y \le -\dfrac{2}{3}x + \dfrac{140}{3} \\[2mm] x \ge 0, \; y \ge 0 \end{cases}$$

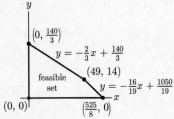

Find the vertices:

y-intercept of $y = -\dfrac{2}{3}x + \dfrac{140}{3} : \left(0, \dfrac{140}{3}\right)$

Intersection of $y = -\dfrac{16}{19}x + \dfrac{1050}{19}$ and $y = -\dfrac{2}{3}x + \dfrac{140}{3}$:

$-\dfrac{16}{19}x + \dfrac{1050}{19} = -\dfrac{2}{3}x + \dfrac{140}{3}$

$\dfrac{10}{57}x = \dfrac{490}{57}$

$x = 49$

$y = -\dfrac{16}{19}x + \dfrac{1050}{19} = -\dfrac{16}{19}(49) + \dfrac{1050}{19} = 14$

$(49, 14)$

x-intercept of $y = -\dfrac{16}{19}x + \dfrac{1050}{19} : \left(\dfrac{525}{8}, 0\right)$

Vertex	Revenue $= 46x + 55y$
$\left(0, \dfrac{140}{3}\right)$	$46(0) + 55(46.67) \approx 2566.67$
$(49, 14)$	$46(49) + 55(14) = 3024$
$\left(\dfrac{525}{8}, 0\right)$	$46(65.625) + 55(0) = 3018.75$

The maximum profit is achieved at $(49, 14)$. Make 49 regular bags and 14 deluxe.

41. Since $x \geq 0$ and $y \geq 0$, $4x + 3y \geq x + y$.
Therefore, the inequality $x + y \geq 6$ implies
that $4x + 3y \geq 6$, which contradicts
$4x + 3y \leq 4$.
The feasible set contains no points.

43.

	A	B
1	12	36
2	0	12
3		12
4		0
5		84
6		
7	**Cell**	**Content**
8	B1	=3*x+2*y
9	B2	=x+4*y
10	B3	=x
11	B4	=y
12	B5	=7*x+4*y

Exercises 3.3

1. a. $21x + 14y = c$

$$y = -\frac{3}{2}x + \frac{c}{14}$$

   b. Up

   c. B

3. The objective function $ax + by$ has

constant value on any line of slope $-\dfrac{a}{b}$.

The slope of the line containing $(8, 3)$ and

$(9, 0)$ is $\dfrac{0 - 3}{9 - 8} = -3$. Therefore, if

$-\dfrac{a}{b} < 0$, we require $-\dfrac{a}{b} \leq -3$, or

$\dfrac{a}{b} \geq 3$, where $a > 0$ and $b > 0$. One

possibility is $a = 5$ and $b = 1$, giving the
objective function
$5x + y$.
(We could also have chosen an objective
function that is constant on lines of
nonnegative or undefined slope of the
form $ax + by$ with
$a \geq 0$ and $b \leq 0$.)

5. The objective function $ax + by$ has

constant value on any line of slope $-\dfrac{a}{b}$.

Since the slope of the line containing $(3,$

$8)$ and $(8, 3)$ is $\dfrac{3 - 8}{8 - 3} = -1$, and the slope

of the line containing $(8, 3)$ and

$(9, 0)$ is $\dfrac{0 - 3}{9 - 8} = -3$, we require

$-1 \geq -\dfrac{a}{b} \geq -3$, or $1 \leq \dfrac{a}{b} \leq 3$.

One possibility is $a = 2$ and $b = 1$, giving
the objective function $2x + y$.

7. The objective function $ax + by$ has constant value on any line of slope $-\dfrac{a}{b}$. The slope of the line containing (6, 1) and (9, 0) is $\dfrac{0-1}{9-6} = -\dfrac{1}{3}$.

   We require $0 \geq -\dfrac{a}{b} \geq -\dfrac{1}{3}$, or $0 \leq \dfrac{a}{b} \leq \dfrac{1}{3}$, where $a \geq 0$ and $b > 0$. One possibility is $a = 1$ and $b = 5$, giving the objective function $x + 5y$.

9. The objective function $ax + by$ has constant value on any line of slope $-\dfrac{a}{b}$. The slope of the line containing (1, 6) and (6, 1) is $\dfrac{1-6}{6-1} = -1$, and the slope of the line containing (6, 1) and (9, 0) is $\dfrac{0-1}{9-6} = -\dfrac{1}{3}$.

   We require $-\dfrac{1}{3} \geq -\dfrac{a}{b} \geq -1$, or $\dfrac{1}{3} \leq \dfrac{a}{b} \leq 1$, where $a > 0$ and $b > 0$. One possibility is $a = 2$ and $b = 3$, giving the objective function $2x + 3y$.

11. The objective function $3x + 2y$ has constant value on any line of slope $-\dfrac{3}{2}$. Since this is between -1 and -4, the objective function is maximized at C.

13. The objective function $10x + 2y$ has constant value on any line of slope -5. Since this is less (steeper) than -4, the objective function is maximized at D.

15. The objective function $2x + 10y$ has constant value on any line of slope $-\dfrac{1}{5}$. Since this is between 0 and $-\dfrac{1}{4}$, the objective function is minimized at D.

17. The objective function $2x + 3y$ has constant value on any line of slope $-\dfrac{2}{3}$. Since this is between $-\dfrac{1}{4}$ and -1, the objective function is minimized at C.

19. The objective function $x + ky$ has constant value on any line of slope $-\dfrac{1}{k}$. The slope of the line containing (0, 5) and (3, 4) is $\dfrac{4-5}{3-0} = -\dfrac{1}{3}$, and the slope of the line containing (3, 4) and (4, 0) is $\dfrac{0-4}{4-3} = -4$.

   We require $-4 \leq -\dfrac{1}{k} \leq -\dfrac{1}{3}$. This is equivalent to $4 \geq \dfrac{1}{k} \geq \dfrac{1}{3}$, or $\dfrac{1}{4} \leq k \leq 3$.

21.

	Brand A	*Brand B*	*Requirement*
Protein	3 units	1 unit	6 units
Carbohydrate	1 unit	1 unit	4 units
Fat	2 units	6 units	12 units
Cost	$.80	$.50	

Let x be the number of units of brand A, and let y be the number of units of brand B.

Protein: $3x + y \geq 6$
Carbohydrate: $x + y \geq 4$
Fat: $2x + 6y \geq 12$

In standard form the inequalities are:

$$\begin{cases} y \geq -3x + 6 \\ y \geq -x + 4 \\ y \geq -\dfrac{1}{3}x + 2 \\ x \geq 0, \ y \geq 0 \end{cases}$$

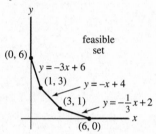

Vertex	*Cost* $= .8x + .5y$
$(0, 6)$	3
$(1, 3)$	2.3
$(3, 1)$	2.9
$(6, 0)$	4.8

The minimum cost of $2.30 is obtained at $(1, 3)$.
Feed 1 can of brand A and 3 cans of brand B.

23. Let the variables represent amounts in thousands of dollars.

	Low-risk	*Medium-risk*	*High-risk*
Yield	.06	.07	.08
Variables	x	y	$9 - x - y$

The required inequalities are:

$$\begin{cases} x \leq y + 1 \\ x + y \geq 5 \\ y + (9 - x - y) \leq 7 \\ x \geq 0, \ y \geq 0 \\ 9 - x - y \geq 0 \end{cases} \text{ or } \begin{cases} y \geq x - 1 \\ y \geq -x + 5 \\ x \geq 2 \\ x \geq 0, \ y \geq 0 \\ y \leq -x + 9 \end{cases}$$

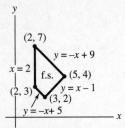

(The inequalities $x \geq 0$ and $y \geq 0$ do not appear in the graph, as they are assured by the other inequalities.)

The objective function for the expected yield is $.06x + .07y + .08(9 - x - y)$, or $.72 - .02x - .01y$.

Vertex	Yield = $.72 - .02x - .01y$
(2, 3)	.65
(2, 7)	.61
(5, 4)	.58
(3, 2)	.64

The maximum yield of $650 is achieved at (2, 3). Then $9 - x - y = 9 - 2 - 3 = 4$.

Invest $2000 in low-risk stocks, $3000 in medium-risk stocks, and $4000 in high-risk stocks.

25.

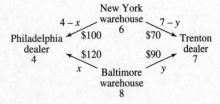

The required inequalities are:

$$\begin{cases} x \geq 0, \ y \geq 0 \\ 4 - x \geq 0 \\ 7 - y \geq 0 \\ x + y \leq 8 \\ (4 - x) + (7 - y) \leq 6 \end{cases} \quad \text{or} \quad \begin{cases} x \geq 0, \ y \geq 0 \\ x \leq 4 \\ y \leq 7 \\ y \leq -x + 8 \\ y \geq -x + 5 \end{cases}$$

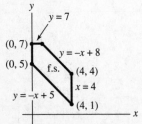

(The inequality $y \geq 0$ does not appear in the graph, as it is assured by the other inequalities.)

The objective function for the cost is $120x + 90y + 100(4 - x) + 70(7 - y)$, or $890 + 20x + 20y$.

Vertex	Cost = 890 + 20x + 20y
(0, 5)	990
(0, 7)	1030
(1, 7)	1050
(4, 4)	1050
(4, 1)	990

The minimum cost of \$990 is achieved at (0, 5) or (4, 1)—or anywhere along the segment connecting these points.

There are several ways to minimize costs, as summarized below.

Baltimore to Philadelphia	Baltimore to Trenton	New York to Philadelphia	New York to Trenton
0	5	4	2
1	4	3	3
2	3	2	4
3	2	1	5
4	1	0	6

27. Let the variables represent the number of thousands of gallons.

	Gasoline	Jet fuel	Diesel fuel
Profit	\$.15	\$.12	\$.10
Variables	x	y	$100 - x - y$

The required inequalities are:

$$\begin{cases} x \geq 5, \ y \geq 5 \\ 100 - x - y \geq 5 \\ x + y \geq 20 \\ x + (100 - x - y) \geq 50 \end{cases} \quad \text{or} \quad \begin{cases} x \geq 5, \ y \geq 5 \\ y \geq -x + 95 \\ y \geq -x + 20 \\ y \leq 50 \end{cases}$$

(Note that the inequalities above have ignored the somewhat subtle issue of whether there will be enough gasoline for *both* the airline and the trucking firm at the same time. This is only a potential issue if both suppliers require gasoline—that is, if $y \leq 20$ and $100 - x - y \leq 50$. In this case, we require that the gasoline requirements of each firm be less than the amount of gasoline actually produced—that is,

$(20 - y) + (50 - (100 - x - y)) \leq x$. This inequality is equivalent to $-30 \leq 0$, so it is always satisfied.)

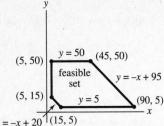

The objective function for the profit is $.15x + .12y + .1(100 - x - y)$, or $10 + .05x + .02y$.

Vertex	Profit = 10 + .05x + .02y
(5, 50)	11.25
(45, 50)	13.25
(90, 5)	14.6
(15, 5)	10.85
(5, 15)	10.55

The maximum profit of $14,600 is achieved at (90, 5). Then $100 - x - y = 100 - 90 - 5 = 5$.
Produce 90,000 gallons of gasoline, 5000 gallons of jet fuel, and 5000 gallons of diesel fuel.

29.

	High-capacity	Low-capacity	Available
Cost ($thousands)	50	30	1080
Drivers	1	1	30
Capacity	320 cases	200 cases	

Let x be the number of high-capacity trucks, and let y be the number of low-capacity trucks. The required inequalities are:

$$\begin{cases} 50x + 30y \le 1080 \\ x + y \le 30 \\ x \le 15 \\ x \ge 0,\ y \ge 0 \end{cases} \quad \text{or} \quad \begin{cases} y \le -\dfrac{5}{3}x + 36 \\ y \le -x + 30 \\ x \le 15 \\ x \ge 0,\ y \ge 0 \end{cases}$$

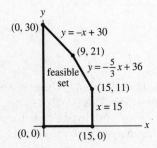

Vertex	Capacity $= 320x + 200y$
(0, 0)	0
(0, 30)	6000
(9, 21)	7080
(15, 11)	7000
(15, 0)	4800

The maximum capacity of 7080 cases is achieved at (9, 21).
Buy 9 high-capacity trucks and 21 low-capacity trucks.

31. Let $x =$ pounds of coffee shipped from San Jose to Salt Lake City, and
$y =$ pounds of coffee shipped from San Jose to Reno.

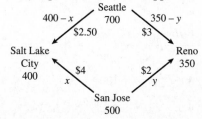

The required inequalities are:

$$\begin{cases} x + y \le 500 \\ (400 - x) + (350 - y) \le 700 \\ x \ge 0,\, y \ge 0 \\ 400 - x \ge 0,\, 350 - y \ge 0 \end{cases} \text{ or } \begin{cases} y \le -x + 500 \\ y \ge -x + 50 \\ x \ge 0,\, y \ge 0 \\ x \le 400,\, y \le 350 \end{cases}$$

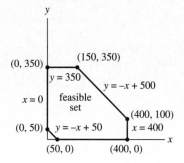

Objective function: [cost] $= 4x + 2y + 2.5(400 - x) + 3(350 - y) = 2050 + 1.5x - y$

Vertex	Cost $= 2050 + 1.5x - y$
(0, 50)	2000
(0, 350)	1700
(150, 350)	1925
(400, 100)	2550
(400, 0)	2650
(50, 0)	2125

The minimum cost of $1700 is at (0, 350).
Ship 400 pounds of coffee from Seattle to Salt Lake City and 350 pounds from San Jose to Reno.

33.

	Kit I	Kit II	Kit III	Available
Filters	1	2	1	54
Gravel (pounds)	2	2	3	100
Fish food	1	0	2	53
Profit	$7	$10	$13	

Let x be the number of Kit I's, and let y be the number of Kit II's. Then $x - y$ is the number of Kit III's. The required inequalities are:

$$\begin{cases} x + 2y + (x - y) \le 54 \\ 2x + 2y + 3(x - y) \le 100 \\ x + 2(x - y) \le 53 \\ x \ge 0, y \ge 0 \end{cases} \quad \text{or} \quad \begin{cases} y \le -2x + 54 \\ y \ge 5x - 100 \\ y \ge \frac{3}{2}x - \frac{53}{2} \\ x \ge 0, y \ge 0 \end{cases}$$

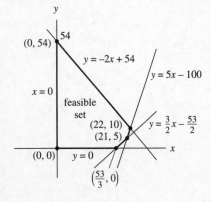

The objective function is $[\text{profit}] = 7x + 10y + 13(x - y)$
$$= 20x - 3y$$

Vertex	Profit = $20x - 3y$
$(0, 0)$	0
$(0, 54)$	-162
$(22, 10)$	410
$(21, 5)$	405
$\left(\frac{53}{3},\ 0\right)$	$353\frac{1}{3}$

The maximum profit of \$410 is acheived at $(22, 10)$. Create 22 Basic I kits, 10 Basic II kits; and 12 Deluxe kits.

35.

Cell	Name	Final Value	Reduced Cost	Objective Coefficient	Allowable Decrease	Allowable Increase
\$A\$1	x	0.857142857	0	21	21	11.66666667
\$A\$2	y	3.428571429	0	14	17.5	7

The range of optimality for the cost of rice is $[21 - 11.67, 21 + 21]$ or $[9.33, 42]$.
The range of optimality for the cost of soybeans is $[14 - 7, 14 + 17.5]$ or $[7, 31.5]$.

Chapter 3 Supplementary Exercises

1.

	Type A	Type B	Required or available
Passengers	50	300	1400
Flight attendants	3	4	42
Cost	\$14,000	\$90,000	

Let x be the number of type A planes, and let y be the number of type B planes. The required inequalities are:

$$\begin{cases} 50x + 300y \geq 1400 \\ 3x + 4y \leq 42 \\ x \geq y \\ x \geq 0,\ y \geq 0 \end{cases} \text{ or } \begin{cases} y \geq -\dfrac{1}{6}x + \dfrac{14}{3} \\ y \leq -\dfrac{3}{4}x + \dfrac{21}{2} \\ y \leq x \\ x \geq 0,\ y \geq 0 \end{cases}$$

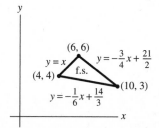

(The inequalities $x \geq 0$ and $y \geq 0$ are not shown in the graph because they are assured by the other inequalities.)

Vertex	$Cost = 14{,}000x + 90{,}000y$
(4, 4)	416,000
(6, 6)	624,000
(10, 3)	410,000

The minimum cost of \$410,000 is achieved at (10, 3).
Use 10 type A planes and 3 type B planes.

2.

	Wheat germ	Enriched oat flour	Required
Niacin	2 milligrams	3 milligrams	7 milligrams
Iron	3 milligrams	3 milligrams	9 milligrams
Thiamin	.5 milligram	.25 milligram	1 milligram
Cost	6 cents	8 cents	

Let x be the number of ounces of wheat germ, and let y be the number of ounces of enriched oat flour.

The required inequalities are:

$$\begin{cases} 2x + 3y \geq 7 \\ 3x + 3y \geq 9 \\ .5x + .25y \geq 1 \\ x \geq 0,\ y \geq 0 \end{cases} \text{ or } \begin{cases} y \geq -\dfrac{2}{3}x + \dfrac{7}{3} \\ y \geq -x + 3 \\ y \geq -2x + 4 \\ x \geq 0,\ y \geq 0 \end{cases}$$

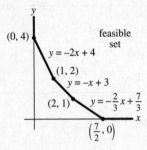

Vertex	$Cost = 6x + 8y$
(0, 4)	32
(1, 2)	22
(2, 1)	20
$\left(\frac{7}{2}, 0\right)$	21

The minimum cost of 20 cents is achieved at (2, 1).
Use 2 ounces of wheat germ and 1 ounce of enriched oat flour.

3.

	Hardtops	*Sports cars*	*Available*
Assemble	8 labor-hours	18 labor-hours	360 labor-hours
Paint	2 labor-hours	2 labor-hours	50 labor-hours
Upholster	2 labor-hours	1 labor-hour	40 labor-hours
Profit	$90	$100	

Let x be the number of hardtops and let y be the number of Sports Cars.
The required inequalities are:

$$\begin{cases} 8x + 18y \le 360 \\ 2x + 2y \le 50 \\ 2x + y \le 40 \\ x \ge 0,\ y \ge 0 \end{cases} \text{ or } \begin{cases} y \le -\dfrac{4}{9}x + 20 \\ y \le -x + 25 \\ y \le -2x + 40 \\ x \ge 0,\ y \ge 0 \end{cases}$$

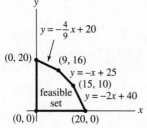

Vertex	Profit = 90x + 100y
(0, 0)	0
(0, 20)	2000
(9, 16)	2410
(15, 10)	2350
(20, 0)	1800

The maximum profit of $2410 is achieved at (9, 16).
Produce 9 hardtops and 16 sports cars.

4.

	Type A	*Type B*	*Available*
Peanuts	6 oz	12 oz	5400 oz
Raisins	1 oz	3 oz	1200 oz
Cashews	4 oz	2 oz	2400 oz
Revenue	$4.25	$6.55	

Let x be the amount of mixture A, and let y be the amount of mixture B.

Peanuts: $6x + 12y \le 5400$

Raisins: $x + 3y \le 1200$

Cashews: $4x + 2y \le 2400$

In standard form the inequalities are:

$$\begin{cases} y \le -\dfrac{1}{2}x + 450 \\[2mm] y \le -\dfrac{1}{3}x + 400 \\[2mm] y \le -2x + 1200 \\[2mm] x \ge 0, \; y \ge 0 \end{cases}$$

Find the vertices.

Lower left corner: (0, 0)

y-intercept of $y = -\dfrac{1}{3}x + 400$: (0, 400)

Intersection of $y = -\dfrac{1}{3}x + 400$ and $y = -\dfrac{1}{2}x + 450$:

$$-\frac{1}{3}x + 400 = -\frac{1}{2}x + 450$$

$$\frac{1}{6}x = 50$$

$$x = 300$$

$$y = -\frac{1}{3}x + 400 = -\frac{1}{3}(300) + 400 = 300$$

(300, 300)

Intersection of $y = -\dfrac{1}{2}x + 450$ and $y = -2x + 1200$:

$$-\frac{1}{2}x + 450 = -2x + 1200$$

$$\frac{3}{2}x = 750$$

$$x = 500$$

$$y = -\frac{1}{2}x + 450 = -\frac{1}{2}(500) + 450 = 200$$

(500, 200)

x-intercept of $y = -2x + 1200$: (600, 0)

Vertex	Revenue = $4.25x + 6.55y$
$(0, 0)$	$4.25(0) + 6.55(0) = 0$
$(0, 400)$	$4.25(0) + 6.55(400) = 2620$
$(300, 300)$	$4.25(300) + 6.55(300) = 3240$
$(500, 200)$	$4.25(500) + 6.55(200) = 3435$
$(600, 0)$	$4.25(600) + 6.55(0) = 2550$

The maximum revenue is achieved at $(500, 200)$.
Make 500 boxes of mixture A and 200 boxes of mixture B.

5.

	Elementary	Intermediate	Advanced
Profit	$8000	$7000	$1000
Variables	x	y	$72 - x - y$

The required inequalities are:

$$\begin{cases} 72 - x - y \geq 4 \\ x \geq 3y \\ y \geq 2(72 - x - y) \\ x \geq 0, \ y \geq 0 \end{cases} \text{ or } \begin{cases} y \leq -x + 68 \\ y \leq \dfrac{1}{3}x \\ y \leq -\dfrac{2}{3}x + 48 \\ x \geq 0, \ y \geq 0 \end{cases}$$

The objective function for the annual profit is $8000x + 7000y + 1000(72 - x - y)$, or $72{,}000 + 7000x + 6000y$.

Vertex	Profit = $72{,}000 + 7000x + 6000y$
$(48, 16)$	504,000
$(51, 17)$	531,000
$(60, 8)$	540,000

The maximum annual profit of $540,000 is achieved at $(60, 8)$.
Then $72 - x - y = 72 - 60 - 8 = 4$.
Publish 60 elementary books, 8 intermediate books, and 4 advanced books.

6.

	Rochester	Queens	Available
Transport time	15 hours	20 hours	2100 hours
Cost	$15	$30	$3000
Profit	$40	$30	

Let x be the number of computers sent from Rochester, and let y be the number of computers sent from Queens. The required inequalities are:

$$\begin{cases} 15x + 20y \le 2100 \\ 15x + 30y \le 3000 \\ x \le 80 \\ y \le 120 \\ x \ge 0, \ y \ge 0 \end{cases} \quad \text{or} \quad \begin{cases} y \le -\dfrac{3}{4}x + 105 \\ y \le -\dfrac{1}{2}x + 100 \\ x \le 80 \\ y \le 120 \\ x \ge 0, \ y \ge 0 \end{cases}$$

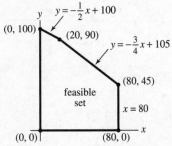

(The inequality $y \le 120$ does not appear in the graph, as it is assured by the other inequalities.)

Vertex	Profit = 40x + 30y
(0, 0)	0
(0, 100)	3000
(20, 90)	3500
(80, 45)	4550
(80, 0)	3200

The maximum profit of $4550 is achieved at (80, 45).
Transport 80 computers from Rochester and 45 computers from Queens.

7.

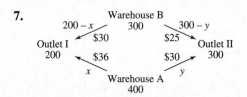

The required inequalities are:

$$\begin{cases} x \ge 0, \; y \ge 0 \\ 200 - x \ge 0 \\ 300 - y \ge 0 \\ x + y \le 400 \\ (200 - x) + (300 - y) \le 300 \end{cases} \quad \text{or} \quad \begin{cases} x \ge 0, \; y \ge 0 \\ x \le 200 \\ y \le 300 \\ y \le -x + 400 \\ y \ge -x + 200 \end{cases}$$

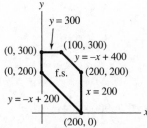

The objective function for the cost is $36x + 30y + 30(200 - x) + 25(300 - y)$, or $13{,}500 + 6x + 5y$.

Vertex	Cost $= 13{,}500 + 6x + 5y$
(0, 200)	14,500
(0, 300)	15,000
(100, 300)	15,600
(200, 200)	15,700
(200, 0)	14,700

The minimum cost of \$14,500 is achieved at (0, 200). Then $200 - x = 200 - 0 = 200$, and $300 - y = 300 - 200 = 100$.

Transport 200 computers from warehouse A to outlet II, 200 computers from warehouse B to outlet I, and 100 computers from warehouse B to outlet II.

8.

	CD	*Mutual fund*	*Stocks*
Yield	.05	.07	.09
Variables	x	y	$10{,}000 - (x + y)$

The required inequalities are:

$$\begin{cases} y \le x + 10{,}000 - (x + y) \\ y + 10{,}000 - (x + y) \le 8000 \\ x \ge 0,\ y \ge 0 \\ x \le 10{,}000,\ y \le 10{,}000 \end{cases} \text{or} \begin{cases} y \le 5000 \\ x \ge 2000 \\ y \ge 0 \\ x \le 10{,}000 \end{cases}$$

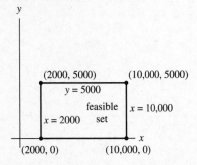

The objective function is [return] = $.05x + .07y + .09[10{,}000 - (x + y)]$ or $900 - .04x - .02y$.

Vertex	*Return* $= 900 - .04x - .02y$
(2000, 0)	820
(2000, 5000)	720
(10,000, 5000)	400
(10,000, 0)	500

The maximum return of $820 is achieved at (2000, 0). Invest $2000 in the CD, $0 in mutual funds; and $8000 in stocks.

9. a. Yes; The added constraint may add to the original cost therefore it may increase the optimal cost.

b. No; The added constraint will not decrease the original cost.

10. a. Yes; The removed constraint may increase the profit therefore it may increase the optimal profit.

b. No; The added constraint will not decrease the original profit.

11. Answers will vary.

12. One of the boundary points will always be a maximum or a minimum of the problem.

13. Every point on a line (or line segment) always has the same objective function value, provided the line (or line segment) has the same slope as a line with constant objective function.

Chapter 3 Chapter Test

1. Step 1: Translate the problem into mathematical language.
   Identify variables and write the inequalities and objective function.
   Step 2: Graph the feasible set.
   Step 3: Determine the vertices of the feasible set.
   Step 4: Evaluate the objective function at each vertex.
   Determine the optimal point.

2. **a.** Let x = hours spent at plant A
   y = hours spent at plant B
   Minimize $70x + 90y$ subject to:
   $$\begin{cases} 20x + 30y \geq 300 \\ 10x + 20y \geq 200 \\ x \geq 0, \ y \geq 0 \end{cases}$$

   b. Let x = amount in mutual funds
   y = amount in bonds
   z = amount in CDs
   Maximize
   $0.07x + 0.06y + 0.045z$ subject to
   $$\begin{cases} y \geq 150{,}000 \\ z \leq 200{,}000 \\ z \leq 500{,}000 - x - y \\ x \geq .5y \\ x \geq 0, \ y \geq 0, \ z \geq 0 \end{cases}$$

3. **a.**

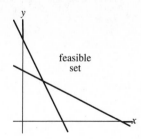

   b. Answers may vary. *Sample answer:* No, this problem does not have a solution. The feasible region is unbounded, and as x and y get larger, $3x + 2y$ will also get larger.

4. A: $\begin{cases} 5x + 7y = 70 \\ \quad x = 0 \end{cases}$
   $$7y = 70$$
   $$y = 10$$
   $(0, 10)$

   B: $\begin{cases} x + y = 12 \\ 5x + 7y = 70 \end{cases}$
   $$\begin{cases} y = 12 - x \\ y = -\dfrac{5}{7}x + 10 \end{cases}$$
   $$12 - x = -\frac{5}{7}x + 10$$
   $$2 = \frac{2}{7}x$$
   $$x = 7$$
   $$y = 12 \ 7 = 5$$
   $(7, 5)$

   C: $\begin{cases} x + y = 12 \\ \quad x = 9 \end{cases}$
   $$9 + y = 12$$
   $$y = 3$$
   $(9, 3)$

   D: $\begin{cases} x = 9 \\ y = 0 \end{cases}$
   $(9, 0)$

   E: $(0, 0)$

5.
Vertex	$20x$ $10y$
(3, 0)	60
(3, 6)	0
(7, 9)	50
(10, 8)	120
(10, 0)	200

   a. $x = 10$, $y = 0$

   b. $x = 3$, $y = 6$

6. a. $8x + 5y = 3200$

$$y = -\frac{8}{5}x + 640$$

b. $8x + 5y = c$

$$y = -\frac{8}{5}x + \frac{c}{5}$$

c. down

d. E

7. $5x + 8y = c$

$$y = -\frac{5}{8}x + \frac{c}{8}$$

The slope of the objective function is

$-\frac{5}{8}$. So this line is steeper than one with

a slope of $-\frac{1}{2}$, but not as steep as one

with a slope of 1. Thus, the objective function is minimized at point C.

8.

	Bears	Wreaths	Available
Prep Time	3 labor-hours	2 labor-hour	36 labor-hours
Assembly	6 labor-hour	5 labor-hours	78 labor-hours
Finishing	2 labor-hours	3 labor-hours	42 labor-hours
Profit	$10	$8	

Let x be the number of bears manufactured each day, and let y be the number of wreaths.

Prep: $3x + 2y \le 36$

Assembly: $6x + 5y \le 78$

Finishing: $2x + 3y \le 42$

In standard form the inequalities are:

$$\begin{cases} y \le -\dfrac{3}{2}x + 18 \\[2mm] y \le -\dfrac{6}{5}x + \dfrac{78}{5} \\[2mm] y \le -\dfrac{2}{3}x + 14 \\[2mm] x \ge 0,\ y \ge 0 \end{cases}$$

Find the vertices.

Intersection of $y = -\dfrac{6}{5}x + \dfrac{78}{5}$ and $y = -\dfrac{2}{3}x + 14$:

$$-\frac{6}{5}x + \frac{78}{5} = -\frac{2}{3}x + 14$$

$$\frac{8}{15}x = \frac{8}{5}$$

$$x = 3$$

$$y = -\frac{6}{5}x + \frac{78}{5} = -\frac{6}{5}(3) + \frac{78}{5} = 12$$

$(3, 12)$

Intersection of $y = -\dfrac{6}{5}x + \dfrac{78}{5}$ and $y = -\dfrac{3}{2}x + 18$:

$$-\frac{6}{5}x + \frac{78}{5} = -\frac{3}{2}x + 18$$

$$\frac{3}{10}x = \frac{12}{5}$$

$$x = 8$$

$$y = -\frac{6}{5}x + \frac{78}{5} = -\frac{6}{5}(8) + \frac{78}{5} = 6$$

(8, 6)

Intersection of $y = -\dfrac{2}{3}x + 14$ and $y = -\dfrac{3}{2}x + 18$:

$$-\frac{2}{3}x + 14 = -\frac{3}{2}x + 18$$

$$\frac{5}{6}x = 4$$

$$x = 4.8$$

$$y = -\frac{2}{3}x + 14 = -\frac{2}{3}(4.8) + 14 = 10.8$$

(4.8, 10.8) We would need to use the point (4, 10) because that is the number of completed bears and wreaths that there would be.

Vertex	$Profit = 10x + 8y$
(3, 12)	$10(3) + 8(12) = 126$
(8, 6)	$10(8) + 8(6) = 128$
(4, 10)	$10(4) + 8(10) = 120$

The maximum profit is 128 and occurs when the owner manufactures 8 bears and 6 wreaths.

Chapter 4

Exercises 4.1

$$1. \begin{cases} 20x + 30y + u & = 3500 \\ 50x + 10y + v & = 5000 \\ -8x - 13y + M = & 0 \end{cases}$$

Maximize M given $x \geq 0, y \geq 0, u \geq 0, v \geq 0$.

$$3. \begin{cases} x + y + z + u & = 100 \\ 3x + z + v & = 200 \\ 5x + 10y + w & = 100 \\ -x - 2y + 3z + M = & 0 \end{cases}$$

Maximize M given $x \geq 0, y \geq 0, z \geq 0, u \geq 0, v \geq 0, w \geq 0$.

$$5. \begin{cases} 4x + 6y - 7z + u & = 16 \\ 3x + 2y + v & = 11 \\ 9y + 3z + w & = 21 \\ -3x - 5y - 12z + M = & 0 \end{cases}$$

Maximize M given $x \geq 0, y \geq 0, z \geq 0, u \geq 0, v \geq 0, w \geq 0$.

7. a.

$$\begin{array}{ccccccc} x & y & u & v & w & t & M \end{array}$$
$$\left[\begin{array}{ccccccc|c} 3 & 2 & 1 & 0 & 0 & 0 & 0 & 10 \\ 1 & 0 & 0 & 1 & 0 & 0 & 0 & 15 \\ 0 & 1 & 0 & 0 & 1 & 0 & 0 & 3 \\ 1 & 1 & 0 & 0 & 0 & 1 & 0 & 5 \\ -1 & -15 & 0 & 0 & 0 & 0 & 1 & 0 \end{array} \right]$$

b. $x = 0, y = 0, u = 10, v = 15, w = 3, t = 5, M = 0$

9. a.

$$\begin{array}{cccccc} x & y & u & v & w & M \end{array}$$
$$\left[\begin{array}{cccccc|c} 1 & 3 & 1 & 0 & 0 & 0 & 24 \\ 0 & 1 & 0 & 1 & 0 & 0 & 5 \\ 1 & 7 & 0 & 0 & 1 & 0 & 10 \\ -2 & -1 & 0 & 0 & 0 & 1 & 50 \end{array} \right]$$

b. $x = 0, y = 0, u = 24, v = 5, w = 10, M = 50$

11. $x = 15, y = 0, u = 10, v = 0, M = 20$

13. $x = 10, y = 0, z = 15, u = 23, v = 0, w = 0,$
$M = -11$

15. a. Divide the first row by 2, then use matrix operations to change the remaining values in column 1 to zeros.

$$\begin{array}{ccccc} & x & y & u & v & M \end{array}$$
$$\begin{array}{c} x \\ v \\ M \end{array} \left[\begin{array}{ccccc|c} 1 & \frac{3}{2} & \frac{1}{2} & 0 & 0 & 6 \\ 0 & -\frac{1}{2} & -\frac{1}{2} & 1 & 0 & 4 \\ 0 & -5 & 5 & 0 & 1 & 60 \end{array} \right]$$
$x = 6, y = 0, u = 0, v = 4, M = 60$

b. Divide the first row by 3, then use matrix operations to change the remaining values in column 2 to zeros.

$$\begin{array}{c} \\ y \\ v \\ M \end{array} \begin{array}{cccccc} x & y & u & v & M \\ \left[\begin{array}{ccccc|c} \frac{2}{3} & 1 & \frac{1}{3} & 0 & 0 & 4 \\ \frac{1}{3} & 0 & -\frac{1}{3} & 1 & 0 & 6 \\ \frac{10}{3} & 0 & \frac{20}{3} & 0 & 1 & 80 \end{array}\right] \end{array}$$

$x = 0$, $y = 4$, $u = 0$, $v = 6$, $M = 80$

c. Use matrix operations to change the remaining values in column 1 to zeros.

$$\begin{array}{c} \\ u \\ x \\ M \end{array} \begin{array}{cccccc} x & y & u & v & M \\ \left[\begin{array}{ccccc|c} 0 & 1 & 1 & -2 & 0 & -8 \\ 1 & 1 & 0 & 1 & 0 & 10 \\ 0 & -10 & 0 & 10 & 1 & 100 \end{array}\right] \end{array}$$

$x = 10$, $y = 0$, $u = -8$, $v = 0$, $M = 100$

d. Use matrix operations to change the remaining values in column 2 to zeros.

$$\begin{array}{c} \\ u \\ y \\ M \end{array} \begin{array}{cccccc} x & y & u & v & M \\ \left[\begin{array}{ccccc|c} -1 & 0 & 1 & -3 & 0 & -18 \\ 1 & 1 & 0 & 1 & 0 & 10 \\ 10 & 0 & 0 & 20 & 1 & 200 \end{array}\right] \end{array}$$

$x = 0$, $y = 10$, $u = -18$, $v = 0$, $M = 200$

17. M becomes greatest after the pivot operation in part (d).

19. a. Group I variables are x and y.

Group II variables are u, v, and M.

b. (i) Divide the first row by 2, then use matrix operations to change the remaining values in column 2 to zeros.

$$\begin{array}{cccccc} x & y & u & v & M \\ \left[\begin{array}{ccccc|c} 1 & \frac{5}{2} & \frac{1}{2} & 0 & 0 & 50 \\ 0 & -\frac{13}{2} & -\frac{3}{2} & 1 & 0 & 150 \\ 0 & 18 & 5 & 0 & 1 & 500 \end{array}\right] \end{array}$$

Feasible: Group I variables: u and y
Group II variables: x, v, M

(ii) Divide the first row by 5, then use matrix operations to change the remaining values in column 2 to zeros.

$$\begin{array}{cccccc} x & y & u & v & M \\ \left[\begin{array}{ccccc|c} \frac{2}{5} & 1 & \frac{1}{5} & 0 & 0 & 20 \\ \frac{13}{5} & 0 & -\frac{1}{5} & 1 & 0 & 280 \\ -\frac{36}{5} & 0 & \frac{7}{5} & 0 & 1 & 140 \end{array}\right] \end{array}$$

Feasible: Group I variables: x and u
Group II variables: y, v, M

(iii) Divide the second row by 3, then use matrix operations to change the remaining values in column 2 to zeros.

$$\begin{array}{cccccc} x & y & u & v & M \\ \left[\begin{array}{ccccc|c} 0 & \frac{13}{3} & 1 & -\frac{2}{3} & 0 & -100 \\ 1 & \frac{1}{3} & 0 & \frac{1}{3} & 0 & 100 \\ 0 & -\frac{11}{3} & 0 & \frac{10}{3} & 1 & 1000 \end{array}\right] \end{array}$$

Not Feasible: Group I variables: y and v
Group II variables: x, u, M

(iv) Use matrix operations to change the remaining value in column 2 to zeros.

$$\begin{array}{cccccc} x & y & u & v & M \\ \left[\begin{array}{ccccc|c} -13 & 0 & 1 & -5 & 0 & -1400 \\ 3 & 1 & 0 & 1 & 0 & 300 \\ 11 & 0 & 0 & 7 & 1 & 2100 \end{array}\right] \end{array}$$

Not Feasible: Group I variables: x and v
Group II variables: y, u, M

c. M becomes greatest after the pivot operation in part (i).

Exercises 4.2

1. a. -12 is the most negative entry in the last row, and $\dfrac{6}{3} < \dfrac{10}{2}$. Pivot about the 3.

b.

$$\begin{array}{c} \\ u \\ y \\ M \end{array} \begin{array}{cccccc} x & y & u & v & M \\ \left[\begin{array}{ccccc|c} \frac{16}{3} & 0 & 1 & -\frac{2}{3} & 0 & 6 \\ \frac{1}{3} & 1 & 0 & \frac{1}{3} & 0 & 2 \\ 0 & 0 & 0 & 4 & 1 & 24 \end{array}\right] \end{array}$$

c. $x = 0$, $y = 2$, $u = 6$, $v = 0$, $M = 24$

3. a. −2 is the only negative entry in the last row, and $\dfrac{5}{10} < \dfrac{12}{12}$. Pivot about the 10.

b.

$$\begin{array}{c}\\u\\y\\M\end{array}\begin{array}{c}x \quad y \quad u \quad v \quad M\\\left[\begin{array}{ccccc|c}-13 & 0 & 1 & -\frac{6}{5} & 0 & 6\\\frac{3}{2} & 1 & 0 & \frac{1}{10} & 0 & \frac{1}{2}\\\hline 7 & 0 & 0 & \frac{1}{5} & 1 & 1\end{array}\right]\end{array}$$

c. $x = 0,\; y = \dfrac{1}{2},\; u = 6,\; v = 0,\; M = 1$

In Exercises 5–25, the pivot elements are underlined.

5.

$$\begin{array}{c}\\u\\v\\M\end{array}\begin{array}{c}x \quad y \quad u \quad v \quad M\\\left[\begin{array}{ccccc|c}1 & 1 & 1 & 0 & 0 & 7\\1 & \underline{2} & 0 & 1 & 0 & 10\\\hline -1 & -3 & 0 & 0 & 1 & 0\end{array}\right]\end{array}$$

$$\begin{array}{c}\\u\\y\\M\end{array}\begin{array}{c}x \quad\; y \quad u \quad\;\; v \quad M\\\left[\begin{array}{ccccc|c}\frac{1}{2} & 0 & 1 & -\frac{1}{2} & 0 & 2\\\frac{1}{2} & 1 & 0 & \frac{1}{2} & 0 & 5\\\hline \frac{1}{2} & 0 & 0 & \frac{3}{2} & 1 & 15\end{array}\right]\end{array}$$

$x = 0,\, y = 5;\, M = 15$

7.

$$\begin{array}{c}\\u\\v\\M\end{array}\begin{array}{c}x \quad y \quad u \quad v \quad M\\\left[\begin{array}{ccccc|c}\underline{5} & 1 & 1 & 0 & 0 & 80\\3 & 2 & 0 & 1 & 0 & 76\\\hline -4 & -2 & 0 & 0 & 1 & 0\end{array}\right]\end{array}$$

$$\begin{array}{c}\\x\\v\\M\end{array}\begin{array}{c}x \quad\; y \quad\;\; u \quad v \quad M\\\left[\begin{array}{ccccc|c}1 & \frac{1}{5} & \frac{1}{5} & 0 & 0 & 16\\0 & \frac{7}{5} & -\frac{3}{5} & 1 & 0 & 28\\\hline 0 & -\frac{6}{5} & \frac{4}{5} & 0 & 1 & 64\end{array}\right]\end{array}$$

$$\begin{array}{c}\\x\\y\\M\end{array}\begin{array}{c}x \quad y \quad\;\; u \quad\;\; v \quad M\\\left[\begin{array}{ccccc|c}1 & 0 & \frac{2}{7} & -\frac{1}{7} & 0 & 12\\0 & 1 & -\frac{3}{7} & \frac{5}{7} & 0 & 20\\\hline 0 & 0 & \frac{2}{7} & \frac{6}{7} & 1 & 88\end{array}\right]\end{array}$$

$x = 12,\, y = 20;\, M = 88$

9.

$$\begin{array}{c}\\u\\v\\M\end{array}\begin{array}{c}x \quad\; y \quad\; z \quad u \quad v \quad M\\\left[\begin{array}{cccccc|c}1 & 0 & \underline{2} & 1 & 0 & 0 & 10\\0 & 3 & 1 & 0 & 1 & 0 & 24\\\hline -1 & -3 & -5 & 0 & 0 & 1 & 0\end{array}\right]\end{array}$$

$$\begin{array}{c}\\z\\v\\M\end{array}\begin{array}{c}x \quad\;\; y \quad z \quad\;\; u \quad v \quad M\\\left[\begin{array}{cccccc|c}\frac{1}{2} & 0 & 1 & \frac{1}{2} & 0 & 0 & 5\\-\frac{1}{2} & \underline{3} & 0 & -\frac{1}{2} & 1 & 0 & 19\\\hline \frac{3}{2} & -3 & 0 & \frac{5}{2} & 0 & 1 & 25\end{array}\right]\end{array}$$

$$\begin{array}{c}\\z\\y\\M\end{array}\begin{array}{c}x \quad\;\; y \quad z \quad\;\; u \quad\; v \quad M\\\left[\begin{array}{cccccc|c}\frac{1}{2} & 0 & 1 & \frac{1}{2} & 0 & 0 & 5\\-\frac{1}{6} & 1 & 0 & -\frac{1}{6} & \frac{1}{3} & 0 & \frac{19}{3}\\\hline 1 & 0 & 0 & 2 & 1 & 1 & 44\end{array}\right]\end{array}$$

$x = 0,\; y = \dfrac{19}{3},\; z = 5;\, M = 44$

11.

$$\begin{array}{c}\\u\\v\\w\\M\end{array}\begin{array}{c}x \quad\; y \quad u \quad v \quad w \quad M\\\left[\begin{array}{cccccc|c}5 & \underline{1} & 1 & 0 & 0 & 0 & 30\\3 & 2 & 0 & 1 & 0 & 0 & 60\\1 & 1 & 0 & 0 & 1 & 0 & 50\\\hline -2 & -3 & 0 & 0 & 0 & 1 & 0\end{array}\right]\end{array}$$

$$\begin{array}{c}\\y\\v\\w\\M\end{array}\begin{array}{c}x \quad\; y \quad\; u \quad v \quad w \quad M\\\left[\begin{array}{cccccc|c}5 & 1 & 1 & 0 & 0 & 0 & 30\\-7 & 0 & -2 & 1 & 0 & 0 & 0\\-4 & 0 & -1 & 0 & 1 & 0 & 20\\\hline 13 & 0 & 3 & 0 & 0 & 1 & 90\end{array}\right]\end{array}$$

$x = 0,\, y = 30;\, M = 90$

Pivoting about the 2 instead gives the same solution although the tableau is different.

13.

$$\begin{array}{c} \begin{array}{ccccc} x & y & u & v & M \end{array} \\ \begin{array}{c} u \\ v \\ M \end{array}\left[\begin{array}{ccccc|c} 2 & \underline{3} & 1 & 0 & 0 & 400 \\ 1 & 1 & 0 & 1 & 0 & 150 \\ \hline -6 & -7 & 0 & 0 & 1 & 300 \end{array}\right] \end{array}$$

$$\begin{array}{c} \begin{array}{ccccc} x & y & u & v & M \end{array} \\ \begin{array}{c} y \\ v \\ M \end{array}\left[\begin{array}{ccccc|c} \frac{2}{3} & 1 & \frac{1}{3} & 0 & 0 & \frac{400}{3} \\ \frac{1}{3} & 0 & -\frac{1}{3} & 1 & 0 & \frac{50}{3} \\ \hline -\frac{4}{3} & 0 & \frac{7}{3} & 0 & 1 & \frac{3700}{3} \end{array}\right] \end{array}$$

$$\begin{array}{c} \begin{array}{ccccc} x & y & u & v & M \end{array} \\ \begin{array}{c} y \\ x \\ M \end{array}\left[\begin{array}{ccccc|c} 0 & 1 & 1 & -2 & 0 & 100 \\ 1 & 0 & -1 & 3 & 0 & 50 \\ \hline 0 & 0 & 1 & 4 & 1 & 1300 \end{array}\right] \end{array}$$

$x = 50, y = 100; M = 1300$

15. Let b be the number of large basketballs and f be the number of footballs manufactured. Maximize $2.50b + 2.00f$ subject to the constraints:

$$\begin{cases} 4b + 3f \le 768 \\ 20b + 30f \le 7200 \\ b \ge 0, f \ge 0 \end{cases}$$

(Note that 48 pounds was converted to ounces and 120 hours to minutes)

$$\begin{array}{c} \begin{array}{ccccc} b & f & u & v & M \end{array} \\ \begin{array}{c} u \\ v \\ M \end{array}\left[\begin{array}{ccccc|c} \underline{4} & 3 & 1 & 0 & 0 & 768 \\ 20 & 30 & 0 & 1 & 0 & 7200 \\ \hline -2.5 & -2 & 0 & 0 & 1 & 0 \end{array}\right] \end{array}$$

$$\begin{array}{c} \begin{array}{ccccc} b & f & u & v & M \end{array} \\ \begin{array}{c} b \\ v \\ M \end{array}\left[\begin{array}{ccccc|c} 1 & \frac{3}{4} & \frac{1}{4} & 0 & 0 & 192 \\ 0 & \underline{15} & -5 & 1 & 0 & 3360 \\ \hline 0 & -\frac{1}{8} & \frac{5}{8} & 0 & 1 & 480 \end{array}\right] \end{array}$$

$$\begin{array}{c} \begin{array}{ccccc} b & f & u & v & M \end{array} \\ \begin{array}{c} b \\ f \\ M \end{array}\left[\begin{array}{ccccc|c} 1 & 0 & \frac{1}{2} & -\frac{1}{20} & 0 & 24 \\ 0 & 1 & -\frac{1}{3} & \frac{1}{15} & 0 & 224 \\ \hline 0 & 0 & \frac{7}{12} & \frac{1}{120} & 1 & 508 \end{array}\right] \end{array}$$

Therefore, 24 basketballs and 224 footballs should be manufactured

17. Let c be the number of chairs, s be the number of sofas, and t be the number of tables manufactured each day. Maximize $80c + 70s + 120t$ subject to the constraints

$$\begin{cases} 6c + 3s + 8t \le 768 \\ c + s + 2t \le 144 \\ 2c + 5s \le 216 \\ c \ge 0, s \ge 0, t \ge 0 \end{cases}$$

$$\begin{array}{c} \begin{array}{ccccccc} c & s & t & u & v & w & M \end{array} \\ \begin{array}{c} u \\ v \\ w \\ M \end{array}\left[\begin{array}{ccccccc|c} 6 & 3 & 8 & 1 & 0 & 0 & 0 & 768 \\ 1 & 1 & \underline{2} & 0 & 1 & 0 & 0 & 144 \\ 2 & 5 & 0 & 0 & 0 & 1 & 0 & 216 \\ \hline -80 & -70 & -120 & 0 & 0 & 0 & 1 & 0 \end{array}\right] \end{array}$$

$$\begin{array}{c} \begin{array}{ccccccc} c & s & t & u & v & w & M \end{array} \\ \begin{array}{c} u \\ t \\ w \\ M \end{array}\left[\begin{array}{ccccccc|c} \underline{2} & -1 & 0 & 1 & -4 & 0 & 0 & 192 \\ \frac{1}{2} & \frac{1}{2} & 1 & 0 & \frac{1}{2} & 0 & 0 & 72 \\ 2 & 5 & 0 & 0 & 0 & 1 & 0 & 216 \\ \hline -20 & -10 & 0 & 0 & 60 & 0 & 1 & 8640 \end{array}\right] \end{array}$$

$$\begin{array}{c} \begin{array}{ccccccc} c & s & t & u & v & w & M \end{array} \\ \begin{array}{c} c \\ t \\ w \\ M \end{array}\left[\begin{array}{ccccccc|c} 1 & -\frac{1}{2} & 0 & \frac{1}{2} & -2 & 0 & 0 & 96 \\ 0 & \frac{3}{4} & 1 & -\frac{1}{4} & \frac{3}{2} & 0 & 0 & 24 \\ 0 & \underline{6} & 0 & -1 & 4 & 1 & 0 & 24 \\ \hline 0 & -20 & 0 & 10 & 20 & 0 & 1 & 10{,}560 \end{array}\right] \end{array}$$

$$\begin{array}{c} \begin{array}{ccccccc} c & s & t & u & v & w & M \end{array} \\ \begin{array}{c} c \\ t \\ s \\ M \end{array}\left[\begin{array}{ccccccc|c} 1 & 0 & 0 & \frac{5}{12} & -\frac{5}{3} & \frac{1}{12} & 0 & 98 \\ 0 & 0 & 1 & -\frac{1}{8} & 1 & -\frac{1}{8} & 0 & 21 \\ 0 & 1 & 0 & -\frac{1}{6} & \frac{2}{3} & \frac{1}{6} & 0 & 4 \\ \hline 0 & 0 & 0 & \frac{20}{3} & \frac{100}{3} & \frac{10}{3} & 1 & 10{,}640 \end{array}\right] \end{array}$$

98 chairs, 21 tables, 4 sofas.

19. Let b be the number of hours spent bicycling, j the number spent jogging, and s the number spent swimming each month. Maximize $200b + 475j + 275s$ subject to the constraints:

$$\begin{cases} b + j + s \le 30 \\ s \le 4 \\ -b + j - s \le 0 \\ b \ge 0,\ j \ge 0,\ s \ge 0 \end{cases}$$

	b	j	s	u	v	w	M	
u	1	1	1	1	0	0	0	30
v	0	0	1	0	1	0	0	4
w	−1	1	−1	0	0	1	0	0
M	−200	−475	−275	0	0	0	1	0

	b	j	s	u	v	w	M	
u	2	0	2	1	0	−1	0	30
v	0	0	1	0	1	0	0	4
j	−1	1	−1	0	0	1	0	0
M	−675	0	−750	0	0	475	1	0

	b	j	s	u	v	w	M	
u	2	0	0	1	−2	−1	0	22
s	0	0	1	0	1	0	0	4
j	−1	1	0	0	1	1	0	4
M	−675	0	0	0	750	475	1	3000

	b	j	s	u	v	w	M	
b	1	0	0	$\frac{1}{2}$	−1	$-\frac{1}{2}$	0	11
s	0	0	1	0	1	0	0	4
j	0	1	0	$\frac{1}{2}$	0	$\frac{1}{2}$	0	15
M	0	0	0	$\frac{675}{2}$	75	$\frac{275}{2}$	1	10,425

11 hours bicycling, 4 hours swimming, 15 hours jogging. He will lose $\dfrac{10,425}{3500} \approx 3$ pounds.

21. Let a be the number of type A restaurants, b the number of type B restaurants, and c the number of type C restaurants. Maximize $40a + 30b + 25c$ subject to the constraints

$$\begin{cases} 600a + 400b + 300c \le 48{,}000 \\ 15a + 9b + 5c \le 1000 \\ a + b + c \le 70 \\ a \ge 0,\ b \ge 0,\ c \ge 0 \end{cases}$$

(dollars in thousands)

	a	b	c	u	v	w	M	
u	600	400	300	1	0	0	0	48,000
v	15	9	5	0	1	0	0	1000
w	1	1	1	0	0	1	0	70
M	−40	−30	−25	0	0	0	1	0

	a	b	c	u	v	w	M	
u	0	40	100	1	−40	0	0	8000
a	1	$\frac{3}{5}$	$\frac{1}{3}$	0	$\frac{1}{15}$	0	0	$\frac{200}{3}$
w	0	$\frac{2}{5}$	$\frac{2}{3}$	0	$-\frac{1}{15}$	1	0	$\frac{10}{3}$
M	0	−6	$-\frac{35}{3}$	0	$\frac{8}{3}$	0	1	$\frac{8000}{3}$

	a	b	c	u	v	w	M	
u	0	−20	0	1	−30	−150	0	7500
a	1	$\frac{2}{5}$	0	0	$\frac{1}{10}$	$-\frac{1}{2}$	0	65
c	0	$\frac{3}{5}$	1	0	$-\frac{1}{10}$	$\frac{3}{2}$	0	5
M	0	1	0	0	$\frac{3}{2}$	$\frac{35}{2}$	1	2725

65 type A restaurants and 5 type C restaurants.

23. Let a be the number of bags of mix A, b the number of bags of mix B, and c the number of bags of mix C. Maximize $3a + 5b + 6c$ subject to the constraints

$$\begin{cases} 12a + 10b + 8c \le 1200 \\ 5a + 6b + 8c \le 800 \\ 3a + 4b + 4c \le 600 \\ a \ge 0, b \ge 0, c \ge 0 \end{cases}$$

	a	b	c	u	v	w	M	
u	12	10	8	1	0	0	0	1200
v	5	6	$\underline{8}$	0	1	0	0	800
w	3	4	4	0	0	1	0	600
M	−3	−5	−6	0	0	0	1	0

	a	b	c	u	v	w	M	
u	7	$\underline{4}$	0	1	−1	0	0	400
c	$\frac{5}{8}$	$\frac{3}{4}$	1	0	$\frac{1}{8}$	0	0	100
w	$\frac{1}{2}$	1	0	0	$-\frac{1}{2}$	1	0	200
M	$\frac{3}{4}$	$-\frac{1}{2}$	0	0	$\frac{3}{4}$	0	1	600

	a	b	c	u	v	w	M	
b	$\frac{7}{4}$	1	0	$\frac{1}{4}$	$-\frac{1}{4}$	0	0	100
c	$-\frac{11}{16}$	0	1	$-\frac{3}{16}$	$\frac{5}{16}$	0	0	25
w	$-\frac{5}{4}$	0	0	$-\frac{1}{4}$	$-\frac{1}{4}$	1	0	100
M	$\frac{13}{8}$	0	0	$\frac{1}{8}$	$\frac{5}{8}$	0	1	650

No bags of Mix A, 100 bags of Mix B, and 25 bags of Mix C.

25.

	x	y	u	v	M	
u	1	$\underline{4}$	1	0	0	300
v	1	2	0	1	0	200
M	−200	−500	0	0	1	0

	x	y	u	v	M	
y	$\frac{1}{4}$	1	$\frac{1}{4}$	0	0	75
v	$\frac{1}{2}$	0	$-\frac{1}{2}$	1	0	50
M	−75	0	125	0	1	37,500

	x	y	u	v	M	
y	0	1	$\frac{1}{2}$	$-\frac{1}{2}$	0	50
x	1	0	−1	2	0	100
M	0	0	50	150	1	45,000

$x = 100$, $y = 50$; $M = 45,000$

27.

	x	y	u	v	M	
u	1	$\underline{4}$	1	0	0	4
v	3	2	0	1	0	6
M	−4	−6	0	0	1	0

	x	y	u	v	M	
y	$\frac{1}{4}$	1	$\frac{1}{4}$	0	0	1
v	$\frac{5}{2}$	0	$-\frac{1}{2}$	1	0	4
M	$-\frac{5}{2}$	0	$\frac{3}{2}$	0	1	6

	x	y	u	v	M	
y	0	1	$\frac{3}{10}$	$-\frac{1}{10}$	0	$\frac{3}{5}$
x	1	0	$-\frac{1}{5}$	$\frac{2}{5}$	0	$\frac{8}{5}$
M	0	0	1	1	1	10

$x = \dfrac{8}{5}$, $y = \dfrac{3}{5}$; $M = 10$

29.

	x	y	z	u	v	w	M	
u	$\underline{4}$	1	5	1	0	0	0	20
v	1	2	4	0	1	0	0	50
w	4	10	1	0	0	1	0	32
M	−16	−4	20	0	0	0	1	0

	x	y	z	u	v	w	M	
x	1	$\frac{1}{4}$	$\frac{5}{4}$	$\frac{1}{4}$	0	0	0	5
v	0	$\frac{7}{4}$	$\frac{11}{4}$	$-\frac{1}{4}$	1	0	0	45
w	0	9	−4	−1	0	1	0	12
M	0	0	40	4	0	0	1	80

$x = 5$, $y = 0$, $z = 0$; $M = 80$

Exercises 4.3

1.

$$
\begin{array}{c}
\\
u\\
v\\
M
\end{array}
\begin{array}{cccccc}
x & y & u & v & M & \\
\left[\begin{array}{ccccc|c}
1 & 1 & 1 & 0 & 0 & 5 \\
2 & \underline{-3} & 0 & 1 & 0 & -12 \\
\hline
-40 & -30 & 0 & 0 & 1 & 0
\end{array}\right]
\end{array}
$$

$$
\begin{array}{c}
\\
u\\
y\\
M
\end{array}
\begin{array}{cccccc}
x & y & u & v & M & \\
\left[\begin{array}{ccccc|c}
\frac{5}{3} & 0 & 1 & \frac{1}{3} & 0 & 1 \\
-\frac{2}{3} & 1 & 0 & -\frac{1}{3} & 0 & 4 \\
\hline
-60 & 0 & 0 & -10 & 1 & 120
\end{array}\right]
\end{array}
$$

$$
\begin{array}{c}
\\
x\\
y\\
M
\end{array}
\begin{array}{cccccc}
x & y & u & v & M & \\
\left[\begin{array}{ccccc|c}
1 & 0 & \frac{3}{5} & \frac{1}{5} & 0 & \frac{3}{5} \\
0 & 1 & \frac{2}{5} & -\frac{1}{5} & 0 & \frac{22}{5} \\
\hline
0 & 0 & 36 & 2 & 1 & 156
\end{array}\right]
\end{array}
$$

$x = \dfrac{3}{5},\ y = \dfrac{22}{5};\ M = 156$

3.

$$
\begin{array}{c}
\\
u\\
v\\
M
\end{array}
\begin{array}{cccccc}
x & y & u & v & M & \\
\left[\begin{array}{ccccc|c}
-1 & \underline{-1} & 1 & 0 & 0 & -3 \\
-2 & 0 & 0 & 1 & 0 & -5 \\
\hline
3 & 1 & 0 & 0 & 1 & 0
\end{array}\right]
\end{array}
$$

$$
\begin{array}{c}
\\
y\\
v\\
M
\end{array}
\begin{array}{cccccc}
x & y & u & v & M & \\
\left[\begin{array}{ccccc|c}
1 & 1 & -1 & 0 & 0 & 3 \\
\underline{-2} & 0 & 0 & 1 & 0 & -5 \\
\hline
2 & 0 & 1 & 0 & 1 & -3
\end{array}\right]
\end{array}
$$

$$
\begin{array}{c}
\\
y\\
x\\
M
\end{array}
\begin{array}{cccccc}
x & y & u & v & M & \\
\left[\begin{array}{ccccc|c}
0 & 1 & -1 & \frac{1}{2} & 0 & \frac{1}{2} \\
1 & 0 & 0 & -\frac{1}{2} & 0 & \frac{5}{2} \\
\hline
0 & 0 & 1 & 1 & 1 & -8
\end{array}\right]
\end{array}
$$

$x = \dfrac{5}{2},\ y = \dfrac{1}{2}$ Since $M = -8$, the minimum is 8. Pivoting about the -2 at the start leads to the same final matrix.

5.

$$
\begin{array}{c}
\\
u\\
v\\
w\\
t\\
M
\end{array}
\begin{array}{cccccccc}
x & y & u & v & w & t & M & \\
\left[\begin{array}{ccccccc|c}
-2 & -1 & 1 & 0 & 0 & 0 & 0 & -11 \\
1 & 1 & 0 & 1 & 0 & 0 & 0 & 10 \\
\frac{1}{3} & 1 & 0 & 0 & 1 & 0 & 0 & 6 \\
-\frac{1}{4} & \underline{-1} & 0 & 0 & 0 & 1 & 0 & -4 \\
\hline
13 & 4 & 0 & 0 & 0 & 0 & 1 & 0
\end{array}\right]
\end{array}
$$

$$
\begin{array}{c}
\\
u\\
v\\
w\\
y\\
M
\end{array}
\begin{array}{cccccccc}
x & y & u & v & w & t & M & \\
\left[\begin{array}{ccccccc|c}
-\frac{7}{4} & 0 & 1 & 0 & 0 & -1 & 0 & -7 \\
\frac{3}{4} & 0 & 0 & 1 & 0 & 1 & 0 & 6 \\
\frac{1}{12} & 0 & 0 & 0 & 1 & 1 & 0 & 2 \\
\frac{1}{4} & 1 & 0 & 0 & 0 & -1 & 0 & 4 \\
\hline
12 & 0 & 0 & 0 & 0 & 4 & 1 & -16
\end{array}\right]
\end{array}
$$

$$
\begin{array}{c}
\\
u\\
v\\
t\\
y\\
M
\end{array}
\begin{array}{cccccccc}
x & y & u & v & w & t & M & \\
\left[\begin{array}{ccccccc|c}
-\frac{5}{3} & 0 & 1 & 0 & 1 & 0 & 0 & -5 \\
\frac{2}{3} & 0 & 0 & 1 & -1 & 0 & 0 & 4 \\
\frac{1}{12} & 0 & 0 & 0 & 1 & 1 & 0 & 2 \\
\frac{1}{3} & 1 & 0 & 0 & 1 & 0 & 0 & 6 \\
\hline
\frac{35}{3} & 0 & 0 & 0 & -4 & 0 & 1 & -24
\end{array}\right]
\end{array}
$$

$$
\begin{array}{c}
\\
x\\
v\\
t\\
y\\
M
\end{array}
\begin{array}{cccccccc}
x & y & u & v & w & t & M & \\
\left[\begin{array}{ccccccc|c}
1 & 0 & -\frac{3}{5} & 0 & -\frac{3}{5} & 0 & 0 & 3 \\
0 & 0 & \frac{2}{5} & 1 & -\frac{3}{5} & 0 & 0 & 2 \\
0 & 0 & \frac{1}{20} & 0 & \frac{21}{20} & 1 & 0 & \frac{7}{4} \\
0 & 1 & \frac{1}{5} & 0 & \frac{6}{5} & 0 & 0 & 5 \\
\hline
0 & 0 & 7 & 0 & 3 & 0 & 1 & -59
\end{array}\right]
\end{array}
$$

$x = 3,\ y = 5$; since $M = -59$, the minimum is 59. Other choices of pivot entries lead to the same final matrix.

7.

$$\begin{array}{c} \\ u \\ v \\ w \\ t \\ M \end{array}\begin{array}{cccccccc} x & y & u & v & w & t & M & \\ \end{array}\left[\begin{array}{ccccccc|c} -2 & -5 & 1 & 0 & 0 & 0 & 0 & -30 \\ 3 & -\underline{5} & 0 & 1 & 0 & 0 & 0 & -5 \\ 8 & 3 & 0 & 0 & 1 & 0 & 0 & 101 \\ -9 & 7 & 0 & 0 & 0 & 1 & 0 & 42 \\ \hline 2 & 7 & 0 & 0 & 0 & 0 & 1 & 0 \end{array}\right]$$

$$\begin{array}{c} \\ u \\ y \\ w \\ t \\ M \end{array}\begin{array}{cccccc} x & y & u & v & w & t & M \\ \end{array}\left[\begin{array}{ccccccc|c} -\underline{5} & 0 & 1 & -1 & 0 & 0 & 0 & -25 \\ -\frac{3}{5} & 1 & 0 & -\frac{1}{5} & 0 & 0 & 0 & 1 \\ \frac{49}{5} & 0 & 0 & \frac{3}{5} & 1 & 0 & 0 & 98 \\ -\frac{24}{5} & 0 & 0 & \frac{7}{5} & 0 & 1 & 0 & 35 \\ \hline \frac{31}{5} & 0 & 0 & \frac{7}{5} & 0 & 0 & 1 & -7 \end{array}\right]$$

$$\begin{array}{c} \\ x \\ y \\ w \\ t \\ M \end{array}\begin{array}{cccccc} x & y & u & v & w & t & M \\ \end{array}\left[\begin{array}{ccccccc|c} 1 & 0 & -\frac{1}{5} & \frac{1}{5} & 0 & 0 & 0 & 5 \\ 0 & 1 & -\frac{3}{25} & -\frac{2}{25} & 0 & 0 & 0 & 4 \\ 0 & 0 & \frac{49}{25} & -\frac{34}{25} & 1 & 0 & 0 & 49 \\ 0 & 0 & -\frac{24}{25} & \frac{59}{25} & 0 & 1 & 0 & 59 \\ \hline 0 & 0 & \frac{31}{25} & \frac{4}{25} & 0 & 0 & 1 & -38 \end{array}\right]$$

$x = 5$, $y = 4$; since $M = -38$, the minimum is 38.

9. Minimize $3a + 1.5b$ subject to the constraints:

$$\begin{cases} 30a + 10b \geq 60 \\ 10a + 10b \geq 40 \\ 20a + 60b \geq 120 \\ a \geq 0, b \geq 0 \end{cases}$$

$$\begin{array}{c} \\ u \\ v \\ w \\ M \end{array}\begin{array}{ccccc} a & b & u & v & w & M \\ \end{array}\left[\begin{array}{cccccc|c} \underline{-30} & -10 & 1 & 0 & 0 & 0 & -60 \\ -10 & -10 & 0 & 1 & 0 & 0 & -40 \\ -20 & -60 & 0 & 0 & 1 & 0 & -120 \\ \hline 3 & \frac{3}{2} & 0 & 0 & 0 & 1 & 0 \end{array}\right]$$

$$\begin{array}{c} \\ a \\ v \\ w \\ M \end{array}\begin{array}{ccccc} a & b & u & v & w & M \\ \end{array}\left[\begin{array}{cccccc|c} 1 & \frac{1}{3} & -\frac{1}{30} & 0 & 0 & 0 & 2 \\ 0 & -\frac{20}{3} & -\frac{1}{3} & 1 & 0 & 0 & -20 \\ 0 & -\frac{160}{3} & -\frac{2}{3} & 0 & 1 & 0 & -80 \\ \hline 0 & \frac{1}{2} & \frac{1}{10} & 0 & 0 & 1 & -6 \end{array}\right]$$

$$\begin{array}{c} \\ a \\ v \\ b \\ M \end{array}\begin{array}{ccccc} a & b & u & v & w & M \\ \end{array}\left[\begin{array}{cccccc|c} 1 & 0 & -\frac{3}{80} & 0 & \frac{1}{160} & 0 & \frac{3}{2} \\ 0 & 0 & -\frac{1}{4} & 1 & -\frac{1}{8} & 0 & -10 \\ 0 & 1 & \frac{1}{80} & 0 & -\frac{3}{160} & 0 & \frac{3}{2} \\ \hline 0 & 0 & \frac{3}{32} & 0 & \frac{3}{320} & 1 & -\frac{27}{4} \end{array}\right]$$

$$\begin{array}{c} \\ a \\ w \\ b \\ M \end{array}\begin{array}{ccccc} a & b & u & v & w & M \\ \end{array}\left[\begin{array}{cccccc|c} 1 & 0 & -\frac{1}{20} & \frac{1}{20} & 0 & 0 & 1 \\ 0 & 0 & 2 & -8 & 1 & 0 & 80 \\ 0 & 1 & \frac{1}{20} & -\frac{3}{20} & 0 & 0 & 3 \\ \hline 0 & 0 & \frac{3}{40} & \frac{3}{40} & 0 & 1 & -\frac{15}{2} \end{array}\right]$$

1 serving of food A, 3 servings of food B

11. Maximize $30a + 50b + 60c$ subject to the constraints:

$$\begin{cases} a + b + c \leq 600 \\ a \geq 100 \\ b \geq 50 \\ b + c \geq 200 \\ a \geq 0, b \geq 0, c \geq 0 \end{cases}$$

$$
\begin{array}{c}
 \\
u \\
v \\
w \\
t \\
M
\end{array}
\begin{array}{c}
\begin{array}{cccccccc}
a & b & c & u & v & w & t & M
\end{array} \\
\left[
\begin{array}{cccccccc|c}
1 & 1 & 1 & 1 & 0 & 0 & 0 & 0 & 600 \\
\underline{-1} & 0 & 0 & 0 & 1 & 0 & 0 & 0 & -100 \\
0 & -1 & 0 & 0 & 0 & 1 & 0 & 0 & -50 \\
0 & -1 & -1 & 0 & 0 & 0 & 1 & 0 & -200 \\
\hline
-30 & -50 & -60 & 0 & 0 & 0 & 0 & 1 & 0
\end{array}
\right]
\end{array}
$$

$$
\begin{array}{c}
 \\
u \\
a \\
w \\
t \\
M
\end{array}
\begin{array}{c}
\begin{array}{cccccccc}
a & b & c & u & v & w & t & M
\end{array} \\
\left[
\begin{array}{cccccccc|c}
0 & 1 & 1 & 1 & 1 & 0 & 0 & 0 & 500 \\
1 & 0 & 0 & 0 & -1 & 0 & 0 & 0 & 100 \\
0 & \underline{-1} & 0 & 0 & 0 & 1 & 0 & 0 & -50 \\
0 & -1 & -1 & 0 & 0 & 0 & 1 & 0 & -200 \\
\hline
0 & -50 & -60 & 0 & -30 & 0 & 0 & 1 & 3000
\end{array}
\right]
\end{array}
$$

$$
\begin{array}{c}
 \\
u \\
a \\
b \\
t \\
M
\end{array}
\begin{array}{c}
\begin{array}{cccccccc}
a & b & c & u & v & w & t & M
\end{array} \\
\left[
\begin{array}{cccccccc|c}
0 & 0 & 1 & 1 & 1 & 1 & 0 & 0 & 450 \\
1 & 0 & 0 & 0 & -1 & 0 & 0 & 0 & 100 \\
0 & 1 & 0 & 0 & 0 & -1 & 0 & 0 & 50 \\
0 & 0 & \underline{-1} & 0 & 0 & -1 & 1 & 0 & -150 \\
\hline
0 & 0 & -60 & 0 & -30 & -50 & 0 & 1 & 5500
\end{array}
\right]
\end{array}
$$

$$
\begin{array}{c}
 \\
u \\
a \\
b \\
c \\
M
\end{array}
\begin{array}{c}
\begin{array}{cccccccc}
a & b & c & u & v & w & t & M
\end{array} \\
\left[
\begin{array}{cccccccc|c}
0 & 0 & 0 & 1 & 1 & 0 & \underline{1} & 0 & 300 \\
1 & 0 & 0 & 0 & -1 & 0 & 0 & 0 & 100 \\
0 & 1 & 0 & 0 & 0 & -1 & 0 & 0 & 50 \\
0 & 0 & 1 & 0 & 0 & 1 & -1 & 0 & 150 \\
\hline
0 & 0 & 0 & 0 & -30 & 10 & -60 & 1 & 14{,}500
\end{array}
\right]
\end{array}
$$

$$
\begin{array}{c}
 \\
t \\
a \\
b \\
c \\
M
\end{array}
\begin{array}{c}
\begin{array}{cccccccc}
a & b & c & u & v & w & t & M
\end{array} \\
\left[
\begin{array}{cccccccc|c}
0 & 0 & 0 & 1 & 1 & 0 & 1 & 0 & 300 \\
1 & 0 & 0 & 0 & -1 & 0 & 0 & 0 & 100 \\
0 & 1 & 0 & 0 & 0 & -1 & 0 & 0 & 50 \\
0 & 0 & 1 & 1 & 1 & 1 & 0 & 0 & 450 \\
\hline
0 & 0 & 0 & 60 & 30 & 10 & 0 & 1 & 32{,}500
\end{array}
\right]
\end{array}
$$

Stock 100 of brand A, 50 of brand B, and 450 of brand C.

13. Let x = number of computers shipped from Chicago to Detroit. Let y = number of computers shipped from Chicago to Fletcher. Then (40 − x) equals the number of computers shipped from Boston to Detroit and (30 − y) represents the number of computers shipped from Boston to Fletcher.

Minimize:

$125(40 − x) + 100x + 180(30 − y) + 160y$

or $10400 − 25x − 20y$

Subject to:

$$\begin{cases} x + y \le 80 \\ x + y \ge 20 \\ x \le 40 \\ y \le 30 \\ x \ge 0, y \ge 0 \end{cases}$$

	x	y	u	v	w	t	M	
u	1	1	1	0	0	0	0	80
v	−1	−1	0	1	0	0	0	−20
w	1	0	0	0	1	0	0	40
t	0	1	0	0	0	1	0	30
M	−25	−20	0	0	0	0	1	−10400

	x	y	u	v	w	t	M	
u	0	0	1	1	0	0	0	60
x	1	1	0	−1	0	0	0	20
w	0	−1	0	1	1	0	0	20
t	0	1	0	0	0	1	0	30
M	0	5	0	−25	0	0	1	−9900

	x	y	u	v	w	t	M	
u	0	1	1	0	−1	0	0	40
x	1	0	0	0	1	0	0	40
v	0	−1	0	1	1	0	0	20
t	0	1	0	0	0	1	0	30
M	0	−20	0	0	25	0	1	−9400

	x	y	u	v	w	t	M	
u	0	0	1	0	−1	−1	0	10
x	1	0	0	0	1	0	0	40
v	0	0	0	1	1	1	1	50
y	0	1	0	0	0	1	0	30
M	0	0	0	0	25	20	1	−8800

Therefore, the minimum cost is $8800 when you ship all the computers from Chicago.

15.

	x	y	u	v	M	
u	4	1	1	0	0	5
v	−1	−3	0	1	0	−4
M	−1	2	0	0	1	0

	x	y	u	v	M	
x	1	$\frac{1}{4}$	$\frac{1}{4}$	0	0	$\frac{5}{4}$
v	0	$-\frac{11}{4}$	$\frac{1}{4}$	1	0	$-\frac{11}{4}$
M	0	$\frac{9}{4}$	$\frac{1}{4}$	0	1	$\frac{5}{4}$

	x	y	u	v	M	
x	1	0	$\frac{3}{11}$	$\frac{1}{11}$	0	1
y	0	1	$-\frac{1}{11}$	$-\frac{4}{11}$	0	1
M	0	0	$\frac{5}{11}$	$\frac{9}{11}$	1	−1

$x = 1, y = 1; M = −1$

Exercises 4.4

1. x (paring knives) goes from 14 to

   $14 + 54\left(-\dfrac{5}{27}\right) = 4$, y (pocket knives)

   goes from 8 to $8 + 54\left(\dfrac{7}{27}\right) = 22$, and

   the profit goes from $82 to

   $82 + 54\left(\dfrac{20}{27}\right) = \$122$.

3. The slack variable involved is w. x (sets to Rockville) goes from 20 to $20 + (50 - 45)(1) = 25$, y (sets to Annapolis) goes from 25 to $25 + (50 - 45)(0) = 25$, and the cost goes from \$260 to $-[-260 + (50 - 45)(2)] = \$250$.

5. $100 + h\left(\dfrac{1}{4}\right) \geq 0$, $25 + h\left(-\dfrac{3}{16}\right) \geq 0$,

   and $100 + h\left(-\dfrac{1}{4}\right) \geq 0$, so

   $h \geq -400$, $h \leq \dfrac{400}{3}$ and $h \leq 400$ or

   $-400 \leq h \leq \dfrac{400}{3}$.

7. $\begin{bmatrix} 9 & 1 & 1 \\ 4 & 8 & -3 \end{bmatrix}$

9. $\begin{bmatrix} 7 \\ 6 \\ 5 \\ 1 \end{bmatrix}$

11. Yes

13. Minimize $\begin{bmatrix} 7 & 5 & 4 \end{bmatrix} \begin{bmatrix} x \\ y \\ z \end{bmatrix}$ subject to the

   constraints $\begin{bmatrix} 3 & 8 & 9 \\ 1 & 2 & 5 \\ 4 & 1 & 7 \end{bmatrix} \begin{bmatrix} x \\ y \\ z \end{bmatrix} \geq \begin{bmatrix} 75 \\ 80 \\ 67 \end{bmatrix}$ and

   $\begin{bmatrix} x \\ y \\ z \end{bmatrix} \geq \begin{bmatrix} 0 \\ 0 \\ 0 \end{bmatrix}$.

15. Maximize $\begin{bmatrix} 3 & 5 \end{bmatrix} \begin{bmatrix} x \\ y \end{bmatrix}$ subject to the

   constraints $\begin{bmatrix} 3 & 6 \\ 7 & 5 \\ 4 & 3 \end{bmatrix} \begin{bmatrix} x \\ y \end{bmatrix} \leq \begin{bmatrix} 90 \\ 138 \\ 120 \end{bmatrix}$ and

   $\begin{bmatrix} x \\ y \end{bmatrix} \geq \begin{bmatrix} 0 \\ 0 \end{bmatrix}$.

17. Minimize $2x + 3y$ subject to the

   constraints $\begin{cases} 7x + 4y \geq 33 \\ 5x + 8y \geq 44 \\ x + 3y \geq 55 \\ x \geq 0,\ y \geq 0 \end{cases}$.

19. Constraints

Cell	Name	Final Value	Shadow Price	Constraint R.H. Side	Allowable Increase	Allowable Decrease
\$B\$1		45	−2	45	10	20
\$B\$2		45	0	15	30	1E + 30
\$B\$3		20	0	30	1E + 30	10
\$B\$4		25	−1	25	20	10
\$B\$5		20	0	0	20	1E + 30
\$B\$6		25	0	0	25	1E + 30

The first line of the table corresponds to the constraint for the number of sets shipped from College Park. The shadow price is –2, and the range feasibility is [45 – 20, 45 + 10] or [25, 55].

Exercises 4.5

1. Primal: Maximize $[4 \ 2]\begin{bmatrix} x \\ y \end{bmatrix}$ subject to

$$\begin{bmatrix} 5 & 1 \\ 3 & 2 \end{bmatrix}\begin{bmatrix} x \\ y \end{bmatrix} \leq \begin{bmatrix} 80 \\ 76 \end{bmatrix} \text{ and } \begin{bmatrix} x \\ y \end{bmatrix} \geq \begin{bmatrix} 0 \\ 0 \end{bmatrix}.$$

Dual: Minimize $[80 \ 76]\begin{bmatrix} u \\ v \end{bmatrix}$ subject to

$$\begin{bmatrix} 5 & 3 \\ 1 & 2 \end{bmatrix}\begin{bmatrix} u \\ v \end{bmatrix} \geq \begin{bmatrix} 4 \\ 2 \end{bmatrix} \text{ and } \begin{bmatrix} u \\ v \end{bmatrix} \geq \begin{bmatrix} 0 \\ 0 \end{bmatrix}.$$

Minimize $80u + 76v$ subject to the

constraints $\begin{cases} 5u + 3v \geq 4 \\ u + 2v \geq 2 \\ u \geq 0, v \geq 0 \end{cases}$.

3. Primal: Minimize $[10 \ 12]\begin{bmatrix} x \\ y \end{bmatrix}$ subject to

$$\begin{bmatrix} 1 & 2 \\ -1 & 1 \\ 2 & 3 \end{bmatrix}\begin{bmatrix} x \\ y \end{bmatrix} \geq \begin{bmatrix} 1 \\ 2 \\ 1 \end{bmatrix} \text{ and } \begin{bmatrix} x \\ y \end{bmatrix} \geq \begin{bmatrix} 0 \\ 0 \end{bmatrix}.$$

Dual: Maximize $[1 \ 2 \ 1]\begin{bmatrix} u \\ v \\ w \end{bmatrix}$ subject to

$$\begin{bmatrix} 1 & -1 & 2 \\ 2 & 1 & 3 \end{bmatrix}\begin{bmatrix} u \\ v \\ w \end{bmatrix} \leq \begin{bmatrix} 10 \\ 12 \end{bmatrix} \text{ and }$$

$$\begin{bmatrix} u \\ v \\ w \end{bmatrix} \geq \begin{bmatrix} 0 \\ 0 \\ 0 \end{bmatrix}.$$

Maximize $u + 2v + w$ subject to the

constraints $\begin{cases} u - v + 2w \leq 10 \\ 2u + v + 3w \leq 12 \\ u \geq 0, v \geq 0, w \geq 0 \end{cases}$.

5. Primal: Minimize $[3 \ 5 \ 1]\begin{bmatrix} x \\ y \\ z \end{bmatrix}$ subject to

$$\begin{bmatrix} -2 & 4 & 6 \\ 8 & 1 & 9 \end{bmatrix}\begin{bmatrix} x \\ y \\ z \end{bmatrix} \geq \begin{bmatrix} -7 \\ 10 \end{bmatrix} \text{ and } \begin{bmatrix} x \\ y \\ z \end{bmatrix} \geq \begin{bmatrix} 0 \\ 0 \\ 0 \end{bmatrix}.$$

Dual: Maximize $[-7 \ 10]\begin{bmatrix} u \\ v \end{bmatrix}$ subject to

$$\begin{bmatrix} -2 & 8 \\ 4 & 1 \\ 6 & 9 \end{bmatrix}\begin{bmatrix} u \\ v \end{bmatrix} \leq \begin{bmatrix} 3 \\ 5 \\ 1 \end{bmatrix} \text{ and } \begin{bmatrix} u \\ v \end{bmatrix} \geq \begin{bmatrix} 0 \\ 0 \end{bmatrix}.$$

Maximize $-7u + 10v$ subject to the

constraints $\begin{cases} -2u + 8v \leq 3 \\ 4u + v \leq 5 \\ 6u + 9v \leq 1 \\ u \geq 0, v \geq 0 \end{cases}$.

7. $x = 12$, $y = 20$, $M = 88$; $u = \dfrac{2}{7}$, $v = \dfrac{6}{7}$, $M = 88$.

9. $x = 0$, $y = 2$, $M = 24$; $u = 0$, $v = 12$, $w = 0$, $M = 24$

11. Maximize $3u + 5v$ subject to the

constraints $\begin{cases} u + 2v \leq 3 \\ u \leq 1 \\ u \geq 0, v \geq 0 \end{cases}$.

Solve the dual.

	u	v	x	y	M	
x	1	$\underline{2}$	1	0	0	3
y	1	0	0	1	0	1
M	-3	-5	0	0	1	0

	u	v	x	y	M	
v	$\frac{1}{2}$	1	$\frac{1}{2}$	0	0	$\frac{3}{2}$
y	$\underline{1}$	0	0	1	0	1
M	$-\frac{1}{2}$	0	$\frac{5}{2}$	0	1	$\frac{15}{2}$

$$
\begin{array}{c}
\quad\; u \quad\; v \quad\; x \quad\; y \quad\; M \\
\begin{array}{c} v \\ u \\ M \end{array}
\left[\begin{array}{ccccc|c}
0 & 1 & \frac{1}{2} & -\frac{1}{2} & 0 & 1 \\
1 & 0 & 0 & 1 & 0 & 1 \\
\hline
0 & 0 & \frac{5}{2} & \frac{1}{2} & 1 & 8
\end{array}\right]
\end{array}
$$

$x = \dfrac{5}{2}, y = \dfrac{1}{2}$, minimum $= 8$; $u = 1$, $v = 1$, maximum $= 8$

13. Minimize $6u + 9v + 12w$ subject to the

constraints $\begin{cases} u + 3v \ge 10 \\ -2u + w \ge 12 \\ v + 3w \ge 10 \\ u \ge 0, v \ge 0, w \ge 0 \end{cases}$.

Solve the primal.

$$
\begin{array}{c}
\quad\; x \quad\;\; y \quad\;\; z \quad\; u \quad\; v \quad\; w \quad\; M \\
\begin{array}{c} u \\ v \\ w \\ M \end{array}
\left[\begin{array}{ccccccc|c}
1 & -2 & 0 & 1 & 0 & 0 & 0 & 6 \\
3 & 0 & 1 & 0 & 1 & 0 & 0 & 9 \\
0 & 1 & 3 & 0 & 0 & 1 & 0 & 12 \\
\hline
-10 & -12 & -10 & 0 & 0 & 0 & 1 & 0
\end{array}\right]
\end{array}
$$

$$
\begin{array}{c}
\quad\; x \;\; y \;\; z \;\; u \;\; v \;\; w \;\; M \\
\begin{array}{c} u \\ v \\ y \\ M \end{array}
\left[\begin{array}{ccccccc|c}
1 & 0 & 6 & 1 & 0 & 2 & 0 & 30 \\
3 & 0 & 1 & 0 & 1 & 0 & 0 & 9 \\
0 & 1 & 3 & 0 & 0 & 1 & 0 & 12 \\
\hline
-10 & 0 & 26 & 0 & 0 & 12 & 1 & 144
\end{array}\right]
\end{array}
$$

$$
\begin{array}{c}
\quad\; x \;\; y \;\; z \;\;\; u \;\;\; v \;\;\; w \;\; M \\
\begin{array}{c} u \\ x \\ y \\ M \end{array}
\left[\begin{array}{ccccccc|c}
0 & 0 & \frac{17}{3} & 1 & -\frac{1}{3} & 2 & 0 & 27 \\
1 & 0 & \frac{1}{3} & 0 & \frac{1}{3} & 0 & 0 & 3 \\
0 & 1 & 3 & 0 & 0 & 1 & 0 & 12 \\
\hline
0 & 0 & \frac{88}{3} & 0 & \frac{10}{3} & 12 & 1 & 174
\end{array}\right]
\end{array}
$$

$x = 3, y = 12, z = 0$, maximum $= 174$; $u = 0, v = \dfrac{10}{3}, w = 12$, minimum $= 174$

15. Suppose we can hire workers out at a profit of u dollars per hour, sell the steel at a profit of v dollars per unit, and sell the wood at a profit of w dollars per unit. To find the minimum profit at which that should be done, minimize $90u + 138v + 120w$ subject to the constraints

$\begin{cases} 3u + 7v + 4w \ge 3 \\ 6u + 5v + 3w \ge 5 \\ u \ge 0, v \ge 0, w \ge 0 \end{cases}$.

17. Suppose we can buy anthracite at u dollars per ton, ordinary coal at v dollars per ton, and bituminous coal at w dollars per ton. To find the maximum cost at which this should be done, maximize $80u + 60v + 75w$ subject to the

constraints $\begin{cases} 4u + 4v + 7w \le 150 \\ 10u + 5v + 5w \le 200. \\ u \ge 0, v \ge 0, w \ge 0 \end{cases}$

19. The new primal problem is to maximize $3x + 5y + pz$, where p is the profit per table knife and z is the number of table knives produced, subject to the constraints

$\begin{cases} 3x + 6y + 4z \le 90 \\ 7x + 5y + 6z \le 138 \\ 4x + 3y + 2z \le 120 \\ x \ge 0, y \ge 0, z \ge 0 \end{cases}$.

The dual is to minimize $90u + 138v + 120w$ subject to the constraints

$\begin{cases} 3u + 7v + 4w \ge 3 \\ 6u + 5v + 3w \ge 5 \\ 4u + 6v + 2w \ge p \\ u \ge 0, v \ge 0, w \ge 0 \end{cases}$.

The original solution, with

$u = \dfrac{20}{7}, v = \dfrac{1}{9}, w = 0$, will still be

optimal if $4\left(\dfrac{20}{27}\right) + 6\left(\dfrac{1}{9}\right) + 2(0) \ge p$.

Since $4\left(\dfrac{20}{27}\right) + 6\left(\dfrac{1}{9}\right) + 2(0) \approx \$3.63$,

that is the minimum profit per table knife that needs to be realized to warrant adding table knives to the product line.

21. Dual: Maximize $5u + 8v$ subject to the

constraints $\begin{cases} u + 2v \le 16 \\ 3u + 4v \le 42. \\ u \ge 0, v \ge 0 \end{cases}$

$$\begin{array}{c} \begin{array}{ccccc} u & v & x & y & M \end{array} \\ \begin{array}{c} x \\ y \\ M \end{array} \left[\begin{array}{ccccc|c} 1 & \underline{2} & 1 & 0 & 0 & 16 \\ 3 & 4 & 0 & 1 & 0 & 42 \\ \hline -5 & -8 & 0 & 0 & 1 & 0 \end{array} \right] \end{array}$$

$$\begin{array}{c} \begin{array}{ccccc} u & v & x & y & M \end{array} \\ \begin{array}{c} v \\ y \\ M \end{array} \left[\begin{array}{ccccc|c} \frac{1}{2} & 1 & \frac{1}{2} & 0 & 0 & 8 \\ \underline{1} & 0 & -2 & 1 & 0 & 10 \\ \hline -1 & 0 & 4 & 0 & 1 & 64 \end{array} \right] \end{array}$$

$$\begin{array}{c} \begin{array}{ccccc} u & v & x & y & M \end{array} \\ \begin{array}{c} v \\ u \\ M \end{array} \left[\begin{array}{ccccc|c} 0 & 1 & \frac{3}{2} & -\frac{1}{2} & 0 & 3 \\ 1 & 0 & -2 & 1 & 0 & 10 \\ \hline 0 & 0 & 2 & 1 & 1 & 74 \end{array} \right] \end{array}$$

$y = 2, y = 1, M = 74$

Chapter 4 Supplementary Exercises

1.

$$\begin{array}{c} \begin{array}{ccccc} x & y & u & v & M \end{array} \\ \begin{array}{c} u \\ v \\ M \end{array} \left[\begin{array}{ccccc|c} 2 & 1 & 1 & 0 & 0 & 7 \\ -1 & \underline{1} & 0 & 1 & 0 & 1 \\ \hline -3 & -4 & 0 & 0 & 1 & 0 \end{array} \right] \end{array}$$

$$\begin{array}{c} \begin{array}{ccccc} x & y & u & v & M \end{array} \\ \begin{array}{c} u \\ y \\ M \end{array} \left[\begin{array}{ccccc|c} \underline{3} & 0 & 1 & -1 & 0 & 6 \\ -1 & 1 & 0 & 1 & 0 & 1 \\ \hline -7 & 0 & 0 & 4 & 1 & 4 \end{array} \right] \end{array}$$

$$\begin{array}{c} \begin{array}{ccccc} x & y & u & v & M \end{array} \\ \begin{array}{c} x \\ y \\ M \end{array} \left[\begin{array}{ccccc|c} 1 & 0 & \frac{1}{3} & -\frac{1}{3} & 0 & 2 \\ 0 & 1 & \frac{1}{3} & \frac{2}{3} & 0 & 3 \\ \hline 0 & 0 & \frac{7}{3} & \frac{5}{3} & 1 & 18 \end{array} \right] \end{array}$$

$x = 2, y = 3, M = 18$

2.

$$\begin{array}{c} \begin{array}{ccccc} x & y & u & v & M \end{array} \\ \begin{array}{c} y \\ v \\ M \end{array} \left[\begin{array}{ccccc|c} 1 & \underline{1} & 1 & 0 & 0 & 7 \\ 4 & 3 & 0 & 1 & 0 & 24 \\ \hline -2 & -5 & 0 & 0 & 1 & 0 \end{array} \right] \end{array}$$

$$\begin{array}{c} \begin{array}{ccccc} x & y & u & v & M \end{array} \\ \begin{array}{c} y \\ v \\ M \end{array} \left[\begin{array}{ccccc|c} 1 & 1 & 1 & 0 & 0 & 7 \\ 1 & 0 & -3 & 1 & 0 & 3 \\ \hline 3 & 0 & 5 & 0 & 1 & 35 \end{array} \right] \end{array}$$

$x = 0, y = 7, M = 35$

3.

$$\begin{array}{c} \begin{array}{cccccc} x & y & u & v & w & M \end{array} \\ \begin{array}{c} u \\ v \\ w \\ M \end{array} \left[\begin{array}{cccccc|c} 1 & \underline{2} & 1 & 0 & 0 & 0 & 14 \\ 1 & 1 & 0 & 1 & 0 & 0 & 9 \\ 3 & 2 & 0 & 0 & 1 & 0 & 24 \\ \hline -2 & -3 & 0 & 0 & 0 & 1 & 0 \end{array} \right] \end{array}$$

$$\begin{array}{c} \begin{array}{cccccc} x & y & u & v & w & M \end{array} \\ \begin{array}{c} y \\ v \\ w \\ M \end{array} \left[\begin{array}{cccccc|c} \frac{1}{2} & 1 & \frac{1}{2} & 0 & 0 & 0 & 7 \\ \underline{\frac{1}{2}} & 0 & -\frac{1}{2} & 1 & 0 & 0 & 2 \\ 2 & 0 & -1 & 0 & 1 & 0 & 10 \\ \hline -\frac{1}{2} & 0 & \frac{3}{2} & 0 & 0 & 1 & 21 \end{array} \right] \end{array}$$

$$\begin{array}{c} \begin{array}{cccccc} x & y & u & v & w & M \end{array} \\ \begin{array}{c} y \\ x \\ w \\ M \end{array} \left[\begin{array}{cccccc|c} 0 & 1 & 1 & -1 & 0 & 0 & 5 \\ 1 & 0 & -1 & 2 & 0 & 0 & 4 \\ 0 & 0 & 1 & -4 & 1 & 0 & 2 \\ \hline 0 & 0 & 1 & 1 & 0 & 1 & 23 \end{array} \right] \end{array}$$

$x = 4, y = 5, M = 23$

4.

$$\begin{array}{c} \\ u \\ v \\ w \\ M \end{array} \begin{array}{cccccc} x & y & u & v & w & M \\ \left[\begin{array}{cccccc|c} 1 & 2 & 1 & 0 & 0 & 0 & 10 \\ 4 & 3 & 0 & 1 & 0 & 0 & 30 \\ -2 & \underline{1} & 0 & 0 & 1 & 0 & 0 \\ \hline -3 & -7 & 0 & 0 & 0 & 1 & 0 \end{array}\right] \end{array}$$

$$\begin{array}{c} \\ u \\ v \\ y \\ M \end{array} \begin{array}{cccccc} x & y & u & v & w & M \\ \left[\begin{array}{cccccc|c} \underline{5} & 0 & 1 & 0 & -2 & 0 & 10 \\ 10 & 0 & 0 & 1 & -3 & 0 & 30 \\ -2 & 1 & 0 & 0 & 1 & 0 & 0 \\ \hline -17 & 0 & 0 & 0 & 7 & 1 & 0 \end{array}\right] \end{array}$$

$$\begin{array}{c} \\ x \\ v \\ y \\ M \end{array} \begin{array}{cccccc} x & y & u & v & w & M \\ \left[\begin{array}{cccccc|c} 1 & 0 & \frac{1}{5} & 0 & -\frac{2}{5} & 0 & 2 \\ 0 & 0 & -2 & 1 & 1 & 0 & 10 \\ 0 & 1 & \frac{2}{5} & 0 & \frac{1}{5} & 0 & 4 \\ \hline 0 & 0 & \frac{17}{5} & 0 & \frac{1}{5} & 1 & 34 \end{array}\right] \end{array}$$

$x = 2, y = 4, M = 34$

5.

$$\begin{array}{c} \\ u \\ v \\ M \end{array} \begin{array}{ccccc} x & y & u & v & M \\ \left[\begin{array}{ccccc|c} \underline{-7} & -5 & 1 & 0 & 0 & -40 \\ -1 & -4 & 0 & 1 & 0 & -9 \\ \hline 1 & 1 & 0 & 0 & 1 & 0 \end{array}\right] \end{array}$$

$$\begin{array}{c} \\ x \\ v \\ M \end{array} \begin{array}{ccccc} x & y & u & v & M \\ \left[\begin{array}{ccccc|c} 1 & \frac{5}{7} & -\frac{1}{7} & 0 & 0 & \frac{40}{7} \\ 0 & -\frac{23}{7} & -\frac{1}{7} & 1 & 0 & -\frac{23}{7} \\ \hline 0 & \frac{2}{7} & \frac{1}{7} & 0 & 1 & -\frac{40}{7} \end{array}\right] \end{array}$$

$$\begin{array}{c} \\ x \\ y \\ M \end{array} \begin{array}{ccccc} x & y & u & v & M \\ \left[\begin{array}{ccccc|c} 1 & 0 & -\frac{4}{23} & \frac{5}{23} & 0 & 5 \\ 0 & 1 & \frac{1}{23} & -\frac{7}{23} & 0 & 1 \\ \hline 0 & 0 & \frac{3}{23} & \frac{2}{23} & 1 & -6 \end{array}\right] \end{array}$$

$x = 5, y = 1, M = -6$; the minimum is 6.

6.

$$\begin{array}{c} \\ u \\ v \\ M \end{array} \begin{array}{ccccc} x & y & u & v & M \\ \left[\begin{array}{ccccc|c} \underline{-1} & -1 & 1 & 0 & 0 & -6 \\ -1 & -2 & 0 & 1 & 0 & 0 \\ \hline 3 & 2 & 0 & 0 & 1 & 0 \end{array}\right] \end{array}$$

$$\begin{array}{c} \\ x \\ v \\ M \end{array} \begin{array}{ccccc} x & y & u & v & M \\ \left[\begin{array}{ccccc|c} 1 & \underline{1} & -1 & 0 & 0 & 6 \\ 0 & -1 & -1 & 1 & 0 & 6 \\ \hline 0 & -1 & 3 & 0 & 1 & -18 \end{array}\right] \end{array}$$

$$\begin{array}{c} \\ y \\ v \\ M \end{array} \begin{array}{ccccc} x & y & u & v & M \\ \left[\begin{array}{ccccc|c} 1 & 1 & -1 & 0 & 0 & 6 \\ 1 & 0 & -2 & 1 & 0 & 12 \\ \hline 1 & 0 & 2 & 0 & 1 & -12 \end{array}\right] \end{array}$$

$x = 0, y = 6, M = -12$; the minimum is 12.

7.

$$\begin{array}{c} \\ u \\ v \\ w \\ M \end{array} \begin{array}{cccccc} x & y & u & v & w & M \\ \left[\begin{array}{cccccc|c} -1 & -4 & 1 & 0 & 0 & 0 & -8 \\ -1 & -1 & 0 & 1 & 0 & 0 & -5 \\ \underline{-2} & -1 & 0 & 0 & 1 & 0 & -7 \\ \hline 20 & 30 & 0 & 0 & 0 & 1 & 0 \end{array}\right] \end{array}$$

$$\begin{array}{c} \\ u \\ v \\ x \\ M \end{array} \begin{array}{cccccc} x & y & u & v & w & M \\ \left[\begin{array}{cccccc|c} 0 & -\frac{7}{2} & 1 & 0 & -\frac{1}{2} & 0 & -\frac{9}{2} \\ 0 & -\frac{1}{2} & 0 & 1 & -\frac{1}{2} & 0 & -\frac{3}{2} \\ 1 & \frac{1}{2} & 0 & 0 & -\frac{1}{2} & 0 & \frac{7}{2} \\ \hline 0 & 20 & 0 & 0 & 10 & 1 & -70 \end{array}\right] \end{array}$$

$$\begin{array}{c} \\ u \\ w \\ x \\ M \end{array} \begin{array}{cccccc} x & y & u & v & w & M \\ \left[\begin{array}{cccccc|c} 0 & \underline{-3} & 1 & -1 & 0 & 0 & -3 \\ 0 & 1 & 0 & -2 & 1 & 0 & 3 \\ 1 & 1 & 0 & -1 & 0 & 0 & 5 \\ \hline 0 & 10 & 0 & 20 & 0 & 1 & -100 \end{array}\right] \end{array}$$

$$\begin{array}{c} \\ y \\ w \\ x \\ M \end{array} \begin{array}{cccccc} x & y & u & v & w & M \\ \left[\begin{array}{cccccc|c} 0 & 1 & -\frac{1}{3} & \frac{1}{3} & 0 & 0 & 1 \\ 0 & 0 & \frac{1}{3} & -\frac{7}{3} & 1 & 0 & 2 \\ 1 & 0 & \frac{1}{3} & -\frac{4}{3} & 0 & 0 & 4 \\ \hline 0 & 0 & \frac{10}{3} & \frac{50}{3} & 0 & 1 & -110 \end{array}\right] \end{array}$$

$x = 4, y = 1, M = -110$; the minimum is 110.

8.

$$\begin{array}{c} \\ u \\ v \\ w \\ M \end{array}\begin{array}{cccccc} x & y & u & v & w & M \\ \left[\begin{array}{cccccc|c} -2 & -1 & 1 & 0 & 0 & 0 & -10 \\ -3 & -2 & 0 & 1 & 0 & 0 & -18 \\ -1 & \underline{-2} & 0 & 0 & 1 & 0 & -10 \\ 5 & 7 & 0 & 0 & 0 & 1 & 0 \end{array}\right] \end{array}$$

$$\begin{array}{c} \\ u \\ v \\ y \\ M \end{array}\begin{array}{ccccccc} x & y & u & v & w & M \\ \left[\begin{array}{cccccc|c} -\frac{3}{2} & 0 & 1 & 0 & -\frac{1}{2} & 0 & -5 \\ -2 & 0 & 0 & 1 & \underline{-1} & 0 & -8 \\ \frac{1}{2} & 1 & 0 & 0 & -\frac{1}{2} & 0 & 5 \\ \frac{3}{2} & 0 & 0 & 0 & \frac{7}{2} & 1 & -35 \end{array}\right] \end{array}$$

$$\begin{array}{c} \\ u \\ w \\ y \\ M \end{array}\begin{array}{ccccccc} x & y & u & v & w & M \\ \left[\begin{array}{cccccc|c} -\frac{1}{2} & 0 & 1 & -\frac{1}{2} & 0 & 0 & -1 \\ \underline{2} & 0 & 0 & -1 & 1 & 0 & 8 \\ \frac{3}{2} & 1 & 0 & -\frac{1}{2} & 0 & 0 & 9 \\ -\frac{11}{2} & 0 & 0 & \frac{7}{2} & 0 & 1 & -63 \end{array}\right] \end{array}$$

$$\begin{array}{c} \\ x \\ w \\ y \\ M \end{array}\begin{array}{cccccc} x & y & u & v & w & M \\ \left[\begin{array}{cccccc|c} 1 & 0 & -2 & 1 & 0 & 0 & 2 \\ 0 & 0 & \underline{4} & -3 & 1 & 0 & 4 \\ 0 & 1 & 3 & -2 & 0 & 0 & 6 \\ 0 & 0 & -11 & 9 & 0 & 1 & -52 \end{array}\right] \end{array}$$

$$\begin{array}{c} \\ x \\ u \\ y \\ M \end{array}\begin{array}{cccccc} x & y & u & v & w & M \\ \left[\begin{array}{cccccc|c} 1 & 0 & 0 & -\frac{1}{2} & \frac{1}{2} & 0 & 4 \\ 0 & 0 & 1 & -\frac{3}{4} & \frac{1}{4} & 0 & 1 \\ 0 & 1 & 0 & \frac{1}{4} & -\frac{3}{4} & 0 & 3 \\ 0 & 0 & 0 & \frac{3}{4} & \frac{11}{4} & 1 & -41 \end{array}\right] \end{array}$$

$x = 4$, $y = 3$, $M = -41$; the minimum is 41.

9.

$$\begin{array}{c} \\ t \\ u \\ v \\ w \\ M \end{array}\begin{array}{ccccccccc} x & y & z & t & u & v & w & M \\ \left[\begin{array}{cccccccc|c} 1 & 0 & 0 & 1 & 0 & 0 & 0 & 0 & 4 \\ 0 & 1 & 0 & 0 & 1 & 0 & 0 & 0 & 6 \\ 0 & 0 & \underline{1} & 0 & 0 & 1 & 0 & 0 & 8 \\ 4 & 3 & 2 & 0 & 0 & 0 & 1 & 0 & 38 \\ -36 & -48 & -70 & 0 & 0 & 0 & 0 & 1 & 0 \end{array}\right] \end{array}$$

$$\begin{array}{c} \\ t \\ u \\ z \\ w \\ M \end{array}\begin{array}{cccccccc} x & y & z & t & u & v & w & M \\ \left[\begin{array}{cccccccc|c} 1 & 0 & 0 & 1 & 0 & 0 & 0 & 0 & 4 \\ 0 & \underline{1} & 0 & 0 & 1 & 0 & 0 & 0 & 6 \\ 0 & 0 & 1 & 0 & 0 & 1 & 0 & 0 & 8 \\ 4 & 3 & 0 & 0 & 0 & -2 & 1 & 0 & 22 \\ -36 & -48 & 0 & 0 & 0 & 70 & 0 & 1 & 560 \end{array}\right] \end{array}$$

$$\begin{array}{c} \\ t \\ y \\ z \\ w \\ M \end{array}\begin{array}{cccccccc} x & y & z & t & u & v & w & M \\ \left[\begin{array}{cccccccc|c} 1 & 0 & 0 & 1 & 0 & 0 & 0 & 0 & 4 \\ 0 & 1 & 0 & 0 & 1 & 0 & 0 & 0 & 6 \\ 0 & 0 & 1 & 0 & 0 & 1 & 0 & 0 & 8 \\ \underline{4} & 0 & 0 & 0 & -3 & -2 & 1 & 0 & 4 \\ -36 & 0 & 0 & 0 & 48 & 70 & 0 & 1 & 848 \end{array}\right] \end{array}$$

$$\begin{array}{c} \\ t \\ y \\ z \\ x \\ M \end{array}\begin{array}{cccccccc} x & y & z & t & u & v & w & M \\ \left[\begin{array}{cccccccc|c} 0 & 0 & 0 & 1 & \frac{3}{4} & \frac{1}{2} & -\frac{1}{4} & 0 & 3 \\ 0 & 1 & 0 & 0 & 1 & 0 & 0 & 0 & 6 \\ 0 & 0 & 1 & 0 & 0 & 1 & 0 & 0 & 8 \\ 1 & 0 & 0 & 0 & -\frac{3}{4} & -\frac{1}{2} & \frac{1}{4} & 0 & 1 \\ 0 & 0 & 0 & 0 & 21 & 52 & 9 & 1 & 884 \end{array}\right] \end{array}$$

$x = 1$, $y = 6$, $z = 8$, $M = 884$

10.

$$\begin{array}{c} \\ t \\ u \\ v \\ M \end{array}\begin{array}{ccccccccc} x & y & z & w & t & u & v & M \\ \left[\begin{array}{cccccccc|c} 6 & 9 & 12 & 15 & 1 & 0 & 0 & 0 & 672 \\ 1 & -1 & \underline{2} & 2 & 0 & 1 & 0 & 0 & 92 \\ 5 & 10 & -5 & 4 & 0 & 0 & 1 & 0 & 280 \\ -3 & -4 & -5 & -4 & 0 & 0 & 0 & 1 & 0 \end{array}\right] \end{array}$$

$$\begin{array}{c} \\ t \\ z \\ v \\ M \end{array}\begin{array}{cccccccc} x & y & z & w & t & u & v & M \\ \left[\begin{array}{cccccccc|c} 0 & \underline{15} & 0 & 3 & 1 & -6 & 0 & 0 & 120 \\ \frac{1}{2} & -\frac{1}{2} & 1 & 1 & 0 & \frac{1}{2} & 0 & 0 & 46 \\ \frac{15}{2} & \frac{15}{2} & 0 & 9 & 0 & \frac{5}{2} & 1 & 0 & 510 \\ -\frac{1}{2} & -\frac{13}{2} & 0 & 1 & 0 & \frac{5}{2} & 0 & 1 & 230 \end{array}\right] \end{array}$$

$$\begin{array}{c} \\ y \\ z \\ v \\ M \end{array}\begin{array}{cccccccc} x & y & z & w & t & u & v & M \\ \left[\begin{array}{cccccccc|c} 0 & 1 & 0 & \frac{1}{5} & \frac{1}{15} & -\frac{2}{5} & 0 & 0 & 8 \\ \frac{1}{2} & 0 & 1 & \frac{11}{10} & \frac{1}{30} & \frac{3}{10} & 0 & 0 & 50 \\ \frac{15}{2} & 0 & 0 & \frac{15}{2} & -\frac{1}{2} & \frac{11}{2} & 1 & 0 & 450 \\ -\frac{1}{2} & 0 & 0 & \frac{23}{10} & \frac{13}{30} & -\frac{1}{10} & 0 & 1 & 282 \end{array}\right] \end{array}$$

$$\begin{array}{c} \\ y \\ z \\ x \\ M \end{array}\begin{array}{cccccccc} x & y & z & w & t & u & v & M \\ \left[\begin{array}{cccccccc|c} 0 & 1 & 0 & \frac{1}{5} & \frac{1}{15} & -\frac{2}{5} & 0 & 0 & 8 \\ 0 & 0 & 1 & \frac{3}{5} & \frac{1}{15} & -\frac{1}{15} & -\frac{1}{15} & 0 & 20 \\ 1 & 0 & 0 & 1 & -\frac{1}{15} & \frac{11}{15} & \frac{2}{15} & 0 & 60 \\ 0 & 0 & 0 & \frac{14}{5} & \frac{2}{5} & \frac{4}{15} & \frac{1}{15} & 1 & 312 \end{array}\right] \end{array}$$

$x = 60$, $y = 8$, $z = 20$, $w = 0$, $M = 312$

11. Minimize $14u + 9v + 24w$ subject to the

constraints $\begin{cases} u + v + 3w \geq 2 \\ 2u + v + 2w \geq 3 \\ u \geq 0, v \geq 0, w \geq 0 \end{cases}$.

12. Maximize $8u + 5v + 7w$ subject to the

constraints $\begin{cases} u + v + 2w \le 20 \\ 4u + v + w \le 30 \\ u \ge 0,\ v \ge 0,\ w \ge 0 \end{cases}$.

13. Primal: $x = 4$, $y = 5$, maximum = 23; Dual: $u = 1$, $v = 1$, $w = 0$, minimum = 23

14. Primal: $x = 4$, $y = 1$, minimum = 110;

Dual: $u = \dfrac{10}{3}$, $v = \dfrac{50}{3}$, $w = 0$, maximum = 110

15. $A = \begin{bmatrix} 1 & 2 \\ 1 & 1 \\ 3 & 2 \end{bmatrix}$, $B = \begin{bmatrix} 14 \\ 9 \\ 24 \end{bmatrix}$,

$C = \begin{bmatrix} 2 & 3 \end{bmatrix}$, $X = \begin{bmatrix} x \\ y \end{bmatrix}$

Primal: Maximize CX subject to $AX \le B$, $X \ge 0$.

Dual: $U = \begin{bmatrix} u \\ v \\ w \end{bmatrix}$

Minimize $B^T U$ subject to $A^T U \ge C^T$, $U \ge 0$.

16. $A = \begin{bmatrix} 1 & 4 \\ 1 & 1 \\ 2 & 1 \end{bmatrix}$, $B = \begin{bmatrix} 8 \\ 5 \\ 7 \end{bmatrix}$,

$C = \begin{bmatrix} 20 & 30 \end{bmatrix}$, $X = \begin{bmatrix} x \\ y \end{bmatrix}$

Primal: Minimize CX subject to $AX \ge B$, $X \ge 0$.

Dual: $U = \begin{bmatrix} u \\ v \\ w \end{bmatrix}$

Maximize $B^T U$ subject to $A^T U \le C^T$, $U \ge 0$

17. a. Let a be the number of attack sticks and b the number of defense sticks. Maximize $16a + 20b$ subject to the constraints

$\begin{cases} 2a + b \le 120 \\ a + 3b \le 150 \\ 2a + 2b \le 140 \\ a \ge 0,\ b \ge 0 \end{cases}$.

	a	b	u	v	w	M	
u	2	1	1	0	0	0	120
v	1	$\underline{3}$	0	1	0	0	150
w	2	2	0	0	1	0	140
M	−16	−20	0	0	0	1	0

	a	b	u	v	w	M	
u	$\frac{5}{3}$	0	1	$-\frac{1}{3}$	0	0	70
b	$\frac{1}{3}$	1	0	$\frac{1}{3}$	0	0	50
w	$\frac{4}{3}$	0	0	$-\frac{2}{3}$	1	0	40
M	$-\frac{28}{3}$	0	0	$\frac{20}{3}$	0	1	1000

	a	b	u	v	w	M	
u	0	0	1	$\frac{1}{2}$	$-\frac{5}{4}$	0	20
b	0	1	0	$\frac{1}{2}$	$-\frac{1}{4}$	0	40
a	1	0	0	$-\frac{1}{2}$	$\frac{3}{4}$	0	30
M	0	0	0	2	7	1	1280

30 attack sticks, 40 defense sticks

b. The new problem is to maximize $16a + 20b + pc$, where c is the number of tennis rackets and p is the profit on each racket, subject to the

constraints: $\begin{cases} 2a + b + c \le 120 \\ a + 3b + 4c \le 150 \\ 2a + 2b + 2c \le 140 \\ c \ge 0,\ b \ge 0,\ c \ge 0 \end{cases}$.

The dual problem is to minimize $120u + 150v + 140w$ subject to the

constraints: $\begin{cases} 2u + v + 2w \ge 8 \\ u + 3v + 2w \ge 10 \\ u + 4v + 2w \ge p \\ u \ge 0,\ v \ge 0,\ w \ge 0 \end{cases}$.

The original solution, with $u = 0$, $v = 2$, $w = 7$ will be optimal if $0 + 4(2) + 2(7) \ge p$. Since $0 + 4(2) + 2(7) = \$22$, that is the profit per tennis racket that needs to be realized to justify the diversification.

18. The new problem is to maximize $70a + 210b + 140c + pd$, where d is the number of brand D stereo systems and p is the profit per system, subject to the constraints:

$$\begin{cases} a + b + c + d \le 100 \\ 5a + 4b + 4c + 3d \le 480 \\ 40a + 20b + 30c + 30d \le 3200 \\ a \ge 0,\, b \ge 0,\, c \ge 0,\, d \ge 0 \end{cases}$$

The dual problem is to minimize $100u + 480v + 3200w$, subject to the

constraints: $\begin{cases} u + 5v + 40w \ge 70 \\ u + 4v + 20w \ge 210 \\ u + 4v + 30w \ge 140 \\ u + 3v + 30w \ge p \\ u \ge 0,\, v \ge 0,\, w \ge 0 \end{cases}$.

The original solution, with $u = 210$, $v = 0$, $w = 0$, will still be optimal if $210 + 3(0) + 30(0) \ge p$. Since $210 + 3(0) + 30(0) = \$210$, that is the required profit per brand D system. The original solution was to sell 100 units of brand B at a profit of $210 each. Brand D units have to be at least this profitable. (Differences in storage space and commission turned out not to matter.)

Chapter 4 Chapter Test

1. $\begin{cases} x + y - 2z + u \qquad\qquad = 10 \\ 2x - 6y + 3z \quad + v \qquad = 18 \\ x + 3y + z \qquad + w \quad = 21 \\ -2x - y + 3z \qquad\qquad + M = 0 \end{cases}$

$$\begin{array}{ccccccc} x & y & z & u & v & w & M \end{array}$$
$$\left[\begin{array}{ccccccc|c} 1 & 1 & -2 & 1 & 0 & 0 & 0 & 10 \\ 2 & -1 & 3 & 0 & 1 & 0 & 0 & 18 \\ 1 & 3 & 1 & 0 & 0 & 1 & 0 & 21 \\ \hline -2 & -1 & 3 & 0 & 0 & 0 & 1 & 0 \end{array}\right]$$

2. $x = 35$, $y = 0$, $z = 30$, $u = 0$, $v = 0$, $w = 42$, $M = 560$;
For the dual, $x = 0$, $y = 14$, $z = 0$, $u = 0$, $v = \dfrac{7}{2}$, $w = 0$, $M = 560$

3. Maximize $3x - 4y$ subject to
$$\begin{cases} 6x + 7y \le 120 \\ 15x + 5y \le 195 \\ x \ge 0,\, y \ge 0 \end{cases}$$

$$\begin{array}{ccccc} x & y & u & v & M \end{array}$$
$$\left[\begin{array}{ccccc|c} 6 & 7 & 1 & 0 & 0 & 120 \\ 15 & 5 & 0 & 1 & 0 & 195 \\ \hline -3 & 4 & 0 & 0 & 1 & 0 \end{array}\right]$$

$$\begin{array}{ccccc} x & y & u & v & M \end{array}$$
$$\left[\begin{array}{ccccc|c} 6 & 7 & 1 & 0 & 0 & 120 \\ 1 & \frac{1}{3} & 0 & \frac{1}{15} & 0 & 13 \\ \hline -3 & 4 & 0 & 0 & 1 & 0 \end{array}\right]$$

$$\begin{array}{ccccc} x & y & u & v & M \end{array}$$
$$\left[\begin{array}{ccccc|c} 0 & 5 & 1 & -\frac{2}{5} & 0 & 42 \\ 1 & \frac{1}{3} & 0 & \frac{1}{15} & 0 & 13 \\ \hline 0 & 5 & 0 & \frac{1}{5} & 1 & 39 \end{array}\right]$$

$x = 13$, $y = 0$, $u = 42$, $v = 0$, $M = 39$

4. Maximize $-12x - 5y$ subject to

$$\begin{cases} -\dfrac{1}{2}x - y \le -5 \\ x + y \le 10 \\ -3x - y \le -10 \\ x \ge 0, y \ge 0 \end{cases}$$

$$\begin{array}{cccccc} x & y & u & v & w & M \\ \end{array}$$
$$\left[\begin{array}{cccccc|c} -\frac{1}{2} & -1 & 1 & 0 & 0 & 0 & -5 \\ 1 & 1 & 0 & 1 & 0 & 0 & 10 \\ -3 & -1 & 0 & 0 & 1 & 0 & -10 \\ \hline 12 & 5 & 0 & 0 & 0 & 1 & 0 \end{array}\right]$$

$$\begin{array}{cccccc} x & y & u & v & w & M \\ \end{array}$$
$$\left[\begin{array}{cccccc|c} 0 & -\frac{5}{6} & 1 & 0 & -\frac{1}{6} & 0 & -\frac{10}{3} \\ 0 & \frac{2}{3} & 0 & 1 & \frac{1}{3} & 0 & \frac{20}{3} \\ 1 & \frac{1}{3} & 0 & 0 & -\frac{1}{3} & 0 & \frac{10}{3} \\ \hline 0 & 1 & 0 & 0 & 4 & 1 & -40 \end{array}\right]$$

$$\begin{array}{cccccc} x & y & u & v & w & M \\ \end{array}$$
$$\left[\begin{array}{cccccc|c} 0 & 5 & -6 & 0 & 1 & 0 & 20 \\ 0 & -1 & 2 & 1 & 0 & 0 & 0 \\ 1 & 2 & -2 & 0 & 0 & 0 & 10 \\ \hline 0 & -19 & 24 & 0 & 0 & 1 & -120 \end{array}\right]$$

$$\begin{array}{cccccc} x & y & u & v & w & M \\ \end{array}$$
$$\left[\begin{array}{cccccc|c} 0 & 1 & -\frac{6}{5} & 0 & \frac{1}{5} & 0 & 4 \\ 0 & 0 & \frac{4}{5} & 1 & \frac{1}{5} & 0 & 4 \\ 1 & 0 & \frac{2}{5} & 0 & -\frac{2}{5} & 0 & 2 \\ \hline 0 & 0 & \frac{6}{5} & 0 & \frac{19}{5} & 1 & -44 \end{array}\right]$$

$x = 2, y = 4, u = 0, v = 4, w = 0, M = 44$

5. Maximize $6u + 3v$ subject to

$$\begin{cases} u + 2v \le 3 \\ u - v \le 2 \\ u \ge 0, v \ge 0 \end{cases}$$

6. a. Let x = number of morning papers
y = number of evening papers

Maximize $\begin{bmatrix} .50 & .35 \end{bmatrix}\begin{bmatrix} x \\ y \end{bmatrix}$ subject to

the constraints

$$\begin{bmatrix} 2 & 1 \\ 2 & 3 \end{bmatrix}\begin{bmatrix} x \\ y \end{bmatrix} \le \begin{bmatrix} 6000 \\ 9600 \end{bmatrix} \text{ and } \begin{bmatrix} x \\ y \end{bmatrix} \ge \begin{bmatrix} 0 \\ 0 \end{bmatrix}$$

b. Minimize $\begin{bmatrix} 6000 & 9600 \end{bmatrix}\begin{bmatrix} u \\ v \end{bmatrix}$

subject to the constraints

$$\begin{bmatrix} 2 & 2 \\ 1 & 3 \end{bmatrix}\begin{bmatrix} u \\ v \end{bmatrix} \ge \begin{bmatrix} .50 \\ .35 \end{bmatrix} \text{ and } \begin{bmatrix} u \\ v \end{bmatrix} \ge \begin{bmatrix} 0 \\ 0 \end{bmatrix}$$

c. Minimize $6000u + 9600v$ subject to the constraints

$$\begin{cases} 2u + 2y \ge .50 \\ u + 3v \ge .35 \\ u \ge 0, v \ge 0 \end{cases}$$

u is a measure of the value of a pound of paper.
v is a measure of the value of a minute of labor.

d. The dual gives the minimum acceptable profit that can be achieved by selling the paper and hiring out the workers.

Chapter 5

Exercises 5.1

1. a. $S' = \{5, 6, 7\}$

   b. $S \cup T = \{1, 2, 3, 4, 5, 7\}$

   c. $S \cap T = \{1, 3\}$

   d. $S' \cap T = \{5, 7\}$

3. a. $R \cup S = \{a, b, c, d, e, f\}$

   b. $R \cap S = \{c\}$

   c. $S \cap T = \varnothing$

5. $\varnothing, \{1\}, \{2\}, \{1, 2\}$

7. a. $M \cap F = \{$all male college students who like football$\}$

   b. $M' = \{$all female college students$\}$

   c. $M' \cap F' = \{$all female college students who don't like football$\}$

   d. $M \cup F = \{$all male college students or all college students who like football$\}$

9. a. $S = \{1983, 1984, 1987, 1999, 2003, 2006\}$

   b. $T = \{1980, 1983, 1985, 1989, 1991, 1995, 1996, 1997, 1998, 1999, 2003\}$

   c. $S \cap T = \{1983, 1999, 2003\}$

   d. $S \cup T = \{1980, 1983, 1984, 1985, 1987, 1989, 1991, 1995, 1996, 1997, 1998, 1999, 2003, 2006\}$

   e. $S' \cap T = \{1980, 1985, 1989, 1991, 1995, 1996, 1997, 1998\}$

   f. $S \cap T' = \{1984, 1987, 2006\}$

11. From 1980 to 2007, during only three years did the Standard and Poor's Index increase by 2% or more during the first 5 days and not increase by 16% or more for that year.

13. a. $R \cup S = \{a, b, c, e\}$
      $(R \cup S)' = \{d, f\}$

   b. $R \cup S \cup T = \{a, b, c, e, f\}$

   c. $R \cap S = \{a, c\}$
      $R \cap S \cap T = (R \cap S) \cap T = \varnothing$

   d. $T' = \{a, b, c, d\}$
      $R \cap S \cap T' = (R \cap S) \cap T' = \{a, c\}$

   e. $R' = \{d, e, f\}$; $S \cap T = \{e\}$
      $R' \cap S \cap T = R' \cap (S \cap T) = \{e\}$

   f. $S \cup T = \{a, c, e, f\}$

   g. $R \cup S = \{a, b, c, e\}$;
      $R \cup T = \{a, b, c, e, f\}$
      $(R \cup S) \cap (R \cup T) = \{a, b, c, e\}$

   h. $R \cap S = \{a, c\}$; $R \cap T = \varnothing$
      $(R \cap S) \cup (R \cap T) = \{a, c\}$

   i. $R' = \{d, e, f\}$; $T' = \{a, b, c, d\}$
      $R' \cap T' = \{d\}$

15. $(S')' = S$

17. $S \cup S' = U$

19. $T \cap S \cap T' = S \cap (T \cap T') = S \cap \varnothing = \varnothing$

21. $\{$divisions that had increases in labor costs or total revenue$\} = L \cup T$

23. $\{$divisions that made a profit despite an increase in labor costs$\} = L \cap P$

25. $\{$profitable divisions with increases in labor costs and total revenue$\} = P \cap L \cap T$

27. $\{$applicants who have not received speeding tickets$\} = S'$

29. $\{$applicants who have received speeding tickets, caused accidents, or were arrested for drunk driving$\}$
    $= S \cup A \cup D$

31. {applicants who have not both caused accidents and received speeding tickets but who have been arrested for drunk driving}
    $= (A \cap S)' \cap D$

33. $A \cap D$ = {male students at Mount College}

35. $A \cap B$ = {people who are both teachers and students at Mount College}

37. $A \cup C' = A \cup D$ = {males or students at Mount College}

39. $D' = C$ = {females at Mount College}

41. {people who don't like strawberry ice cream} = S'

43. {people who like vanilla or chocolate but not strawberry ice cream} = $(V \cup C) \cap S'$

45. {people who like neither chocolate nor vanilla ice cream} = $(V \cup C)'$

47. a. $R = \{B, C, D, E\}$

    b. $S = \{C, D, E, F\}$

    c. $T = \{A, D, E, F\}$

    d. $R' = \{A, F\}$
        $R' \cup S = \{A, C, D, E, F\}$

    e. $R' \cap T = \{A, F\}$

    f. $R \cap S = \{C, D, E\}$
        $R \cap S \cap T = (R \cap S) \cap T = \{D, E\}$

49. Any subset with 2 as an element is an example. Possible answer: {2}

51. If S is a subset of T, then $S \cup T = T$.

53. True; 5 is an element of the set {3, 5, 7}.

55. True; {b} is a subset of the set {b, c}.

57. False; 0 is not an element of the empty set $\varnothing$.

59. True; any set is a subset of itself.

Exercises 5.2

1. $n(S \cup T) = n(S) + n(T) - n(S \cap T)$
   $= 5 + 4 - 2 = 7$

3. $n(S \cup T) = n(S) + n(T) - n(S \cap T)$
   $15 = 7 + 8 - n(S \cap T)$
   $n(S \cap T) = 7 + 8 - 15 = 0$

5. $n(S \cup T) = n(S) + n(T) - n(S \cap T)$
   $13 = n(S) + 7 - 5$
   $n(S) = 13 - 7 + 5 = 11$

7. S is a subset of T.

9. Let P = {adults in South America fluent in Portuguese} and
   S = {adults in South America fluent in Spanish}.
   Then $P \cup S$ = {adults in South America fluent in Portuguese or Spanish} and
   $P \cap S$ = {adults in South America fluent in Portuguese and Spanish}.
   $n(P) = 134, n(S) = 130$,
   $n(P \cup S) = 245$ (numbers in millions)
   $n(P \cup S) = n(P) + n(S) - n(P \cap S)$
   $245 = 134 + 130 - n(P \cap S)$
   $n(P \cap S) = 134 + 130 - 245 = 19$
   19 million are fluent in both languages.

11. Let U = {all letters of the alphabet},
    let V = {letters with vertical symmetry}, and
    H = {letters with horizontal symmetry}.
    Then $V \cup H$ = {letters with vertical or horizontal symmetry} and
    $V \cap H$ = {letters with both vertical and horizontal symmetry}.
    $n(V) = 11, n(H) = 9, n(V \cap H) = 4$
    $n(V \cup H) = n(V) + n(H) - n(V \cap H)$
    $n(V \cup H) = 11 + 9 - 4 = 16$
    $n((V \cup H)') = n(U) - n(V \cup H) = 26 - 16 = 10$
    There are 10 letters with no symmetry.

13. Let A = {cars with automatic transmission} and
    P = {cars with power steering}.
    Then $A \cup P$ = {cars with automatic transmission or power steering} and
    $A \cap P$ = {cars with both automatic transmission and power steering},
    $n(A) = 325, n(P) = 216, n(A \cap P) = 89$
    $n(A \cup P) = n(A) + n(P) - n(A \cap P)$
    $\qquad\qquad = 325 + 216 - 89$
    $\qquad\qquad = 452$
    452 cars were manufactured with at least one of the two options.

15. Consists of points not in *S* but in *T*.

17. Consists of points in *S* or not in *T*.

19. $(S' \cap T)' = S \cup T'$
Consists of points in *S* or not in *T*.

21. Consists of points in *S* but not in *T* or points in *T* but not in *S*.

23. $S \cup (S \cap T) = S$
Consists of points in *S*.

25. $S \cup S' = U$
Consists of all points.

27. Consists of points in *R* and *S* but not in *T*.

29. Consists of points in *R* or points in both *S* and *T*.

31. Consists of points in *R* but not in *S* or points in both *R* and *T*.

33. Consists of points in both *R* and *T*.

35. Consists of points not in *R*, *S*, or *T*.

37. Consists of points in *R* and *T* or points in *S* but not *T*.

39. $S' \cup (S \cap T)' = S' \cup S' \cup T' = S' \cup T'$

41. $(S' \cup T)' = S \cap T'$

43. $T \cup (S \cap T)' = T \cup S' \cup T'$
$\qquad\qquad\quad = (T \cup T') \cup S'$
$\qquad\qquad\quad = U$

45. S'

47. $R \cap T$

49. $R' \cap S \cap T$

51.

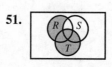

$T \cup (R \cap S')$

53. First draw a Venn diagram for
$(R \cap S') \cup (S \cap T') \cup (T \cap R')$.

The set consists of the complement.
$(R \cap S \cap T) \cup (R' \cap S' \cap T')$

55. People who are not illegal aliens or everyone over the age of 18 who is employed

57. Everyone over the age of 18 who is unemployed

59. *B* is a subset of A', so $A' \cup B = A'$.
Noncitizens who are unemployed

Exercises 5.3

1. $5 + 6 = 11$

3. $6 + 5 + 15 + 20 = 46$

5. 11

7. $19 + 10 + 5 + 6 + 15 + 20 = 75$

9. $10 + 5 + 15 = 30$

11. $n(S \cap T') = n(S) - n(S \cap T) = 5 - 2 = 3$
    $n(S' \cap T) = n(T) - n(S \cap T) = 6 - 2 = 4$
    $n(S \cup T) = n(S) + n(T) - n(S \cap T)$
    $\qquad = 5 + 6 - 2 = 9$
    $n(S' \cap T') = n(U) - n(S \cup T) = 14 - 9 = 5$

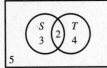

13. $n(S \cap T) = n(S) + n(T) - n(S \cup T)$
    $\qquad = 12 + 14 - 18 = 8$
    $n(S \cap T') = n(S) - n(S \cap T) = 12 - 8 = 4$
    $n(S' \cap T) = n(T) - n(S \cap T) = 14 - 8 = 6$
    $n(S' \cap T') = n(U) - n(S \cup T) = 20 - 18 = 2$

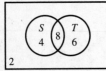

15. $n(S \cup T) = n(U) - n(S' \cap T') = 75 - 40 = 35$
    $n(S \cap T) = n(S) + n(T) - n(S \cup T) = 15 + 25 - 35 = 5$
    $n(S \cap T') = n(S) - n(S \cap T) = 15 - 5 = 10$
    $n(S' \cap T) = n(T) - n(S \cap T) = 25 - 5 = 20$

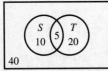

17. $n(S \cap T) = n(S) + n(T) - n(S \cup T) = 3 + 4 - 6 = 1$
    $n(U) = n(S' \cup T') + n(S \cap T) = 9 + 1 = 10$
    $n(S \cap T') = n(S) - n(S \cap T) = 3 - 1 = 2$
    $n(S' \cap T) = n(T) - n(S \cap T) = 4 - 1 = 3$

$n(S' \cap T') = n(U) - n(S \cup T) = 10 - 6 = 4$

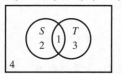

19. $n(R \cap S \cap T') = n(R \cap S) - n(R \cap S \cap T) = 7 - 2 = 5$

$n(R \cap S' \cap T) = n(R \cap T) - n(R \cap S \cap T) = 6 - 2 = 4$

$n(R' \cap S \cap T) = n(S \cap T) - n(R \cap S \cap T) = 5 - 2 = 3$

$n(R \cap S' \cap T') = n(R) - n(R \cap S \cap T') - n(R \cap S' \cap T) - n(R \cap S \cap T) = 17 - 5 - 4 - 2 = 6$

$n(R' \cap S \cap T') = n(S) - n(R \cap S \cap T') - n(R' \cap S \cap T) - n(R \cap S \cap T) = 17 - 5 - 3 - 2 = 7$

$n(R' \cap S' \cap T) = n(T) - n(R \cap S' \cap T) - n(R' \cap S \cap T) - n(R \cap S \cap T) = 17 - 4 - 3 - 2 = 8$

$n(R \cup S \cup T) = 6 + 7 + 8 + 5 + 4 + 3 + 2 = 35$

$n(R' \cap S' \cap T') = n(U) - n(R \cup S \cup T) = 44 - 35 = 9$

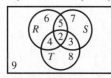

21. $n(R) = n(R \cup S) + n(R \cap S) - n(S) = 21 + 7 - 14 = 14$

$n(U) = n(R) + n(R') = 22 + 14 = 36$

$n(R \cap S \cap T') = n(R \cap S) - n(R \cap S \cap T) = 7 - 5 = 2$

$n(R \cap S' \cap T) = n(R \cap T) - n(R \cap S \cap T) = 11 - 5 = 6$

$n(R' \cap S \cap T) = n(S \cap T) - n(R \cap S \cap T) = 9 - 5 = 4$

$n(R \cap S' \cap T') = n(R) - n(R \cap S \cap T') - n(R \cap S' \cap T) - n(R \cap S \cap T) = 14 - 2 - 6 - 5 = 1$

$n(R' \cap S \cap T') = n(S) - n(R \cap S \cap T') - n(R' \cap S \cap T) - n(R \cap S \cap T) = 14 - 2 - 4 - 5 = 3$

$n(R' \cap S' \cap T) = n(T) - n(R \cap S' \cap T) - n(R' \cap S \cap T) - n(R \cap S \cap T) = 22 - 6 - 4 - 5 = 7$

$n(R \cup S \cup T) = 1 + 3 + 7 + 2 + 6 + 4 + 5 = 28$

$n(R' \cap S' \cap T') = n(U) - n(R \cup S \cup T) = 36 - 28 = 8$

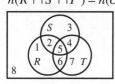

23. Let $U = \{$high school students surveyed$\}$, $R = \{$students who like rock music$\}$, and $H = \{$students who like hip-hop music$\}$.

$n(U) = 70$; $n(R) = 35$; $n(H) = 15$; $n(R \cap H) = 5$

$n(R \cup H) = n(R) + n(H) - n(R \cap H) = 35 + 15 - 5 = 45$

$n((R \cup H)') = n(U) - n(R \cup H) = 70 - 45 = 25$

25 students do not like either rock or hip-hop music.

25. Let $U = \{$lines$\}$, $V = \{$lines with verbs$\}$, $A = \{$lines with adjectives$\}$

$n(U) = 14$; $n(V) = 11$; $n(A) = 9$;

$n(V \cap A) = 7$

$n(V \cap A') = n(V) - n(V \cap A) = 11 - 7 = 4$

Four lines have a verb with no adjective.

$n(V' \cap A) = n(A) - n(V \cap A) = 9 - 7 = 2$

Two lines have an adjective but no verb.

$n(V \cup A) = n(V) + n(A) - n(V \cap A)$

$\qquad = 11 + 9 - 7 = 13$

$n((V \cup A)') = n(U) - n(V \cap A) = 14 - 13 = 1$

One line has neither an adjective nor a verb.

For Exercises 27–29, let $U = \{$students who took the exam$\}$, $F = \{$students who correctly answered the first question$\}$, $S = \{$students who correctly answered the second question$\}$. Then $n(U) = 130$, $n(F) = 90$, $n(S) = 62$, $n(F \cap S) = 50$. Draw and complete the Venn diagram as follows.

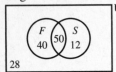

27. $n((F \cup S)') = 28$

29. $n(S \cap F') = 12$

31. Let $U = \{$students in finite math$\}$, $M = \{$male students$\}$, $B = \{$students who are business majors$\}$, and $F = \{$first-year students$\}$. $n(U) = 35$; $n(M) = 22$; $n(B) = 19$; $n(F) = 27$; $n(M \cap B) = 14$; $n(M \cap F) = 17$; $n(B \cap F) = 15$; $n(M \cap B \cap F) = 11$

   a. Draw a Venn diagram as shown.

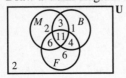

   b. $n(F' \cap M' \cap B') = 2$

      There are two upperclass women nonbusiness majors.

   c. $n(M' \cap B) = 1 + 4 = 5$

      There are five women business majors.

For Exercises 33–37, let $U = \{$people surveyed$\}$, $I = \{$people who learned from the Internet$\}$, $T = \{$people who learned from television$\}$, $N = \{$people who learned from newspapers$\}$. Then $n(U) = 400$, $n(I) = 180$, $n(T) = 190$, $n(N) = 190$, $n(I \cap T) = 80$, $n(I \cap N) = 90$, $n(T \cap N) = 50$, $n(I \cap T \cap N) = 30$. Draw and complete

the Venn diagram as follows.

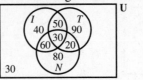

33. $n((I \cap N') \cup (I' \cap N)) = 40 + 50 + 80 + 20 = 190$

35. $n((I \cup T) \cap N') = 40 + 90 + 50 = 180$

37. $n((I \cap T' \cap N') \cup (I' \cap T \cap N') \cup (I' \cap T' \cap N))$
$\qquad = 40 + 90 + 80$
$\qquad = 210$

39. Draw and complete the Venn diagram as follows:

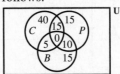

$40 + 15 + 15 + 15 + 5 + 10 + 0 = 100$
The correct answer is (d).

For Exercises 41–43, $n(U) = 4000$, $n(F) = 2000$, $n(S) = 3000$, $n(L) = 500$, $n(F \cap S) = 1500$, $n(F \cap L) = 300$, $n(S \cap L) = 200$, $n(F \cap S \cap L) = 50$. Draw and complete the Venn diagram as follows.

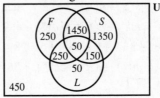

41. $(L \cup F \cup S)' = 450$

43. $L \cup S \cup F' = 4000 - 250 = 3750$

45. Let $U = \{$college students surveyed$\}$, $F = \{$first-year students$\}$, $D = \{$voted Democratic$\}$, $n(U) = 100$; $n(F) = 50$; $n(D) = 55$ $n(F' \cap D') = n((F \cup D)') = 25$ $n(F \cup D) = n(U) - n((F \cup D)') = 100 - 25 = 75$ $n(F \cap D) = n(F) + n(D) - n(F \cup D)$
$\qquad\qquad = 50 + 55 - 75$
$\qquad\qquad = 30$
30 freshmen voted Democratic.

47. Let $U = \{\text{students}\}$, $D = \{\text{students who passed the diagnostic test}\}$, $C = \{\text{students who passed the course}\}$.
$n(U) = 30$; $n(D) = 21$; $n(C) = 23$; $n(D \cap C') = 2$

$n(D \cap C) = n(D) - n(D \cap C') = 21 - 2 = 19$

$n(D' \cap C) = n(C) - n(D \cap C) = 23 - 19 = 4$

Four students passed the course even though they failed the diagnostic test.

For Exercises 49–53, let $U = \{\text{students}\}$,
$M = \{\text{male students}\}$, $B = \{\text{biology majors}\}$.
Then $n(U) = 52$, $n(M \cap B) = 5$, $n(M' \cap B) = 15$, and $n(M' \cap B') = 12$. Therefore $n(B) = 5 + 15 = 20$,

$n(M \cup B) = 52 - 12 = 40$, so
$40 = n(M) + 20 - 5$ (inclusion-exclusion) and
$n(M) = 40 - 20 + 5 = 25$.
Draw and complete the Venn diagram as follows.

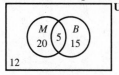

49. 40

51. 27

53. 20

For Exercises 55–61, let $U = \{\text{surveyed students}\}$,
$R = \{\text{students who like rock}\}$,
$C = \{\text{students who like country}\}$,
$J = \{\text{students who like jazz}\}$. Then $n(U) = 190$,
$n(R) = 114$, $n(C) = 50$, $n(R \cap J) = 15$, $n(C \cap J) = 11$, $n(R' \cap C' \cap J) = 20$, $n(R \cap C' \cap J) = 10$, $n(R \cap C \cap J') = 9$ and
$n((R \cup C \cup J)') = 20$. Draw and complete the Venn diagram as follows.

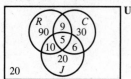

55. $n(R \cap C' \cap J') = 90$

57. $n(R' \cap C \cap J) = 6$

59. $n((R \cap C' \cap J') \cup (R' \cap C \cap J') \cup (R' \cap C' \cap J)) = 90 + 30 + 20 = 140$

61. $n((R \cap C) \cup (R \cap J) \cup (C \cap J)) = 9 + 10 + 6 + 5$
$$= 30$$

63. Let $U = \{\text{executives}\}$, $F = \{\text{executives who read } Fortune\}$, $T = \{\text{executives who read } Time\}$,
$M = \{\text{executives who read } Money\}$.
$n(U) = 180$, $n(F) = 75$, $n(T) = 70$, $n(M) = 55$, $n(F \cap T) = 25$, $n(T \cap M) = 25$, $n(M \cap T \cap N) = 5$
Draw and complete the Venn diagram as follows.

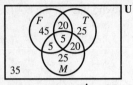

$n((F \cap T \cap M)') = 35$

65. Let $U = \{\text{students}\}$, $P = \{\text{students who play piano}\}$, $V = \{\text{students who play violin}\}$ and $C = \{\text{students who play clarinet}\}$. Then let $x = n(P \cap V \cap C)$ and complete the Venn diagram as follows.

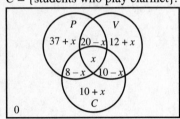

$37 + x + 12 + x + 10 + x + (20 - x) + (8 - x) + (10 - x) + x = 97 + x = 100$, so $x = 3$.

Exercises 5.4

1. $3 \cdot 5 = 15$ routes

3. $26 \cdot 26 = 676$ words

5. $20 \cdot 19 = 380$ tickets

7. $5 \cdot 4 = 20$ ways

9. $5 \cdot 4 \cdot 3 \cdot 2 \cdot 1 = 120$ ways

11. $2 \cdot 2 \cdot 2 \cdot 2 \cdot 2 \cdot 2 = 2^6 = 64$ sequences

13. $20 \cdot 19 \cdot 18 = 6840$ possibilities

15. $30 \cdot 29 = 870$ ways

17. $4 \cdot 3 \cdot 2 \cdot 1 = 24$ words

19. $2 \cdot 2 \cdot 2 \cdot 2 \cdot 2 = 2^5 = 32$ ways

21. $3 \cdot 12 \cdot 10 \cdot 10 \cdot 10 \cdot 10 = 360{,}000$ serial numbers

23. **a.** $7 \cdot 7 \cdot 7 \cdot 7 = 2401$ words

   b. $7 \cdot 6 \cdot 5 \cdot 4 = 840$ words

   c. $1 \cdot 7 \cdot 7 \cdot 7 = 343$ words

   d. Let operation 1 be choosing the letter at the end.
   $2 \cdot 6 \cdot 5 \cdot 4 = 240$ words

25. **a.** $9 \cdot 8 \cdot 7 \cdot 6 \cdot 5 \cdot 4 \cdot 3 \cdot 2 \cdot 1 = 362{,}880$ batting orders

   b. $8 \cdot 7 \cdot 6 \cdot 5 \cdot 4 \cdot 3 \cdot 2 \cdot 1 \cdot 1 = 40{,}320$ batting orders

c. $1 \cdot 6 \cdot 5 \cdot 4 \cdot 3 \cdot 2 \cdot 1 \cdot 1 \cdot 1$
$= 720$ batting orders

27. A Venn diagram with three circles has 8 regions that can be shaded. Let each region be an operation. Each operation can be performed two ways, shaded or not shaded.
$2 \cdot 2 \cdot 2 \cdot 2 \cdot 2 \cdot 2 \cdot 2 \cdot 2 = 256$ ways

29. Each statement has three options.
$3 \cdot 3 \cdot 3 \cdot 3 \cdot 3 \cdot 3 = 3^6 = 729$ ways

31. $6 \cdot 7 \cdot 4 = 168$ days

33. The sequence is DLLLDDD. The number of license plates is $(10)(26)^3(10)^3 = 175,760,000$.

35. The number of different area codes is $(8)(2)(10) = 160$ area codes.

37. Let the winning team (I) consist of the players A,B,C,D,E,F,G,H, I, and J. Then let AB represent player A shaking hands with player B. Then the number of ways all that player A can shake hands is AB, AC, AD, AE, AF, AG, AH, AI, AJ = 9. Player B: BC, BD, BE, BF, BG, BH, BI, BJ = 8 ways (BA is the same handshake as AB and players cannot shake hands with themselves so that BB is not a valid handshake). Player C can shake hands 7 different ways , 6 handshakes for player D, 5 for player E, and so on. So the total number of handshakes with all the players from the winning team is
$9 + 8 + 7 + 6 + 5 + 4 + 3 + 2 + 1 = 45$
$= (10(9))/2 = 45$. Now, represent the players from team II (losing team) as K, L, M, N, O, P, Q, R, S, T. Then A has 10 handshakes: AK, AL, AM, AN, AO, AP, AQ, AR, AS, AT. So does B, C, D, ... J. So the number of different handshakes between the winning team members and losing team members is
$10 + 10 + 10 + 10 + 10 + 10 + 10 + 10 + 10 + 10$
$= 100$.
So the total number of handshakes is
$100 + 45 = 145$.

39. The first house has four choices of color. Each house after that has only three choices.
$4 \cdot 3 \cdot 3 \cdot 3 \cdot 3 \cdot 3 = 972$ ways

41. $4 \cdot 4 \cdot 4 \cdot 4 \cdot 4 \cdot 4 \cdot 4 \cdot 4 \cdot 4 \cdot 4 = 4^{10}$
$= 1,048,576$ ways

43. $4 \cdot 4 = 16$ ways

45. The first song can be any one of the three waltzes, the second song any of the three tangos, the third any of the two remaining waltzes, and so on. Altogether, there are
$3 \cdot 3 \cdot 2 \cdot 2 \cdot 1 \cdot 1 = 36$ different arrangements of the three waltzes and three tangos. The correct answer is (d).

47. $4 \cdot 2 \cdot 3 = 24$ computers

49. Let operation 1 be choosing a box. There are 5 choices.
Let operation 2 be choosing a wrapping paper. There are 11 choices (including not choosing a wrapping paper).
Let operation 3 be choosing a ribbon with a special item for the bow. There are
$7 \cdot 2 \cdot 10 + 1 = 141$ choices (including not choosing a special item and not choosing a ribbon).
There are $5 \cdot 11 \cdot 141 = 7755$ ways to gift wrap.

51. $3 \cdot 5 + 3 \cdot 6 = 33$ segments

53. $6 \cdot 9 \cdot 6 = 324$

55. $7 \cdot 6 \cdot 6 \cdot 6 = 1512$ numbers

57. $10^9 - 1 = 999,999,999$ numbers

59. $2 \cdot 2 \cdot 2 \cdot 2 = 16$ paths

61. $6 \cdot 1 \cdot 4 \cdot 1 \cdot 2 \cdot 1 = 48$ ways

63. The proponents have a choice to start with the leftmost seat or to start with the rightmost seat. There are $4 \cdot 3 \cdot 2 \cdot 1 = 24$ ways to order the proponents. Thus the proponents can be seated $2 \cdot 24 = 48$ ways. Once the proponents have been seated there are $4 \cdot 3 \cdot 2 \cdot 1 = 24$ ways to order the opponents. Thus all 8 panelists can be seated in $48 \cdot 24 = 1152$ ways.

65. A five-digit number must begin with the digits 1–9; the next two digits can be chosen freely and the final two are determined by matching the first two. Hence there are
$9 \cdot 10 \cdot 10 \cdot 1 \cdot 1 = 900$ five-digit palindromes.

67. A restaurant has 5 appetizers and 6 salads on its menu. How many ways can a diner select an appetizer and salad combination?

Exercises 5.5

1. $P(4, 2) = 4 \cdot 3 = 12$

3. $P(6, 3) = 6 \cdot 5 \cdot 4 = 120$

5. $C(10, 3) = \dfrac{P(10, 3)}{3!} = \dfrac{10 \cdot 9 \cdot 8}{3 \cdot 2 \cdot 1} = 120$

7. $C(5, 4) = \dfrac{P(5, 4)}{4!} = \dfrac{5 \cdot 4 \cdot 3 \cdot 2}{4 \cdot 3 \cdot 2 \cdot 1} = 5$

9. $P(7, 1) = 7$

11. $P(n, 1) = n$

13. $C(4, 4) = \dfrac{P(4, 4)}{4!} = \dfrac{4 \cdot 3 \cdot 2 \cdot 1}{4 \cdot 3 \cdot 2 \cdot 1} = 1$

15. $C(n, n-2) = \dfrac{P(n, n-2)}{(n-2)!}$

$\qquad = \dfrac{n \cdot (n-1) \cdots 4 \cdot 3}{(n-2) \cdot (n-3) \cdots 2 \cdot 1}$

$\qquad = \dfrac{n \cdot (n-1)}{2 \cdot 1}$

$\qquad = \dfrac{n(n-1)}{2}$

17. $6! = 6 \cdot 5 \cdot 4 \cdot 3 \cdot 2 \cdot 1 = 720$

19. $\dfrac{9!}{7!} = \dfrac{9 \cdot 8 \cdot 7 \cdot 6 \cdot 5 \cdot 4 \cdot 3 \cdot 2 \cdot 1}{7 \cdot 6 \cdot 5 \cdot 4 \cdot 3 \cdot 2 \cdot 1} = 9 \cdot 8 = 72$

21. $P(4, 4) = 4 \cdot 3 \cdot 2 \cdot 1 = 24$ ways

23. $C(9, 2) = \dfrac{P(9, 2)}{2!} = \dfrac{9 \cdot 8}{2 \cdot 1} = 36$ selections

25. $P(15, 3) = 15 \cdot 14 \cdot 13 = 2730$ ways

27. $C(10, 4) = \dfrac{P(10, 4)}{4!} = \dfrac{10 \cdot 9 \cdot 8 \cdot 7}{4 \cdot 3 \cdot 2 \cdot 1}$

$\qquad = 210$ deluxe chocolate banana splits

29. $C(10, 6) = \dfrac{P(10, 6)}{6!} = \dfrac{10 \cdot 9 \cdot 8 \cdot 7 \cdot 6 \cdot 5}{6 \cdot 5 \cdot 4 \cdot 3 \cdot 2 \cdot 1}$

$\qquad = 210$ ways

31. $P(5, 5) = 5 \cdot 4 \cdot 3 \cdot 2 \cdot 1 = 120$ ways

33. $C(8, 2) = \dfrac{P(8, 2)}{2!} = \dfrac{8 \cdot 7}{2 \cdot 1} = 28$ games

35. $P(10, 5) = 10 \cdot 9 \cdot 8 \cdot 7 \cdot 6 = 30{,}240$ ways

37. $C(10, 5) = \dfrac{P(10, 5)}{5!} = \dfrac{10 \cdot 9 \cdot 8 \cdot 7 \cdot 6}{5 \cdot 4 \cdot 3 \cdot 2 \cdot 1}$

$\qquad = 252$ ways

39. $P(35, 5) = 35 \cdot 34 \cdot 33 \cdot 32 \cdot 31$

$\qquad = 38{,}955{,}840$ ways

41. $C(100, 3) = \dfrac{P(100, 3)}{3!} = \dfrac{100 \cdot 99 \cdot 98}{3 \cdot 2 \cdot 1}$

$\qquad = 161{,}700$ samples are possible

$C(7, 3) = \dfrac{P(7, 3)}{3!} = \dfrac{7 \cdot 6 \cdot 5}{3 \cdot 2 \cdot 1}$

$\qquad = 35$ samples consist of all defective diskettes

43. $P(26, 3) = 26 \cdot 25 \cdot 24 = 15{,}600$ words

45. $C(100, 5) = \dfrac{P(100, 5)}{5!}$

$\qquad = \dfrac{100 \cdot 99 \cdot 98 \cdot 97 \cdot 96}{5 \cdot 4 \cdot 3 \cdot 2 \cdot 1}$

$\qquad = 75{,}287{,}520$ ways

47. $C(52, 5) = \dfrac{P(52, 5)}{5!} = \dfrac{52 \cdot 51 \cdot 50 \cdot 49 \cdot 48}{5 \cdot 4 \cdot 3 \cdot 2 \cdot 1}$

$\qquad = 2{,}598{,}960$ hands

49. There are 13 clubs in a deck.

$C(13, 5) = \dfrac{P(13, 5)}{5!} = \dfrac{13 \cdot 12 \cdot 11 \cdot 10 \cdot 9}{5 \cdot 4 \cdot 3 \cdot 2 \cdot 1}$

$\qquad = 1287$ hands

51. $C(20, 3) = \dfrac{P(20, 3)}{3!} = \dfrac{20 \cdot 19 \cdot 18}{3 \cdot 2 \cdot 1}$

$\qquad = 1140$ ways

53. $P(5, 5) = 5 \cdot 4 \cdot 3 \cdot 2 \cdot 1 = 120$ ways

55. $C(8, 4) = \dfrac{P(8, 4)}{4!} = \dfrac{8 \cdot 7 \cdot 6 \cdot 5}{4 \cdot 3 \cdot 2 \cdot 1} = 70$ ways

57. Moe has $C(9, 2) = \dfrac{P(9, 2)}{2!} = \dfrac{9 \cdot 8}{2 \cdot 1} = 36$ choices.

Joe has $C(7, 3) = \dfrac{P(7, 3)}{3!} = \dfrac{7 \cdot 6 \cdot 5}{3 \cdot 2 \cdot 1} = 35$ choices.

Thus Joe is correct.

59. This is the same as ordering 12 different pictures. Then there are
$P(12, 12) = 12!$
$\qquad = 479{,}001{,}600$ arrangements.

61. There are $P(5, 5) = 5 \cdot 4 \cdot 3 \cdot 2 \cdot 1 = 120$ ways to order the mathematics books. Counting the group of mathematics books as one item, there are $P(5, 5) = 120$ ways to order the novels and the set of mathematics books. All together, there are $120 \cdot 120 = 14,400$ ways to arrange these books.

63. $P(7, 4) = 7 \cdot 6 \cdot 5 \cdot 4 = 840$ arrangements

65. There is only one possibility for the pitcher and there are two possibilities for the first baseman. Of the remaining seven players, there are $P(7, 7) = 7! = 5040$ possibilities.
Thus there are $1 \cdot 2 \cdot 5040 = 10,080$ batting orders.

67. $P(5, 5) = 5! = 120$ ways

69. Of cards from two suits, there are
$$C(26, 5) = \frac{P(26, 5)}{5!} = \frac{26 \cdot 25 \cdot 24 \cdot 23 \cdot 22}{5 \cdot 4 \cdot 3 \cdot 2 \cdot 1}$$
$$= 65,780 \text{ five-card combinations. Of those,}$$
$$C(13, 5) = \frac{P(13, 5)}{5!} = \frac{13 \cdot 12 \cdot 11 \cdot 10 \cdot 9}{5 \cdot 4 \cdot 3 \cdot 2 \cdot 1}$$
$$= 1287 \text{ consist of cards from one of the two suits.}$$
Since there are $C(4, 2) = \dfrac{P(4, 2)}{2!} = \dfrac{4 \cdot 3}{2 \cdot 1} = 6$ possible two-suit combinations, there are
$6 \cdot (65,780 - 2 \cdot 1287) = 379,236$ five-card combinations from exactly two suits.

71. The product is even if at least one of the digits is even. There are $10^5 = 100,000$ possible ZIP codes altogether.

There are $5^5 = 3125$ possible ZIP codes consisting of all odd digits. Thus there are $100,000 - 3125 = 96,875$ ZIP codes in which the product of the digits is even.

73. If there are n people at the party, the number of possible handshakes is $\dfrac{n(n-1)}{2}$. So if there were 45 handshakes,

then $\dfrac{n(n-1)}{2} = 45$ or $n(n - 1) = 90$. The solution is $n = 10$.

75. There are $C(15, 3) = 455$ different side dishes in which all sides are different.
There are $2 \cdot C(15, 2) = 2(105) = 210$ different ways to choose two same side dishes and one different side. (For example if you choose the two sides A and B, there are $C(15,2)$ ways to choose which sides you want. There are two ways to choose which side will be duplicated- for instance AAB or ABB. Finally, there are $C(15,1) = 15$ ways to choose 3 servings of one side dish. The total is $455 + 210 + 15 = 680$.

77. The number of possible combinations of six numbers chosen from 1 to 49 is
$C(49, 6) = 13,983,816$; if the numbers are chosen from 1 to 59 this increases to
$C(59, 6) = 45,057,474$. The ratio is $\dfrac{C(59, 6)}{C(49, 6)} \doteq 3.22$, so the correct answer is (b).

79. A cube has 8 vertices, so one can construct $C(8, 3) = 56$ distinct triangles.

81. Number of programs for a student on the semester system:
$C(752, 5) \cdot C(747, 5) \cdot C(742, 5) \cdot C(737, 5) \cdot C(732, 5) \cdot C(727, 5) \cdot C(722, 5) \cdot C(717, 5)$
Number of programs for a student on the trimester system:
$C(937, 3) \cdot C(934, 3) \cdot C(931, 3) \cdot C(928, 3) \cdot C(925, 3) \cdot C(922, 3) \cdot C(919, 3) \cdot C(916, 3) \cdot C(913, 3) \cdot C(910, 3)$
$\cdot C(907, 3) \cdot C(904, 3)$
Using a computer to compare the two quantities, semester system students have a greater number of different programs.

83. a. $C(45, 5) = 1,221,759$ tickets

b. $C(100, 4) = 3,921,225$ tickets

c. First lottery

d. Number of possible tickets for the first lottery: $P(45, 5) = 146,611,080$
Number of possible tickets for the second lottery: $P(100, 4) = 94,109,400$
Thus the second lottery is better if order matters.

Exercises 5.6

1. a. $2^6 = 64$ outcomes

b. $C(6, 3) = \dfrac{6 \cdot 5 \cdot 4}{3 \cdot 2 \cdot 1} = 20$ outcomes

c. These are the outcomes with exactly no tails, one tail, or two tails.
$C(6, 0) + C(6, 1) + C(6, 2)$
$= 1 + \dfrac{6}{1} + \dfrac{6 \cdot 5}{2 \cdot 1} = 22$ outcomes

d. These are all of the outcomes minus the outcomes with exactly no heads or one head.
$2^6 - C(6, 0) - C(6, 1)$
$= 64 - 1 - \dfrac{6}{1} = 57$ outcomes

3. Four of the nine blocks in the route must be "south."
$C(9, 4) = \dfrac{9 \cdot 8 \cdot 7 \cdot 6}{4 \cdot 3 \cdot 2 \cdot 1} = 126$ routes

5. a. $C(10, 3) = \dfrac{10 \cdot 9 \cdot 8}{3 \cdot 2 \cdot 1} = 120$ samples

b. $C(8, 3) = \dfrac{8 \cdot 7 \cdot 6}{3 \cdot 2 \cdot 1} = 56$ samples

c. From part (b) there are 56 samples with no rotten apples.
Samples that contain at least one rotten apple: $120 - 56 = 64$ samples

7. Consider this a sequence of three operations.
Assign 25 students to dorm A: $C(100, 25)$
Assign 40 of the remaining 75 students to dorm B: $C(75, 40)$
Assign 35 of the remaining 35 students to dorm C: $C(35, 35) = 1$
Ways to assign all 100 students:
$C(100, 25) \cdot C(75, 40) \approx 7.14 \times 10^{44}$

9. Two of the five blocks from A to C must be "south."
Routes from A to C:
$C(5, 2) = \dfrac{5 \cdot 4}{2 \cdot 1} = 10$ routes
Two of the four blocks from C to B must be "south."
Routes from C to B:
$C(4, 2) = \dfrac{4 \cdot 3}{2 \cdot 1} = 6$ routes
Routes from A to B passing through C:
$10 \cdot 6 = 60$ routes

11. First select the states from which each senator comes. There are
$C(50, 5) = \dfrac{50 \cdot 49 \cdot 48 \cdot 47 \cdot 46}{5 \cdot 4 \cdot 3 \cdot 2 \cdot 1}$
$= 2,118,760$ ways to do this.
Once the states have been selected, each state has two possible choices of senators, so there are
$2^5 = 32$ possible committees once the states have been chosen. Ways a committee of five senators can be selected so that no two members are from the same state:
$2,118,760 \cdot 32 = 67,800,320$ ways

13. $C(4, 3) \cdot C(6, 3) = \dfrac{4 \cdot 3 \cdot 2}{3 \cdot 2 \cdot 1} \cdot \dfrac{6 \cdot 5 \cdot 4}{3 \cdot 2 \cdot 1}$
$= 4 \cdot 20$
$= 80$ ways

15. $C(20, 12) \cdot C(8, 2) = 125,970 \cdot 28$
$= 3,527,160$ ways

17. $C(9, 5) \cdot C(8, 6) = 126 \cdot 28 = 3528$ ways

19. $C(15, 3) \cdot C(12, 8) \cdot C(4, 4) = 455 \cdot 495 \cdot 1$
$= 225,225$ ways

21. $P(26, 6) = 26 \cdot 25 \cdot 24 \cdot 23 \cdot 22 \cdot 21$
$= 165,765,600$ plates

23. $P(7, 3) = 7 \cdot 6 \cdot 5 = 210$ words

25. Step 1) There are 10 letters in the word "ABSTEMIOUS."
Step 2) Choose 5 slots to the put the letters "AEIOU" in that order. There are C(10,5) ways to do this.
Step 3) There are P(5,5) number of ways to arrange the five remaining letters.
Total: $C(10, 5) \cdot P(5, 5) = 252(120) = 30,240$.

27. First select the lineup of the four classes. There are 4! = 24 ways to do this. Then arrange three of the four freshmen. There are $P(4, 3) = 24$ ways to do this. Then arrange three of the five sophomores. There are $P(5, 3) = 60$ ways to do this. Then arrange three of the six juniors. There are $P(6, 3) = 120$ ways to do this. Then arrange three of the seven seniors. There are $P(7, 3) = 210$ ways to do this.
Number of different pictures:
$24 \cdot 24 \cdot 60 \cdot 120 \cdot 210 = 870,912,000$ pictures

29. First select the order of the towns. There are 2! = 2 ways to do this. Then select the order in the larger town. There are 3! = 6 ways to do this. Then select the order in the smaller town. There are 2! = 2 ways to do this.
Ways to schedule his stops: $2 \cdot 6 \cdot 2 = 24$ ways

31. $C(4, 3) \cdot C(4, 2) = 4 \cdot 6 = 24$ hands

33. First select the first denomination. There are 13 ways to do this. Then select three of the denomination. There are $C(4, 3) = 4$ ways to do this. Then select the second denomination. There are 12 ways to do this. Then select two of the denomination. There are $C(4, 2) = 6$ ways to do this.
Number of poker hands:
$13 \cdot 4 \cdot 12 \cdot 6 = 3744$ hands

35. The digits that can be used in a detour-prone ZIP code are 0, 1, 6, 8 and 9. There are
$5^5 = 3125$ ZIP codes that use only these digits. However, some of these are the same when read upside down and so are not detour-prone. For this to happen, the middle digit must be 0, 1 or 8 and the last two digits read upside down must match the first two, as in for example 98186. There are
$5 \cdot 5 \cdot 3 = 75$ ways this can happen, so
$3125 - 75 = 3050$ ZIP codes are detour prone.

37. Committees with no restrictions:
$C(12, 4) = 495$ committees
Committees with no seniors:
$C(9, 4) = 126$ committees
Committees with at least one senior:
$495 - 126 = 369$ committees

39. Consider the case with 3 tap, 1 ballet, and 1 modern routine. There are
$C(8, 3) \cdot C(5, 1) \cdot C(2, 1) = 56 \cdot 5 \cdot 2 = 560$ ways to do this.
Consider the case with 1 tap, 3 ballet, and 1 modern routine. There are
$C(8, 1) \cdot C(5, 3) \cdot C(2, 1) = 8 \cdot 10 \cdot 2 = 160$ ways to do this.
Consider the case with 2 tap, 2 ballet, and 1 modern routine. There are
$C(8, 2) \cdot C(5, 2) \cdot C(2, 1) = 28 \cdot 10 \cdot 2$
$= 560$ ways to do this.
Consider the case with 2 tap, 1 ballet, and 2 modern routines. There are
$C(8, 2) \cdot C(5, 1) \cdot C(2, 2) = 28 \cdot 5 \cdot 1 = 140$ ways to do this.
Consider the case with 1 tap, 2 ballet, and 2 modern routines. There are
$C(8, 1) \cdot C(5, 2) \cdot C(2, 2) = 8 \cdot 10 \cdot 1 = 80$ ways to do this.
Thus the manager can choose
$560 + 160 + 560 + 140 + 80 = 1500$ ways.

41. First order the men. There are 6! = 720 ways to do this. Then order the women. There are 6! = 720 ways to do this. Since no two women can sit next to each other, there must be at least one man between each two. This is a total of five men. The sixth man can be at the end or with one of the other men. There are 7 ways to do this. Thus there are a total of
$720 \cdot 720 \cdot 7 = 3,628,800$ ways to seat them.

43. Each family member can decide to come or not come to dinner. Thus there are $2^6 = 64$ different groups that can come to dinner. (This includes no one coming to dinner.)

45. 3 blue marbles and no white marbles chosen:
$C(3, 3) \cdot C(5, 0) = 1 \cdot 1 = 1$ way
2 blue marbles and 1 white marble chosen:
$C(3, 2) \cdot C(5, 1) = 3 \cdot 5 = 15$ ways
number of blue marbles chosen exceeds the number of white marbles chosen:
$1 + 15 = 16$ ways

47. There are $26^4 = 456,976$ possible four-letter sequences; of these, $21^4 = 194,481$ contain no vowels. Hence $456,976 - 194,481 = 262,495$ are possible words.

49. Total number of possible bridge hands: $C(52, 13)$
Bridge hands with all four aces: $C(48, 9)$
Percentage of bridge hands with all four aces:
$$\frac{C(48, 9)}{C(52, 13)} \approx .00264 = .264\%$$

51. Bridge hands with all four aces:
$C(48, 9) = 1,677,106,640$
Bridge hands with the two red kings, the two red queens, and no other kings or queens:
$C(44, 9) = 708,930,508$
A bridge hand with all four aces is more likely.

Exercises 5.7

1. $\binom{6}{2} = C(6, 2) = \frac{6 \cdot 5}{2 \cdot 1} = 15$

3. $\binom{8}{1} = C(8, 1) = \frac{8}{1} = 8$

5. $\binom{18}{16} = C(18, 16) = C(18, 2) = \frac{18 \cdot 17}{2 \cdot 1} = 153$

7. $\binom{7}{0} = C(7, 0) = 1$

9. $\binom{8}{8} = C(8, 8) = 1$

11. $\binom{n}{n-1} = C(n, n-1) = C(n, 1) = \frac{n}{1} = n$

13. $0! = 1$

15. $n \cdot (n-1)! = n!$

17. $\binom{6}{0} + \binom{6}{1} + \binom{6}{2} + \binom{6}{3} + \binom{6}{4} + \binom{6}{5} + \binom{6}{6}$
$= 2^6$
$= 64$

19. $\binom{10}{0} x^{10} + \binom{10}{1} x^9 y + \binom{10}{2} x^8 y^2$
$= x^{10} + 10x^9 y + 45x^8 y^2$

21. $\binom{15}{13} x^2 y^{13} + \binom{15}{14} xy^{14} + \binom{15}{15} y^{15}$
$= 105x^2 y^{13} + 15xy^{14} + y^{15}$

23. $\binom{20}{10} x^{10} y^{10} = 184,756 x^{10} y^{10}$

25. $\binom{11}{7} = 330$

27. $2^6 = 64$ subsets

29. $2^4 = 16$ tips

31. $2^5 = 32$ options (The number of subsets of any size taken from a set of five elements.)

33. $2^8 - 1 = 255$ ways

35. $2 \cdot 3 \cdot 2^{15} = 196,608$ types

37. $2^7 - C(7, 6) - C(7, 7) = 128 - 7 - 1$
$\qquad\qquad\qquad\qquad\qquad\quad = 120$ ways

39. Find the number of subsets of the set $\{a, b, d, e\}$.
$2^4 = 16$ subsets

41. $2^8 - C(8, 0) - C(8, 1) = 256 - 1 - 8$
$\qquad\qquad\qquad\qquad\qquad\quad = 247$ ways

43. $2^{12} = 4096$ groups

45. Each region contains either elements in no sets, exactly one set, exactly two sets, or all three sets. The number of regions containing elements in no set is $\binom{3}{0}$. The number of regions containing elements in exactly one set is $\binom{3}{1}$. The number of regions containing elements in exactly two sets is $\binom{3}{2}$. The number of regions containing elements in all three sets is $\binom{3}{3}$. Thus, there are
$\binom{3}{0} + \binom{3}{1} + \binom{3}{2} + \binom{3}{3} = 1 + 3 + 3 + 1 = 8$ regions.

47. $C(8, 3) \cdot C(12, 2) = 56 \cdot 66 = 3696$ ways

49. $(x - 3y)^7 = [x + (-3y)]^7$

The term with y^4 is

$\binom{7}{4} x^3 (-3y)^4 = 35x^3 \cdot 81y^4 = 2835x^3 y^4$. The

coefficient of y^4 is $2835x^3$.

51. Find the number of subsets of size 1 or more from a set of 15 elements. The total number of subsets of a set of 15 elements (including the null set) is 2^{15}. Thus, if the empty set is not included, the total number of sets of one or more elements is $2^{15} - 1 = 32,767$.

Exercises 5.8

1. $\dfrac{5!}{3!1!1!} = 20$

3. $\dfrac{6!}{2!1!2!1!} = 180$

5. $\dfrac{7!}{3!2!2!} = 210$

7. $\dfrac{12!}{4!4!4!} = 34,650$

9. $\dfrac{12!}{5!3!2!2!} = 166,320$

11. $\dfrac{1}{5!} \cdot \dfrac{15!}{(3!)^5} = 1,401,400$

13. $\dfrac{1}{3!} \cdot \dfrac{18!}{(6!)^3} = 2,858,856$

15. $\binom{20}{7, 5, 8} = \dfrac{20!}{7!5!8!} = 99,768,240$ reports

17. $\dfrac{1}{5!} \cdot \dfrac{20!}{(4!)^5} = 2,546,168,625$ ways

19. $\dfrac{1}{4!} \cdot \dfrac{20!}{(5!)^4} = 488,864,376$ ways

21. $\binom{4}{1, 1, 2} = \dfrac{4!}{1!1!2!} = 12$ ways

23. $\dfrac{1}{7!} \cdot \dfrac{14!}{(2!)^7} = 135,135$ ways

25. The number of ways ten students are to be divided into two five member teams for a basketball game is $\dfrac{10!}{(2)!(5!)^2} = 126$.

27. $\binom{38}{10, 12, 10, 6}$

$= \dfrac{38!}{10!12!10!6!}$

$= 115,166,175,166,136,334,240$ ways

Chapter 5 Supplementary Exercises

1. $\varnothing, \{a\}, \{b\}, \{a, b\}$

2. $(S \cup T')' = S' \cap T$

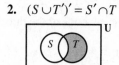

3. $C(16, 2) = \dfrac{16!}{2!14!} = 120$ possibilities

4. $2 \cdot 5! = 240$ ways

5.

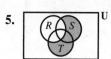

6. $\binom{12}{0} x^{12} + \binom{12}{1} x^{11} y + \binom{12}{2} x^{10} y^2$

$= x^{12} + 12x^{11} y + 66x^{10} y^2$

7. $C(8, 3) \cdot C(6, 2) = \dfrac{8!}{3!5!} \cdot \dfrac{6!}{2!4!} = 56 \cdot 15 = 840$

8. Let U = {people}, P = {people who received placebos}, I = {people who showed improvement}.

$n(U) = 60$; $n(P) = 15$; $n(I) = 40$; $n(P' \cap I) = 30$

Draw and complete a Venn diagram as shown.

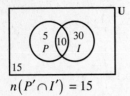

$n(P' \cap I') = 15$

Fifteen of the people who received the drug showed no improvement.

9. $7 \cdot 5 = 35$ combinations

10. $\binom{12}{2,\,4,\,6} = \dfrac{12!}{2!\,4!\,6!} = 13,860$

11. Let $U = \{$applicants$\}$, $F = \{$applicants who speak French$\}$, $S = \{$applicants who speak Spanish$\}$, and $G = \{$applicants who speak German$\}$.

$n(U) = 115;\ n(F) = 70;\ n(S) = 65;\ n(G) = 65;\ n(F \cap S) = 45;\ n(S \cap G) = 35;\ n(F \cap G) = 40;$

$n(F \cap S \cap G) = 35$. Draw and complete a Venn diagram.

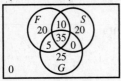

$n((F \cup S \cup G)') = 0$

None of the people speak none of the three languages.

12. $\binom{17}{15} = \binom{17}{2} = \dfrac{17 \cdot 16}{2 \cdot 1} = 136$

For Exercises 13–20, let $U = \{$members of the Earth Club$\}$, $W = \{$members who thought the priority is clean water$\}$, $A = \{$members who thought the priority is clean air$\}$, $R = \{$members who thought the priority is recycling$\}$. Then $n(U) = 100$, $n(W) = 45$, $n(A) = 30$, $n(R) = 42$, $n(W \cap A) = 13$, $n(A \cap R) = 20$, $n(W \cap R) = 16$, $n(W \cap A \cap R) = 9$. Draw and complete the Venn diagram as follows.

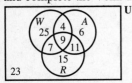

13. $n(A \cap W' \cap R') = 6$

14. $n((W \cap A') \cup (W' \cap A)) = (25 + 7) + (6 + 11) = 32 + 17 = 49$

15. $n((W \cup R) \cap A') = 25 + 15 + 7 = 47$

16. $n(A \cap R \cap W') = 11$

17. $n((W \cap A' \cap R') \cup (W' \cap A \cap R') \cup (W' \cap A' \cap R)) = 25 + 6 + 15 = 46$

18. $n(R') = 23 + 25 + 6 + 4 = 58$

19. $n(R \cap A') = 15 + 7 = 22$

20. $n((W \cup A \cup R)') = 23$

21. $C(9, 4) = C(9, 5) = 126$

22. 2^{40}

23. Let S = {students who ski},
H = {students who play ice hockey}. Then
$$n(S \cup H) = n(S) + n(H) - n(S \cap H)$$
$$= 400 + 300 - 150$$
$$= 550.$$

24. $6 \cdot 10 \cdot 8 = 480$ meals

25. $\begin{pmatrix} 5 \\ 1, 3, 1 \end{pmatrix} = \dfrac{5!}{1!3!1!} = 20$ ways

26. The first digit can be anything but 0, the hundreds digit must be 3, and the last digit must be even.
$$9 \cdot 1 \cdot 5 \cdot 10^4 = 450,000$$

27. $9^2 \cdot 10^8 = 8,100,000,000$

28. $26 + 26^2 + 26^3 + \cdots + 26^{10} = 146,813,779,479,510$

29. Strings of length 8 formed from the symbols a, b, c, d, e:
$5^8 = 390,625$ strings
Strings of length 8 formed from the symbols a, b, c, d:
$4^8 = 65,536$ strings
Strings with at least one e:
$390,625 - 65,536 = 325,089$ strings

30. $C(12, 5) = \dfrac{12!}{5!7!} = 792$

31. $C(100, 14) = \dfrac{100!}{14!86!}$
$= 44,186,942,677,323,600$ groups

32. U = {households}
F = {households that get *Fancy Diet Magazine*}
C = {households that get *Clean Living Journal*}
$$n(F \cup C) = n(F) + n(C) - n(F \cap C)$$
$$= 4000 + 10,000 - 1500$$
$$= 12,500$$
$$n((F \cup C)') = n(U) - n(F \cup C)$$
$$= 40,000 - 12,500$$
$$= 27,500$$
27,500 households get neither.

33. $3^{10} = 59,049$ paths

34. $C(60, 10) = \dfrac{60!}{10!50!} = 75,394,027,566$ ways

35. $5^{10} = 9,765,625$ tests

36. $21^6 = 85,766,121$ strings

37. $C(10, 4) = \dfrac{10!}{4!6!} = 210$ ways

38. $\dfrac{1}{3!} \cdot \dfrac{21!}{(7!)^3} = 66,512,160$ ways

39. $14! = 87,178,291,200$ ways

40. $\dfrac{1}{5!} \cdot \dfrac{100!}{(20!)^5}$ ways

41. Suppose one of the senators from New York has been assigned to a group. There are $C(98, 19)$ ways to complete this group.
$$C(98, 19) = \dfrac{98!}{19!79!} = \dfrac{20 \cdot 80 \cdot 98!}{20!80!}$$
To assign the remaining 80 senators into groups of 20 each, there are $\dfrac{1}{4!} \cdot \dfrac{80!}{(20!)^4}$ ways.

Thus the number of ways to divide the senators if the 2 senators from the state of New York cannot be in the same group is as follows.

$$C(98, 19) \cdot \frac{1}{4!} \cdot \frac{80!}{(20!)^4} = \left(\frac{20 \cdot 80 \cdot 98!}{20!80!}\right)\left(\frac{1}{4!} \cdot \frac{80!}{(20!)^4}\right)$$

$$= \frac{20 \cdot 80 \cdot 98!}{4!(20!)^5}$$

$$= \frac{5 \cdot 20 \cdot 80 \cdot 98!}{5 \cdot 4!(20!)^5}$$

$$= \frac{100 \cdot 80 \cdot 98!}{5!(20!)^5}$$

$$= \frac{80 \cdot 100 \cdot 99 \cdot 98!}{99 \cdot 5!(20!)^5}$$

$$= \frac{80 \cdot 100!}{99 \cdot 5!(20!)^5}$$

$$= \frac{80}{99} \cdot \frac{1}{5!} \cdot \frac{100!}{(20!)^5}$$

Thus there are $\dfrac{80}{99} \cdot \dfrac{1}{5!} \cdot \dfrac{100!}{(20!)^5}$ ways to divide the senators.

$\left(\text{This is } \dfrac{80}{99} \text{ of the answer in Exercise 40.}\right)$

42. $3 \cdot C(10, 2) = 135$

43. A diagonal corresponds to a pair of vertices, except that adjacent pairs must be excluded. Hence there are

$$C(n, 2) - n = \frac{n(n-1)}{2} - n = \frac{n(n-3)}{2} \text{ diagonals.}$$

44. $8 \cdot 6 = 48$

45. $5! \cdot 4! \cdot 3! \cdot 2! \cdot 1! = 120 \cdot 24 \cdot 6 \cdot 2 \cdot 1 = 34,560$

46. $P(12, 5) = 12 \cdot 11 \cdot 10 \cdot 9 \cdot 8 = 95,040$

47. There are four suits. Once a suit is chosen, select 5 of the 13 cards.
Poker hands with cards of the same suit:
$4 \cdot C(13, 5) = 4 \cdot 1287 = 5148$ hands

48. $9 \cdot 9 \cdot 8 = 648$ numbers

49. Exactly the first two digits alike: $9 \cdot 1 \cdot 9 = 81$
First and last digit alike: $9 \cdot 9 \cdot 1 = 81$
Last two digits alike: $9 \cdot 9 \cdot 1 = 81$
Numbers with exactly two digits alike:
$81 + 81 + 81 = 243$

50. $C(4, 3) \cdot C(48, 2) = 4 \cdot 1128 = 4512$ hands

51. $24 \cdot 23 + 24 \cdot 23 \cdot 22 = 552 + 12,144$
$ = 12,696$ names

52. $3! = 6$ pairings

53. $2 \cdot 3! \cdot 3! = 72$ ways

54. $3 \cdot 2 \cdot 2 \cdot 1 = 12$ ways

55. Any two lines will intersect, so the number of intersections is $C(10, 2) = 45$.
Since each of the ten lines is a boundary there are ten sides to the feasible set and there are ten vertices. The number of intersections that occur outside the feasible set is $45 - 10 = 35$.

56. First teacher: $\dfrac{1}{4!} \cdot \dfrac{24!}{(6!)^4} = 96,197,645,544$ ways

Second teacher:
$\dfrac{1}{6!} \cdot \dfrac{24!}{(4!)^6} = 4,509,264,634,875$ ways

The second teacher has more options.

57. $\begin{pmatrix} 10 \\ 3, 4, 3 \end{pmatrix} = \dfrac{10!}{3!4!3!} = 4200$

58. Let n be the number of books.
$n! = 120$, so $n = 5$. There are five books.

59. $7 \cdot 6 \cdot 5 \cdot 1 \cdot 4 \cdot 3 \cdot 2 \cdot 1 \cdot 1 = 5040$ orders

60. a. $C(12, 2) \cdot C(12, 3) = 66 \cdot 220 = 14,520$

b. Select five out of twelve couples. Then determine the spouse from each couple.
$C(12, 5) \cdot 2^5 = 25,344$

61. $C(7, 2) + C(7, 1) + C(7, 0) = 29$ ways

62. There is one possible path from A at the top to the first B on the left, and 1 possible path from there to the final C. Similarly, for the last B on the right. For the second and fourth B's, there are 4 incoming and 4 outgoing paths for the third B, there are 6 incoming and 6 outgoing paths. so the total number of paths from A at the top of C or the bottom is $1^2 + 4^2 + 6^2 + 4^2 + 1^2 = 70$.

63. There are $2 \cdot 26^2$ 3-letter call letters and $2 \cdot 26^3$ 4-letter call letters, so $2 \cdot 26^2 + 2 \cdot 26^3 = 36,504$ call letters are possible.

64. Find n such that $n! = 479,001,600$.
Testing numbers on computer, $n = 12$.

65. $\begin{pmatrix} 25 \\ 10, 9, 6 \end{pmatrix} = 16,360,143,800$ ways

66. a. First choose the three letters. There are $C(26, 3)$ ways to do this. Then choose the three numbers. There are $C(10, 3)$ ways to do this. Then order the six. There are $6!$ ways to do this.
The number of license plates is
$C(26, 3) \cdot C(10, 3) \cdot 6! = 224,640,000$.

b. The number of license plates consisting of three distinct letters followed by three distinct numbers is
$P(26, 3)P(10, 3) = C(26, 3) \cdot 3! \cdot C(10, 3) \cdot 3!$.
Thus there are
$\dfrac{C(26, 3) \cdot C(10, 3) \cdot 6!}{C(26, 3) \cdot 3! C(10, 3) \cdot 3!} = 20$ times as
many plates in (a).

67. If $A \cap B = \varnothing$ then A and B have no elements in common.

Conceptual Exercises

68. If $n(A \cup B) = n(A) + n(B)$, then A and B are disjoint sets and have no elements in common.

69. $n(n - 1)! = n(n - 1)(n - 2)(n - 3)...1 = n!$
For example,
$5 \cdot 4! = 5(4 \cdot 3 \cdot 2 \cdot 1) = 5 \cdot 4 \cdot 3 \cdot 2 \cdot 1 = 5!$.

70. Let $n = 1$. We have $1(1-1)! = 1(0)! = 1! = 1$ So, since 1 times any number is the number, we have $0! = 1! = 1$.

71. A permutation is an arrangement of items in which order is important. A combination is a subset of objects taken from a larger set in which the order of the objects does not make a difference.

72. The number of committees of size 6 is the same as the number of committees of size 4. For each committee of size 6, there is a committee consisting of the four people who were not chosen. For example, use letters A, B,C, D, E, F, G, H, I, J to represent the people. For the committee
{A, B, C, D, E, F} the complementary committee is {G, H, I, J}.

73. $C(10, 3) = C(10, 7)$. The number of subsets of size 3 taken from a 10 member set is the same as the number of subsets of size 7. For example, consider the set $N = \{1, 2, 3, 4, 5, 6, 7, 8, 9, 10\}$. For the subset $A = \{1, 2, 3\}$, there is the corresponding subset $B = \{4, 5, 6, 7, 8, 9, 10\}$.

74. A committee of size 5 either includes or excludes John Doe. If the committee includes him, there are $C(10, 4)$ ways to choose the other 4 members. If the committee excludes him, there are $C(10, 5)$ ways to choose the 5 members. Hence $C(10, 4) + C(10, 5) = C(11, 5)$.

Chapter 5 Chapter Test

1. a. $4! = 4 \cdot 3 \cdot 2 \cdot 1 = 24$

b. $P(7, 3) = 7 \cdot 6 \cdot 5 = 210$

c. $C(18, 16) = \dfrac{18 \cdot 17}{2 \cdot 1} = 153$

d. $\begin{pmatrix} 6 \\ 0 \end{pmatrix} = \dfrac{6!}{0!6!} = 1$

e. $\begin{pmatrix} 5 \\ 2, 1, 2 \end{pmatrix} = \dfrac{5!}{2!1!2!} = 30$

2. a. True

b. True

c. False

3. a. $S' \cap T = \{a, e\}$

b. $(S \cup T)' = \varnothing$

4. $C \cap E$ represents the set of certified public accountants who are self-employed.
$C \cup E'$ Represents the set of people who are either certified public accountants or are not self-employed.

5.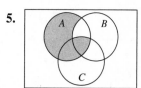

6. Let $C = \{$members of Choral Society$\}$ and $D = \{$members of Drama Club$\}$. Then
$C \cup D = \{$member of Choral Society or Drama Club$\}$ and
$C \cap D = \{$members of Choral Society and Drama Club$\}$.

$$n(C) = 40,\, n(D) = 30,\, n(C \cup D) = 60$$

$$n(C \cup D) = n(C) + n(D) - n(C \cap D)$$

$$60 = 40 + 30 - n(C \cap D)$$

$$n(C \cap D) = 40 + 30 - 60 = 10$$

10 students are members of both groups.

7. Let $E = \{$education majors$\}$, $M = \{$middle-of-the-road politically$\}$, and $V = \{$those who did volunteer work$\}$.
$n(E) = 9,\, n(M) = 43,\, n(V) = 82$
$n(E \cap M) = 4,\, n(E \cap V) = 6,\, n(M \cap V) = 32,\, n(E \cap M \cap V) = 3$
$$n(E \cup M \cup V) = n(E) + n(M) + n(V) - n(E \cap M) - n(E \cap V) - n(M \cap V) + n(E \cap M \cap V)$$
$$= 9 + 43 + 82 - 4 - 6 - 32 + 3$$
$$= 95$$
So $100 - 95 = 5$ freshman did not meet any of the three criteria.

8. $3 \cdot 5 \cdot 4 = 60$ ways

9. a. $\displaystyle C(20, 5) \cdot C(30, 5) = \frac{20!}{15!5!} \cdot \frac{30!}{25!5!}$
$$= (15,504)(142,506)$$
$$= 2,209,413,024 \text{ ways}$$

b. $\displaystyle C(10, 6) = \frac{10!}{6!4!} = 210$ ways
Note that this is the same as $C(10, 4)$.

10. $7 \cdot 6 \cdot 5 = 210$ numbers

11. Consider the children as a set, so there are 3 items to arrange. This can be done in $3! = 6$ ways. For each of those arrangements, the children can be arranged in $4! = 24$ ways. The total number of arrangements is $6 \cdot 24 = 144$.

12. $\displaystyle C(10, 3) \cdot C(20, 3) = \frac{10!}{7!3!} \cdot \frac{20!}{17!3!}$
$$= 120 \cdot 1140$$
$$= 136,800 \text{ ways}$$

13. $\displaystyle \binom{12}{5} = \frac{12!}{7!5!} = 792$

14. $2^7 = 128$

15. $\displaystyle \binom{8}{2,\,2,\,2,\,2} = \frac{8!}{2!2!2!2!} = 2520$ ways

Chapter 6

Exercises 6.2

1. a. The set of all possible pairs: {RS, RT, RU, RV, ST, SU, SV, TU, TV, UV}

  b. The set of pairs containing R: {RS, RT, RU, RV}

  c. The set of pairs containing neither R nor S: {TU, TV, UV}

3. a. {HH, HT, TH, TT}

  b. {HH, HT}

5. a. {(I, red), (I, white), (II, red), (II, white)}

  b. All combinations with I: {(I, red), (I, white)}

7. a. S = {all positive numbers of minutes}

  b. $E \cap F$ = "more than 5 but less than 8 minutes"
$E \cap G = \varnothing$ (There's no time longer than 5 minutes but less than 4 minutes.)
E' = "5 minutes or less"
F' = "8 minutes or more"
$E' \cap F = E'$ = "5 minutes or less"
$E' \cap F \cap G = G$ = "less than 4 minutes"
$E \cup F = S$

9. a. Eight possible combinations: {(Fr, Lib), (Fr, Con), (So, Lib), (So, Con), (Jr, Lib), (Jr, Con), (Sr, Lib), (Sr, Con)}

  b. All combinations with Con: {(Fr, Con), (So, Con), (Jr, Con), (Sr, Con)}

  c. {(Jr, Lib)}

  d. All combinations with neither Fr nor Con: {(So, Lib), (Jr, Lib), (Sr, Lib)}

11. a. No; $E \cap F$ = {2}

  b. Yes; $F \cap G = \varnothing$

13. All combinations of members of S: $\varnothing$, {a}, {b}, {c}, {a, b}, {a, c}, {b, c}, S

15. Yes; $(E \cup F) \cap (E' \cap F') = \{1, 2, 3\} \cap \{4\} = \varnothing$

17. a. {0, 1, 2, 3, 4, 5, 6, 7, 8, 9, 10}

  b. More than half heads: {6, 7, 8, 9, 10}

19. a. No; there are blue-eyed males.

  b. Yes; a brown-eyed female doesn't have blue eyes.

  c. Yes; a brown-eyed female isn't male.

21. {0, 1, 2, 3, 4, 5, 6, 7, 8}

23. A possible outcome is (7, 4). There are $9 \cdot 9 = 81$ outcomes in the sample space.

25. A possible outcome is {2, 6, 9, 10}. There are $C(14, 4) - 1 = 1000$ possible combinations, so $\dfrac{17}{1000} = 0.017 = 1.7\%$ were assigned to the Chicago Bulls.

27. a. 6 suspects $\times$ 6 weapons $\times$ 9 rooms = 324 outcomes

  b. $E \cap F$ = "The murder occurred in the library with a gun."

  c. $E \cup F$ = "Either the murder occurred in the library, or it was perpetrated with a gun."

Exercises 6.3

1. a. $\dfrac{46,277}{774,746}$

  b. $\dfrac{46,277 + 1855}{774,746} = \dfrac{48,132}{774,746}$

  c. $\dfrac{774,746 - 48,132}{774,746} = \dfrac{726,614}{774,746}$

3. a. E = "the numbers add up to 8" = {(2, 6), (3, 5), (4, 4), (5, 3), (6, 2)}
$\Pr(E) = \dfrac{5}{36}$

b. $\Pr(\text{sum is } 2) = \Pr((1, 1)) = \dfrac{1}{36}$; $\Pr(\text{sum is } 3) = \Pr((1, 2)) + \Pr((2, 1)) = \dfrac{2}{36}$

$\Pr(\text{sum is } 4) = \Pr((1, 3)) + \Pr((2, 2)) + \Pr((3, 1)) = \dfrac{3}{36}$

The probability that the sum is less than 5 is $\dfrac{1}{36} + \dfrac{2}{36} + \dfrac{3}{36} = \dfrac{1}{6}$.

5. $\dfrac{1}{38} + \dfrac{1}{38} = \dfrac{2}{38} = \dfrac{1}{19}$

7. a. $1 - \left(\dfrac{1}{3} + \dfrac{1}{2}\right) = \dfrac{1}{6}$

b. The probability is $\dfrac{1}{6}$. Thus $a = 1$ and $a + b = 6$, so $b = 5$. The odds are 1 to 5.

9. a. $.1 + .6 = .7$

b. $.6 + .1 = .7$

11. a. $\dfrac{10}{10+1} = \dfrac{10}{11}$

b. $\dfrac{1}{1+2} = \dfrac{1}{3}$

c. $\dfrac{4}{4+5} = \dfrac{4}{9}$

13. $.09 = \dfrac{9}{100}$

Thus $a = 9$ and $a + b = 100$, so $b = 91$. The odds are 9 to 91.

15. Win: $\dfrac{11}{11+7} = \dfrac{11}{18}$

Lose: $\dfrac{7}{11+7} = \dfrac{7}{18}$

17. a. $\Pr(E \cup F) = \Pr(E) + \Pr(E) - \Pr(E \cap F) = .7$

b.

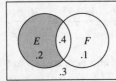

$\Pr(E \cap F') = \Pr(E) - \Pr(E \cap F) = .2$

19. $\Pr(E) = 0.4$
$\Pr(F) = 0.5$
$\Pr(E \cup F) = \Pr(E) + \Pr(F) - \Pr(E \cap F)$
$\Pr(E \cup F) = 0.4 + 0.5 - 0 = 0.9$

21. Pr (A in history) = 0.7 . Pr(A in psychology) = 0.8. Pr(A in history and A in psychology) = 0.9.
The probability that he will get an A in **either** subject
= Pr(A in history or A in psychology)
= 0.7 + 0.8 − 0.9 = 0.6.

23. a. .20 + .25 + .25 = .7

    b. .70 · 10,000 = 7000

25.

Number of Colleges Applied to	Probability
1	.17
2	Pr(2) = Pr(2 or less) − Pr(1) = .29 − .17 = .12
3	Pr(3) = Pr(3 or less) − Pr(2 or less) = .43 − .29 = .14
4	Pr(4) = Pr(4 or less) − Pr(3 or less) = .59 − .43 = .16
5–20	Pr(5 −20) = Pr(20 or less) − Pr(4 or less) = 1 − .59 = .41

27. a. Pr(20–34) = .15
Pr(35–49) = Pr(20–49) − Pr(20–34) = .55
Pr(50–64) = Pr(20–64) − Pr(20–49) = .20
Pr(65–79) = Pr(20–79) − Pr(20–64) = .10

    b. Pr(50–79) = Pr(20–79) − Pr(20–49) = .30

29. a. Some categories are left out—people who use a computer for both school and work, for example.

    b. 100% − 17% = 83% = 0.83

31. The probability that a student attended public school is $.82 = \dfrac{82}{100} = \dfrac{41}{50}$; $a = 41$ and $a + b = 50$, so $b = 9$. The odds are 41 to 9.

33. a. Odds for Sparks = 5:3; Pr(Sparks) = 3/8. Odds for Meteors = 3:1; Pr(Meteors) = 1/4.
Odds for Asteroids = 3:2; Pr(Asteroids) = 2/5
Odds for Suns = 4:1; Pr(Suns) = 1/5

    b. The sum of 3/8 + 1/4 + 2/5 + 1/5 = 49/40 > 1.

    c. Bookies have to make a living so the payoffs are lower.

35. There are more members (13) than Zodiac signs (12) so two or more members will always have the same Zodiac sign; thus the probability is 1.

37. This event never occurs; if 19 of the people receive the correct coat then so must the remaining person. Thus the probability is 0.

Exercises 6.4

1. a. There are $3 \times 3 \times 3 = 27$ combinations of class selections.
$$\frac{3}{27} = \frac{1}{9}$$

b. $\dfrac{3 \times 2 \times 1}{27} = \dfrac{2}{9}$

3. a. $\dfrac{7}{13}$

b. $\dfrac{6}{13}$

c. $\text{Pr}(\{3, 6, 9, 12\}) = \dfrac{4}{13}$

d. $\text{Pr}(\{1, 3, 5, 6, 7, 9, 11, 12, 13\}) = \dfrac{9}{13}$

5. a. $\dfrac{C(6, 5)}{C(13, 5)} = \dfrac{6}{1287} = \dfrac{2}{429}$

b. $\dfrac{C(7, 5)}{C(13, 5)} = \dfrac{21}{1287} = \dfrac{7}{429}$

c. $1 - \dfrac{2}{429} = \dfrac{427}{429}$

7. $1 - \dfrac{C(5, 3)}{C(10, 3)} = 1 - \dfrac{10}{120} = \dfrac{11}{12}$

9. $1 - \dfrac{C(4, 2)}{C(9, 2)} = 1 - \dfrac{6}{36} = \dfrac{5}{6}$

11. Ways for no girls to be chosen: $C(12, 7)$
Ways for exactly 1 girl to be chosen:
$C(12, 6) \times C(10, 1)$
$$1 - \frac{C(12, 7) + C(12, 6) \times C(10, 1)}{C(22, 7)}$$
$$= 1 - \frac{792 + 924 \times 10}{170,544} = \frac{16}{17}$$

13. $\dfrac{C(10, 7)}{C(22, 7)} = \dfrac{5}{7106}$

15. $1 - \dfrac{30 \times 29 \times 28 \times 27}{30^4} = \dfrac{47}{250}$

17. $1 - \dfrac{6 \cdot 6}{7 \cdot 7} = \dfrac{13}{49}$

19. a. $.2 + .15 - .1 = .25$

b. $1 - .25 = .75$

c. $1 - .2 = .8$

21. $\text{Pr}(F) = 1 - \text{Pr}(F') = .4$
$\text{Pr}(E \cap F) = \text{Pr}(E) + \text{Pr}(F) - \text{Pr}(E \cup F)$
$\qquad\qquad = .3 + .4 - .7$
$\qquad\qquad = 0$

23. $\text{Pr}(\text{second sock matches first}) = \dfrac{12 \cdot 1}{12 \cdot 11} = \dfrac{1}{11}$

25. There are $5! = 120$ ways to arrange the family. First determine where the parents will stand. There are two ways with the man at one end. If the man doesn't stand at one end, there are $3 \cdot 2 = 6$ ways for the couple to stand together. Next determine the order of the children. There are $3! = 6$ ways to order the children. Thus there are $(2 + 6) \cdot 6 = 48$ ways to stand with the parents together.

The probability is $\dfrac{48}{120} = \dfrac{2}{5}$.

27. The tourist must travel 8 blocks of which 3 are south. Thus he has $C(8, 3) = 56$ ways to get to B from A.

a. To get from A to B through C there are $C(3, 1) \cdot C(5, 2) = 30$ ways.
The probability is $\dfrac{30}{56} = \dfrac{15}{28}$.

b. To get from A to B through D there are $C(5, 1) \cdot C(3, 2) = 15$ ways.
The probability is $\dfrac{15}{56}$.

c. To get from A to B through C and D there are $C(3, 1) \cdot C(2, 0) \cdot C(3, 2) = 9$ ways.
The probability is $\dfrac{9}{56}$.

d. The number of ways to get from A to B through C or D is $30 + 15 - 9 = 36$.

The probability is $\dfrac{36}{56} = \dfrac{9}{14}$.

29. $1 - \dfrac{C(6, 3)}{C(10, 3)} = 1 - \dfrac{20}{120} = \dfrac{5}{6}$

31. $\dfrac{C(6, 3)}{2^6} = \dfrac{5}{16}$

33. $\dfrac{2}{C(40, 6)} = \dfrac{1}{1,919,190}$

35. The total number of ways to select 5 senators is $C(100, 5) = \dfrac{100!}{95!5!} = 75,287,520$.

The total number of ways to select senators from different states is $\dfrac{100 \cdot 98 \cdot 96 \cdot 94 \cdot 92}{5!} = 67,800,320$.

$\begin{aligned} \text{Pr(no two members from same state)} &= \dfrac{67,800,320}{75,287,520} \\ &\approx .90055 \end{aligned}$

37. $\text{Pr(at least one birthday on June 13)} = 1 - \left(\dfrac{364}{365}\right)^{25} \approx .066$

Because in Table 1 no particular date is being matched. Any two (or more) identical birthdays count as a success.

39. a. $\begin{aligned} \text{Pr(Mary and Laura)} &= \dfrac{C(26, 8)}{C(28, 10)} \\ &= \dfrac{1,562,275}{13,123,110} \\ &\approx .119 \end{aligned}$

b. $\begin{aligned} \text{Pr(Mary and Laura)} &= \dfrac{C(16, 5) \cdot C(10, 3)}{C(16, 5) \cdot C(12, 5)} \\ &= \dfrac{C(10, 3)}{C(12, 5)} \\ &= \dfrac{120}{792} \\ &\approx .152 \end{aligned}$

41. $\begin{aligned} \text{Pr(at least one winner)} &= 1 - \text{Pr(no winners)} \\ &= 1 - \dfrac{3 \cdot 3 \cdot 2}{4 \cdot 4 \cdot 3} \\ &= 1 - \dfrac{18}{48} \\ &= .625 \end{aligned}$

43. a. There are $C(55, 5) \cdot 42 = 146,107,962$ possible picks of which only one is correct, so the odds of winning are 1:146,107,961.

b. The probability of winning is $\dfrac{1}{146,107,962}$.

45. 1) Select two distinct numbers from the set $S = \{\ 1, 2, 3, 4, 5, 6, 7\ \}$.
2) The total number of selections is $7(6) = 42$.
3) Prob(product is even) = 1 - Prob(product is odd)
4) The number of different selections of two numbers in which the product is odd is: 12
They are {1,3}, {1,5}, {1,7}, {3,1}, {3,5}, {3,7}, {5,1}, {5,3}, {5,7}, {7,1}, {7,3}, {7,5}
5) $1 - 12/42 = 30/42 = 5/7$.

47. 1) Find the total number of ways to select 6 numbers from 49 numbers: $C(49,6) = 13,983,816$
2) Find the number of ways to match 3 of the 6 numbers drawn: $C(6,3) = 20$
3) Find the number of ways to choose the remaining 3 numbers: $C(43,3) = 12,341$

4) Prob(choose three of six numbers) = $\dfrac{20 \cdot 12341}{13,983,816} = 0.01765$

49. 1) The number of ways to choose a dinner plate, salad plate and a bowl is $(5)(5)(5) = 125$.
2) The number of ways to choose a different color: $(5)(4)(3) = 60$

3) Pr(they all have different colors) $= \dfrac{60}{125}$
$= 0.48$

51. $0.9^3 = 0.729$

53. a. There are $13 \cdot 12 = 156$ ways to choose the two denominations, $C(4, 3) = 4$ ways to choose the suits for the three of a kind and $C(4, 2) = 6$ ways to choose the suits for the pair; thus there are $156 \cdot 4 \cdot 6 = 3744$ possible full house hands, so the probability is $\dfrac{3744}{C(52, 5)} \approx 0.00144$.

b. There are 13 choices for the denomination and $C(4, 3) = 4$ choices for the suits for the three-of-a-kind; there are $C(12, 2) = 66$ choices for the denominations and $4 \cdot 4 = 16$ choices for the suits of the remaining two cards, for a total of $13 \cdot 4 \cdot 66 \cdot 16 = 54,912$ possible three-of-a-kind poker hands; so the probability is $\dfrac{54,912}{C(52, 5)} \approx 0.0211$.

55. a. There are 4 ways to select the suit of the 4 card group, $C(13, 4)$ ways to select their denominations, and $C(13, 3)^3$ ways to select 3 cards each from the remaining 3 suits. The probability of a 4-3-3-3 bridge hand is $\dfrac{4 \cdot C(13, 4) \cdot C(13, 3)^3}{C(52, 13)} \approx 0.1054$.

b. There are $C(4, 2) = 6$ ways to choose the suits of the 4-card groups and 2 ways to choose the suit of the 3-card group (then the suit of the 2-card group is uniquely determined). The denominations can then be chosen in $C(13, 4)^2 \cdot C(13, 3) \cdot C(13, 2)$ ways, so the probability of a 4-4-3-2 bridge hand is $\dfrac{6 \cdot 2 \cdot C(13, 4)^2 \cdot C(13, 3) \cdot C(13, 2)}{C(52, 13)} \approx 0.2155$.

57. Pr(first one is red) · Pr(second one is red) = .1
Let r = number of red balls.

$$\frac{r}{40} \cdot \frac{r-1}{39} = .1$$
$$r(r-1) = (40)(39)(.1)$$
$$r^2 - r - 156 = 0$$
$$(r+12)(r-13) = 0$$
$$r = 13$$

59. a. $\dfrac{1}{6} \times 24 = 4$

b. $\dfrac{C(24,\,4) \times 5^{20}}{6^{24}} \approx .2139$

61. a. The probability of at least two people having the same birthday in a group of size r is
$$P_r = 1 - \frac{P(365,\, r)}{365^r}, \text{ and } P_r \text{ increases with } r.$$
$P_{22} \approx 0.476$ and $P_{23} \approx 0.507,$ so $r = 23$ is the smallest value of r for which $P_r > 0.5.$

b. $P_{26} \approx 0.598$ and $P_{27} \approx 0.627,$ so $r = 27$ is the smallest value of r for which $P_r > 0.6.$

63. The probability of at least one good picture in a string of n pictures is
$$P_n = 1 - (1 - 0.729)^n = 1 - 0.271^n.$$
$P_5 \approx 0.9985 = 99.85\%$ and
$P_6 \approx 0.9996 = 99.96\%,$ so the photographer must take $n = 6$ pictures to insure that the probability $P_n > 0.999.$

Exercises 6.5

1. $\Pr\left(E|F\right) = \dfrac{.1}{.3} = \dfrac{1}{3}, \Pr\left(F|E\right) = \dfrac{.1}{.5} = \dfrac{1}{5}$

3. $\Pr\left(E|F'\right) = \dfrac{\Pr(E \cap F')}{\Pr(F')}$
$$= \frac{P(E) - \Pr(E \cap F)}{1 - \Pr(F)}$$
$$= \frac{.4}{.7}$$
$$= \frac{4}{7}$$

5. No;
$$\Pr\left(\text{has cancer} \,\middle|\, \text{works for Ajax}\right) \neq \Pr(\text{has cancer})$$

7. a. $.80 \cdot .75 \cdot .60 = .36$

b. $.36 + (1 - .80)(.75)(.60) + (.80)(1 - .75)(.60)$
$+ (.80)(.75)(1 - .60) = .81$

9. $(1 - .01)^5 (1 - .02)^5 (1 - .025)^3$
$= (.99)^5 (.98)^5 (.975)^3$
$\approx .7967$

11. $\Pr\left(E|F\right) = \dfrac{\frac{1}{4}}{\frac{1}{3}} = \dfrac{3}{4}$

$\Pr\left(F|E\right) = \dfrac{\frac{1}{4}}{\frac{1}{2}} = \dfrac{1}{2}$

13. a. $1 - \Pr(\text{Dem. or favors})$
$= 1 - [\Pr(\text{Dem.}) + \Pr(\text{favors})$
$\quad - \Pr(\text{Dem. and favors})] = .4$

b. $\dfrac{\Pr(\text{Dem. and favors})}{\Pr(\text{Dem.})} = .6$

c. $\dfrac{\Pr(\text{Dem. and favors})}{\Pr(\text{favors})} = .75$

15. $\dfrac{\Pr(\{HHH\})}{\Pr(\{HHH,\ HHT,\ HTH,\ THH\})} = \dfrac{1}{4}$

17. $1 - \Pr(\text{Neither } A \text{ nor } B \text{ lives 15 more years})$
$= 1 - (1 - .8)(1 - .7) = .94$

19. Let E = "the sample contains at least one white ball"
and F = "the sample contains balls of both colors."
Then $\Pr(E) = \dfrac{C(2,\,2) + C(2,\,1) \cdot C(3,\,1)}{C(5,\,2)}$
$$= \frac{1 + 2 \cdot 3}{10}$$
$$= \frac{7}{10}$$
and $\Pr(F) = \dfrac{C(2,\,1) \cdot C(3,\,1)}{C(5,\,2)} = \dfrac{2 \cdot 3}{10} = \dfrac{3}{5}.$
$E \cap F = F$ since $F \subset E,$ and so
$\Pr(E \cap F) = \dfrac{3}{5} \neq \dfrac{7}{10} \cdot \dfrac{3}{5}$
and thus the events are not independent.

21. $1 - (1 - .005)^2 = .009975$

23. $(.7)^4 = .2401$

25. $\Pr(E \cap F) = \Pr(E) \cdot \Pr(F)$ Definition of "independent"
$\Pr(E) + \Pr(F) - \Pr(E \cup F) = \Pr(E) \cdot \Pr(F)$ Substitution
$[1 - \Pr(E')] + [1 - \Pr(F')] - [1 - \Pr(E' \cap F')] = [1 - \Pr(E')] \cdot [1 - \Pr(F')]$ Substitution
$\Pr(E' \cap F') = \Pr(E') \cdot \Pr(F')$ Simplification

27. 0 points: $1 - .6 = .4$
1 point: $.6 \cdot .4 = .24$
2 points: $.6 \cdot .6 = .36$

29. $\Pr\left(E \cup F \big| G\right) = \dfrac{\Pr[(E \cup F) \cap G]}{\Pr(G)}$

$= \dfrac{\Pr[(E \cap G) \cup (F \cap G)]}{\Pr(G)}$

$= \dfrac{\Pr(E \cap G) + \Pr(F \cap G) - \Pr[(E \cap G) \cap (F \cap G)]}{\Pr(G)}$

$= \dfrac{\Pr(E \cap G)}{\Pr(G)} + \dfrac{\Pr(F \cap G)}{\Pr(G)} - \dfrac{\Pr[(E \cap F) \cap G]}{\Pr(G)}$

$= \Pr(E|G) + \Pr(F|G) - \Pr(E \cap F|G)$

31. a. $\Pr\left(\text{death} \big| A\right) = \dfrac{12,000}{120,000} = \dfrac{1}{10}$

b. $1000 \times \dfrac{1}{10} = 100$ per 1000

c. $\Pr\left(\text{death} \big| B\right) = \dfrac{4500}{90,000} = \dfrac{1}{20}$

33. a. $\Pr(B) = 80\% \cdot (100\% - 65\%) = .28$

b. $\Pr(C) = 20\% \cdot (100\% - 65\%) = .07$

c. $\Pr(H) = (65\% \cdot 75\%) + (28\% \cdot 40\%)$
$+ (7\% \cdot 10\%) = .6065$

35. a. $\dfrac{250 - (120 + 140 - 50)}{250} = \dfrac{40}{250} = \dfrac{4}{25} = .16$

b. $\dfrac{(120 - 50) + (140 - 50)}{250} = \dfrac{16}{25} = .64$

c. $\dfrac{140 - 50}{250} = \dfrac{9}{25} = .36$

d. $\dfrac{50}{120} = \dfrac{5}{12} \approx .42$

e. $\dfrac{120 - 50}{250 - 140} = \dfrac{7}{11} \approx .64$

f. $\dfrac{40}{250 - 120} = \dfrac{4}{13} \approx .31$

37. Let M = "male," F = M′ = "female,"
Ba = "Bachelor's," Ma = "Master's" and
Do = "Doctor's."

a. $\Pr(\text{Ma}) = \dfrac{211 + 301}{1906} \approx 0.2686$

b. $\Pr(\text{M}) = \dfrac{573 + 211 + 24}{1906} \approx 0.4239$

c. $\Pr(\text{F}|\text{Ba}) = \dfrac{775}{775 + 573} \approx 0.5749$

d. $\Pr(\text{Do}|\text{M}) = \dfrac{24}{24 + 211 + 573} \approx 0.0297$

39. Total voters = $400 + 700 + \cdots + 200 = 2500$

a. $\dfrac{400 + 600}{2500} = .40$

b. $\dfrac{400 + 700 + 300}{2500} = .56$

c. $\dfrac{200}{600 + 300 + 200} \approx .18$

d. $\dfrac{700}{700 + 300} = .70$

e. $\dfrac{600 + 300}{400 + 700 + 600 + 300} = .45$

f. $\dfrac{400 + 600}{400 + 600 + 300 + 200} \approx .67$

41. a. Pr(at least one hit) = 1 − Pr(no hits)
$\qquad\qquad = 1 - (.7)^4 = .7599$

b. Pr(at least one hit in each of first 10 games)
$\qquad = (.7599)^{10} \approx .0642$

c. Pr(at least one has 10-game hitting streak)
= 1 − Pr(no 10-game hitting streaks)
= $1 - (1 - .0642)^{20}$
$\approx .7348$

43. $p(R_2) = p(R_1 \cap R_2) + p(W_1 \cap R_2)$
$\qquad = p(R_1)p(R_2|R_1) + p(W_1)p(R_2|W_1)$
$\qquad = \dfrac{10}{25} \cdot \dfrac{9}{24} + \dfrac{15}{25} \cdot \dfrac{10}{24}$
$\qquad = \dfrac{240}{600}$
$\qquad = \dfrac{2}{5}$

$p(R_1) = \dfrac{10}{25} = \dfrac{2}{5}$

45. $\Pr(\text{all red}) = \dfrac{C(6, 3)}{C(13, 3)} = \dfrac{20}{286} = \dfrac{10}{143}$

$\Pr(\text{all blue}) = \dfrac{C(4, 3)}{C(13, 3)} = \dfrac{4}{286} = \dfrac{2}{143}$

$\Pr(\text{all white}) = \dfrac{C(3, 3)}{C(13, 3)} = \dfrac{1}{286}$

Pr(all same color)
= Pr(all red) + Pr(all blue) + Pr(all white)
$= \dfrac{20 + 4 + 1}{286} = \dfrac{25}{286} \approx 0.0874$

47. Pr(both roast beef) = Pr(both ham)
$\qquad\qquad = \dfrac{C(2, 2)}{C(4, 2)}$
$\qquad\qquad = \dfrac{1}{6}$

Pr(both roast beef) + Pr(both ham) $= \dfrac{1}{6} + \dfrac{1}{6} = \dfrac{1}{3}$

49. Pr(all white|one white)
$= \dfrac{\text{Pr(all white and one white)}}{\text{Pr(one white)}}$
$= \dfrac{\frac{8}{13} \cdot \frac{7}{12} \cdot \frac{6}{11} \cdot \frac{5}{10}}{1 - \frac{5}{13} \cdot \frac{4}{12} \cdot \frac{3}{11} \cdot \frac{2}{10}} \approx .0986$

51. $\dfrac{C(3, 2)}{C(102, 2)} = \dfrac{3}{5151} = \dfrac{1}{1717}$

Note: Since Fred put in 2 additional cards, the total number of cards is 102.

53. If E = {2, 4, 6} and F = {3, 6} we must show that that Pr(E and F) = Pr(E) · pr(F)
Pr(E and F) = 1/6; Pr(E) · Pr(F) = 3/6 · 2/6 = 1/6.
So, events E and F are independent.

55. The probability that the couple will have at least five children is the sum of the events
BBBBG + BBBBBG + BBBBBBG + …
The complement of this event is the event that a girl is born before five children: The events are
G , BG, BBG, BBBG
The probability of each of these events is
1/2 + 1/4 + 1/8 + 1/16 = 15/16 . So the probability the couple will have at least five children is 1 - 15/16 = 1/16.

57. a. Since the probability of winning the lottery is *p*, then if you bet 1000 different combinations the probability is
$p + p + p + ... + p = 1000p$

b. The probability of winning the jackpot at least once when you bet a single combination in 1000 successive weekly lotteries is
1 – Pr(you do not win in 1000 successive lotteries)
$= 1 - (1-p)^{1000}$
$\approx 1 - (1 - 1000p + 499,500p^2)$
$= 1000p - 499,500p^2 < 1000p$
Therefore, you have a higher probability of winning the lottery when you bet 1000 different combinations.

59. Pr(HT) = 0.6 · 0.4 = 0.24;
Pr(TH) = 0.4 · 0.6 = 0.24.
Yes; toss the coin twice. If HT comes up, let the home team kick off. If TH comes up, let the visiting team kick off. If HH or TT comes up, repeat the coin toss until one of each face occurs.

61. No; both outcomes have probability $\left(\dfrac{1}{2}\right)^5 = \dfrac{1}{32}$.

63. $Pr(W) = \dfrac{68,446}{147,482} \approx 0.4641$

$Pr(U)\dfrac{7246}{147,482} \approx 0.0491$

$Pr(W \cap U) = \dfrac{3208}{147,482} \approx 0.0218$

Gender and unemployment status are not independent because $Pr(W \cap U) \neq Pr(W) \cdot Pr(U)$ (although they are very close).

65. a. $\dfrac{4}{52} \cdot \dfrac{3}{51} \cdot \dfrac{2}{50} \cdot \dfrac{1}{49} \approx 3.69 \times 10^{-6}$

b. $\dfrac{2}{52} \cdot \dfrac{1}{51} \cdot \dfrac{2}{50} \cdot \dfrac{1}{49} \approx 6.16 \times 10^{-7}$

c. (a)

Exercises 6.6

1.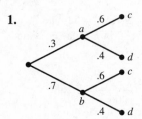

3.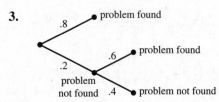

5. $.2 \times .4 = .08$

7. $.10 + .30 \times .05 + .60 \times .30 = .295$

9. Pr(white, then red) + Pr(red, then red)
$= \dfrac{2}{3} \times \dfrac{1}{2} + \dfrac{1}{3} \times \dfrac{3}{4} = \dfrac{7}{12}$

11.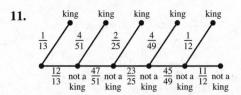

Pr(king on 1st draw) + Pr(king on 2nd draw) + Pr(king on 3rd draw)
= 1 – Pr(not a king on 3rd draw)
$= 1 - \dfrac{12}{13} \times \dfrac{47}{51} \times \dfrac{23}{25} = 1 - \dfrac{4324}{5525} = \dfrac{1201}{5525} \approx .22$

13.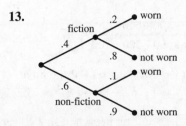

$.40 \times .20 + .60 \times .10 = .14$

15. $\Pr(\text{male}|\text{color-blind}) = \dfrac{\frac{1}{2} \times .05}{\frac{1}{2} \times .05 + \frac{1}{2} \times .004} = \dfrac{25}{27}$

17. $(.5)(.9)(.9) + (.5)(.1)(.7) + (.5)(.7)(.9) + (.5)(.3)(.7) = .86$

19. $\Pr\left(\text{fake}|\text{HH}\right) = \dfrac{\frac{1}{4} \times 1}{\frac{3}{4} \times \frac{1}{4} + \frac{1}{4} \times 1} = \dfrac{4}{7}$

21. a. $\Pr(\text{wins the point}) = .60 \times .75 + .40 \times .75 \times .50 = .60$

 b. $\Pr(\text{first serve good}|\text{wins service point}) = \dfrac{\Pr\left(\text{first serve good and wins service point}\right)}{\Pr\left(\text{wins service point}\right)} = \dfrac{.60 \times .75}{.60}$
$= .75$

23. $1 - (.9999)^n$

25. Same shape

$\Pr(\text{winning}) = \dfrac{3}{6} + \dfrac{3}{6} \times \dfrac{3}{5} \times \dfrac{2}{4} + \dfrac{3}{6} \times \dfrac{3}{5} \times \dfrac{2}{4} \times \dfrac{2}{3} \times \dfrac{1}{2} + \dfrac{3}{6} \times \dfrac{2}{5} \times \dfrac{3}{4} \times \dfrac{2}{3} \times \dfrac{1}{2} = \dfrac{3}{4}$

Probability of winning card game is greater.

27. a. $\Pr(\text{white}) = \dfrac{1}{2} \times \dfrac{1}{2} = \dfrac{1}{4}$

 $\Pr(\text{red}) = 1 - \dfrac{1}{4} = \dfrac{3}{4}$

 b. $\Pr(\text{red}) = .6 \times .5 + .4 \times 1 = .7$

29. $\Pr(\text{night}|\text{part-timer}) = \dfrac{.60 \times 2}{.25 \times 4 + .60 \times 2} = \dfrac{6}{11}$

31. $\Pr(\text{not TB}|\text{NEG}) = \dfrac{\Pr(\text{not TB} \cap \text{NEG})}{\text{PR}(\text{NEG})}$

 $= \dfrac{.989802}{.000004 + .989802}$
 $\approx .999996$

33. $p(W) = p(R_1 \cap W_2) + p(W_1 \cap W_2)$
 $= p(R_1)\, p\left(W_2|R_1\right) + p(W_1)\, p\left(W_2|W_1\right)$
 $= \left(\dfrac{5}{10}\right) \cdot \left(\dfrac{12}{13}\right) + \left(\dfrac{5}{10}\right) \cdot 1$
 $= \dfrac{25}{26}$

35. $\Pr(\text{get same number of heads})$
 $= \Pr(\text{2 heads}) + \Pr(\text{1 head}) + \Pr(\text{0 heads})$
 $= \dfrac{1}{4} \cdot \dfrac{1}{4} + \dfrac{1}{2} \cdot \dfrac{1}{2} + \dfrac{1}{4} \cdot \dfrac{1}{4} = \dfrac{3}{8}$

37.

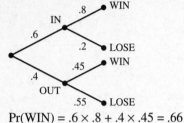

$\Pr(\text{WIN}) = .6 \times .8 + .4 \times .45 = .66$

39. a. Since Stan always scores 400 points, Oliver wins $.99 = 99\%$ of the time.

b. For Oliver to score the most points he must win all 400 games. The probability that this occurs is $0.99^{400} \approx 0.0180$.

41. $\Pr(\text{POS}) = 0.1 \times 1.00 + 0.9 \times 0.01 = 0.109$
$\Pr(\text{POS} \cap \text{USED}) = 0.1 \times 1 = 0.1$

$$\Pr\left(\text{USED} \mid \text{POS}\right) = \frac{0.1}{0.109} \approx 0.92$$

Exercises 6.7

1. $\Pr\left(\text{over } 60 \mid \text{accident}\right) = \dfrac{.10 \times .04}{(.05 \times .06) + (.10 \times .04) + (.25 \times .02) + (.20 \times .015) + (.30 \times .025) + (.10 \times .04)}$

$$= \frac{.004}{.0265}$$

$$= \frac{8}{53}$$

3. $\Pr\left(\text{sophomore} \mid A\right) = \dfrac{.30 \times .4}{(.10 \times .2) + (.30 \times .4) + (.40 \times .3) + (.20 \times .1)} = \dfrac{.12}{.28} = \dfrac{3}{7}$

5. $\Pr\left(\geq \$75,000 \mid 2 \text{ or more cars}\right) = \dfrac{.05 \times .9}{(.10 \times .2) + (.20 \times .5) + (.35 \times .6) + (.30 \times .75) + (.05 \times .9)}$

$$= \frac{.045}{.6}$$

$$= \frac{3}{40}$$

$$= .075$$

7. $\Pr\left(\text{passed exam} \mid A\right) = \dfrac{.80 \times .40}{.80 \times .40 + .20 \times .20}$

$$= \frac{.32}{.36}$$

$$= \frac{8}{9}$$

$$\approx .89$$

9. a. $(.20 \times .20) + (.15 \times .15) + (.25 \times .12) + (.30 \times .10) + (.10 \times .10) = .1325$

b. $\Pr\left(\text{division C}\middle|\text{bilingual}\right) = \dfrac{.25 \times .12}{.1325} = \dfrac{.03}{.1325} = \dfrac{12}{53} \approx .23$

11. $\Pr\left(\text{cancer}\middle|\text{positive}\right) = \dfrac{\Pr(\text{cancer}) \times \Pr\left(\text{positive}\middle|\text{cancer}\right)}{\Pr(\text{cancer}) \times \Pr\left(\text{positive}\middle|\text{cancer}\right) + \Pr(\text{no cancer}) \times \Pr\left(\text{positive}\middle|\text{no cancer}\right)}$

$= \dfrac{.02 \times .75}{(.02 \times .75) + (.98 \times .30)} = \dfrac{5}{103} \approx .049$

13. a. $1 - .99 = .01$

b. $\Pr\left(\text{pregnant}\middle|\text{positive}\right)$

$= \dfrac{\Pr(\text{pregnant}) \times \Pr\left(\text{positive}\middle|\text{pregnant}\right)}{\Pr(\text{pregnant}) \times \Pr\left(\text{positive}\middle|\text{pregnant}\right) + \Pr(\text{not pregnant}) \times \Pr\left(\text{positive}\middle|\text{not pregnant}\right)}$

$= \dfrac{.40 \times .99}{(.40 \times .99) + (.60 \times .02)} = \dfrac{33}{34} \approx .971$

15. $\Pr\left(\text{steroids}\middle|\text{positive}\right) = \dfrac{\Pr(\text{steroids}) \times \Pr\left(\text{positive}\middle|\text{steroids}\right)}{\Pr(\text{steroids}) \times \Pr\left(\text{positive}\middle|\text{steroids}\right) + \Pr(\text{no steroids}) \times \Pr\left(\text{positive}\middle|\text{no steroids}\right)}$

$= \dfrac{.10 \times .93}{(.10 \times .93) + (.90 \times .02)} = \dfrac{31}{37} \approx .838$

17. a. $\Pr(\text{one is}) = \dfrac{13}{52} = \dfrac{1}{4}$

b. $\Pr\left(\text{none is}\middle|\text{random one isn't}\right)$

$= \dfrac{\Pr(\text{none is}) \times \left(\text{random one isn't}\middle|\text{none is}\right)}{\Pr(\text{none is}) \times \Pr\left(\text{random one isn't}\middle|\text{none is}\right) + \Pr(\text{one is}) \times \Pr\left(\text{random one isn't}\middle|\text{one is}\right)}$

$= \dfrac{\frac{3}{4} \times 1}{\left(\frac{3}{4} \times 1\right) + \left(\frac{1}{4} \times \frac{12}{13}\right)} = \dfrac{13}{17} \approx .765$

c. $\Pr(\text{one is}|\text{10 randoms aren't})$

$= \dfrac{\Pr(\text{one is}) \times \Pr\left(\text{10 randoms aren't}\middle|\text{one is}\right)}{\Pr(\text{one is}) \times \Pr\left(\text{10 randoms aren't}\middle|\text{one is}\right) + \Pr(\text{none is}) \times \Pr\left(\text{10 randoms aren't}\middle|\text{none is}\right)}$

$= \dfrac{\frac{1}{4} \times \left(\frac{12}{13}\right)^{10}}{\left[\frac{1}{4} \times \left(\frac{12}{13}\right)^{10}\right] + \left(\frac{3}{4} \times 1\right)}$

$\approx .130$

19. **a.** $\Pr\left(\text{Lakeside}\middle|\text{winner}\right)$

$$= \frac{\Pr(\text{Lakeside}) \times \Pr\left(\text{winner}\middle|\text{Lakeside}\right)}{\Pr(\text{Lakeside}) \times \Pr\left(\text{winner}\middle|\text{Lakeside}\right) + \Pr(\text{Pylesville}) \times \Pr\left(\text{winner}\middle|\text{Pylesville}\right)} \\ {\qquad + \Pr(\text{Millerville}) \times \Pr\left(\text{winner}\middle|\text{Millerville}\right)}$$

$$= \frac{.40 \times .05}{(.40 \times .05) + (.20 \times .02) + (.40 \times .03)} = \frac{5}{9}$$

b. $\dfrac{.20 \times .02}{(.40 \times .05) \times (.20 \times .02) + (.40 \times .03)} = \dfrac{1}{9} \approx 11\%$

Exercises 6.8

1. Use seq(int(6*rand)+1,X,1,36,1)$\to$ L$_1$.

   Theoretical probabilities: $\dfrac{1}{6}$ for each fall.

3. Use seq(int(1000*rand)+1,X,1,10,1)$\to$L$_1$ where a number from 1–810 represents a successful freethrow and 811–1000 represents a miss.

5. Use seq(int(4*rand)+1,X,1,10,1)$\to$L$_1$, where a = 1, b = 2, c = 3, and d = 4.

7. Answers will vary (see Example 4).

9. Answers will vary.

Chapter 6 Supplementary Exercises

1. $1 - \left(\dfrac{1}{2}\right)^5 = \dfrac{31}{32}$

2. No; $\Pr(F|E) = \dfrac{3}{4} \neq \Pr(F) = \dfrac{1}{2}$

3. $\left(\dfrac{1}{3} \times \dfrac{2}{3}\right) + \left(\dfrac{2}{3} \times \dfrac{1}{3}\right) = \dfrac{4}{9}$

4. $\dfrac{5}{10} \times \dfrac{4}{9} \times \dfrac{3}{8} = \dfrac{1}{12}$

5. $\dfrac{\text{number of public colleges offering engineering majors}}{\text{number of public colleges}} = \dfrac{15-5}{50-25} = \dfrac{2}{5}$

6. $\Pr\left(\text{correct}\middle|\text{rejected}\right) = \dfrac{\Pr(\text{correct}) \times \Pr\left(\text{rejected}\middle|\text{correct}\right)}{\Pr(\text{correct}) \times \Pr\left(\text{rejected}\middle|\text{correct}\right) + \Pr(\text{incorrect}) \times \Pr\left(\text{rejected}\middle|\text{incorrect}\right)}$

   $$= \dfrac{.80 \times .05}{(.80 \times .05) + (.20 \times .90)} = \dfrac{2}{11}$$

7. $\dfrac{20}{100} = \dfrac{1}{5}$

8. $\dfrac{2000}{10,000} = \dfrac{1}{5}$

9. Pr(div. by 3 or 5) = Pr(div. by 3) + Pr(div. by 5) – Pr(div. by 15) $= \dfrac{3333}{10,000} + \dfrac{1}{5} - \dfrac{666}{10,000} = .4667$

10. Pr(div. by 3 or 12) = Pr(div. by 3) $= \dfrac{3333}{10,000} = .3333$

11. $\dfrac{7}{8+7} = \dfrac{7}{15}$

12. a. $\dfrac{2 \cdot 1}{5 \cdot 4} = \dfrac{1}{10}$

  b. $4 \times \dfrac{1}{10} = \dfrac{2}{5}$

13. a. Pr(prepared every question) $= \dfrac{C(8,6)}{C(10,6)} = \dfrac{28}{210} = \dfrac{2}{15}$

  b. Pr(not prepared on test) $= \dfrac{C(8,4)}{C(10,6)} = \dfrac{70}{210} = \dfrac{1}{3}$

14. a. $\left(\dfrac{1}{36}\right)^3$

  b. $\left(\dfrac{1}{10}\right)^4$

  c. $\dfrac{3334}{10,000} = \dfrac{1667}{5000}$

  d. $\left(\dfrac{26}{36}\right)^3 \left(\dfrac{1}{2}\right)^4 = \left(\dfrac{13}{18}\right)^3 \left(\dfrac{1}{2}\right)^4$

15. a. $\dfrac{1}{4} \times \dfrac{1}{3} = \dfrac{1}{12}$

  b. $\dfrac{1}{4} + \dfrac{1}{3} - \dfrac{1}{12} = \dfrac{1}{2}$

16. $\dfrac{7}{7+5} = \dfrac{7}{12}$

17. $\dfrac{2}{7} \times \dfrac{1}{6} = \dfrac{1}{21}$

18. $\text{Pr(3 is drawn on 1st or 2nd draw)} = \dfrac{1}{3} + \left(\dfrac{1}{3}\right)\left(\dfrac{1}{2}\right) + \left(\dfrac{1}{3}\right)\left(\dfrac{1}{2}\right) = \dfrac{2}{3}$

19. No; $\text{Pr}\left(F|E\right) = \dfrac{1}{6} > \text{Pr}(F) = \dfrac{5}{36}$

20. $\text{Pr}\left(E|F\right) = \dfrac{\text{Pr}(E \cap F)}{\text{Pr}(F)} = \dfrac{\text{Pr}(E) + \text{Pr}(F) - \text{Pr}(E \cup F)}{\text{Pr}(F)} = \dfrac{.4 + .3 - .5}{.3} = \dfrac{.2}{.3} = \dfrac{2}{3}$

21. $\text{Pr}\left(C|\text{wrong}\right) = \dfrac{\text{Pr}(C) \times \text{Pr}\left(\text{wrong}|C\right)}{\text{Pr}(C) \times \text{Pr}\left(\text{wrong}|C\right) + \text{Pr}(A) \times \text{Pr}\left(\text{wrong}|A\right) + \text{Pr}(B) \times \text{Pr}\left(\text{wrong}|B\right)}$

$= \dfrac{.20 \times .05}{(.20 \times .05) + (.40 \times .02) + (.40 \times .03)}$

$= \dfrac{.01}{.03}$

$= \dfrac{1}{3}$

22. $1 - \left(1 \times \dfrac{6}{7} \times \dfrac{5}{7}\right) = \dfrac{19}{49}$

23. $\text{Pr}(A \cup B) = \dfrac{1}{2}; \ \text{Pr}(A' \cap B) = \dfrac{1}{3}$

$\text{Pr}(A) + \text{Pr}(A' \cap B) = \text{Pr}(A \cup B)$

$\text{Pr}(A) = \dfrac{1}{2} - \dfrac{1}{3} = \dfrac{1}{6}$

24. [number of ways to draw 7 balls] $= \dfrac{10!}{3!} = 604{,}800$

[number of ways to draw 3 odd-numbered balls on odd-numbered draws] $= C(4, 3) \cdot 5 \cdot 4 \cdot 3 \cdot 5 \cdot 4 \cdot 3 \cdot 2$
$= 28{,}800$

$\text{Pr(draw 3 odd-numbered balls on odd-numbered draws)} = \dfrac{28{,}800}{604{,}800} = \dfrac{1}{21}$

25. $\text{Pr}\left(\text{other is \$5}|\text{\$5}\right) = \dfrac{\text{Pr(both are \$5)}}{\text{Pr(\$5)}} = \dfrac{\frac{1}{3}}{\frac{1}{2}} = \dfrac{2}{3}$

26. You should switch. If you stay with your original choice your probability of winning remains $\dfrac{1}{3}$, whereas if you switch you lose only if your original choice was correct, so your probability of winning is $\dfrac{2}{3}$.

27. $\dfrac{13}{13 + 12} = \dfrac{13}{25}$

28. $26\% = \dfrac{13}{50}$; $a = 13$ and $a + b = 50$, so $b = 37$. The odds are 13 to 37.

29. $\dfrac{3}{9!} = \dfrac{1}{120,960}$

30. If n is the number of dragons, then $\dfrac{\text{\# of heads on 1-headed dragons}}{\text{\# of heads}} = \dfrac{\frac{n}{3}}{\frac{n}{3} + 2 \cdot \frac{n}{3} + 3 \cdot \frac{n}{3}} = \dfrac{1}{6}$

31. $\text{Pr(no tails)} + \text{Pr(1 tail each)} + \text{Pr(2 tails each)} + \text{Pr(3 tails each)} = \left(\dfrac{1}{8}\right)^2 + \left(\dfrac{3}{8}\right)^2 + \left(\dfrac{3}{8}\right)^2 + \left(\dfrac{1}{8}\right)^2 = \dfrac{5}{16}$

32. $\text{Pr}\left(\text{one 3}\,\middle|\,\text{no doubles}\right) = \dfrac{5+5}{30} = \dfrac{1}{3}$

33. $\text{Pr}\left(\text{both parents left-handed}\,\middle|\,\text{child left-handed}\right)$

$= \dfrac{\text{Pr(all three left-handed)}}{\text{Pr(child left-handed)}}$

$= \dfrac{.4 \times .25 \times .25}{(.4 \times .25 \times .25) + (.2 \times .25 \times .75) + (.2 \times .75 \times .25) + (.1 \times .75 \times .75)}$

$= \dfrac{4}{25}$

34. $\dfrac{120 - (\text{speak Chinese or Spanish}) - (\text{speak French only})}{120} = \dfrac{120 - (30 + 50 - 12) - (75 - 30 - 15 + 7)}{120} = \dfrac{1}{8}$

35. There are 100 numbers (from 200–299) with a two in the hundreds digit.
There are 10 numbers (from 120–129) with a two in the tens digit.
There are 10 numbers (from 320–329) with a two in the tens digit.
There are 9 numbers (102, 112, 132, 142, …, 192) with a two in the ones digit. There are 9 numbers (302, 312, 332, 342, …, 392) with a two in the ones digit.

The probability is $\dfrac{100 + 10 + 10 + 9 + 9}{301} = \dfrac{138}{301}$.

36. $\dfrac{w+3}{(w+3)+(m+2)} = \dfrac{w+3}{w+m+5}$

(e)

37. $(.60 \times .90) + (.40 \times .05) = .56 = 56\%$

(a)

38. Pr(at least one tail appears in three coins given that at least one head appeared)
1) Look at the sample space: HHH, HHT, HTH, HTT, THH, THT, TTH, TTT.
2) The first seven outcomes have at least one head.
3) Pr (one or more tails in the first seven outcomes) = 6/7

39. Record the six consecutive outcomes; there are 6^6 possibilities, of which 6! contain each number exactly once. Hence the probability is
$$\frac{6!}{6^6} = \frac{5}{324}.$$

40. $\Pr(R_1 \cap G_2 \cap G_3 \cap R_4) = \dfrac{10}{30} \cdot \dfrac{20}{29} \cdot \dfrac{19}{28} \cdot \dfrac{9}{27}$
$$= \frac{95}{1827}$$

41. a. Pr(all three cards are aces)
$$= p(A_1 \cap A_2 \cap A_3) = \frac{4}{52} \cdot \frac{4}{52} \cdot \frac{4}{52} = \frac{1}{2197}$$

    b. Pr (at least one ace) = 1 − Pr(no aces)
$$= 1 - \left(\frac{48}{52}\right)^3 = 1 - \left(\frac{12}{13}\right)^3 = \frac{469}{2197}$$

42. $\Pr(\text{select 4 winning teams}) = \left(\dfrac{1}{16}\right)^4 = \dfrac{1}{65{,}536}$

Odds against $= 1 - \dfrac{1}{65{,}536} : \dfrac{1}{65{,}536} = 65{,}535 : 1$

43. There are $C(4, 2) = 6$ ways to choose a "pair" of socks from the drawer; of these, 2 have the same color, so the probability is $\dfrac{2}{6} = \dfrac{1}{3}.$

44. There are $15 - 5 = 10$ who like only bicycling and $20 - 5 = 15$ who like only jogging, so the probability is $\dfrac{10+15}{50} = 0.5.$

45. $\Pr(\text{4 different numbers}) = \dfrac{6 \cdot 5 \cdot 4 \cdot 3}{6 \cdot 6 \cdot 6 \cdot 6} = \dfrac{5}{18}$, so the odds in favor of getting four different numbers are $\dfrac{15}{18} : 1 - \dfrac{5}{18} = 5$ to 13.

46. There are $C(5, 2) = 10$ ways to choose the pair of people with the same birthday, 365 ways to choose their birthdays, and $364 \times 363 \times 362$ ways to choose the birthdays of the remaining three people. So the probability that exactly two have the same birthday is
$$\frac{10 \cdot 365 \cdot 364 \cdot 363 \cdot 362}{365^5} \approx 0.02695.$$

47. n = number sick
Solve: $0.96n + 0.02(12{,}735 - n) = 650$
$n \approx 421$

48. Use seq(int(6*rand)+1,X,1,15,1)

49. Use seq(int(52*rand)+1,X,1,3,1), with spades = 1 through 13.
Theoretical probability:
Pr = Pr(2 spades) + Pr(3 spades)
$$= 3\left(\frac{1}{4} \cdot \frac{1}{4} \cdot \frac{3}{4}\right) + \left(\frac{1}{4} \cdot \frac{1}{4} \cdot \frac{1}{4}\right) = \frac{5}{32}$$

Conceptual Exercises

50. Two events which are mutually exclusive: A = being a male and B = being a female. Two sets which are not mutually exclusive: M = students who take math and E = students who take English.

51. Two independent events: rolling a die and then tossing a coin. Drawing six balls from a container of 49 balls one week and drawing six balls from a container of 49 balls a second week. (After drawing the first set of six balls, they are replaced).

52. Two events are disjoint if they have no elements in common. Thus $\Pr(A \cap B) = 0.$
Independent events are two events in which the outcome of the first event does not affect the outcome of the second event. The $\Pr(A \cap B) = \Pr(A) \cdot \Pr(B)$ and is not necessarily zero.

53. Suppose A and B are two mutually exclusive events with nonzero probabilities. Then $\Pr(A \cap B) = 0$ and $\Pr(A)\Pr(B) \neq 0$, so A and B are not independent.

54. Drawing two cards from a standard deck of 52 cards without replacement and getting a jack on the first card and an ace on the second card. The events are not independent since the drawing is done without replacement. The events are mutually exclusive since no card can be both an ace and a jack.

55. If E and F are independent events, then the outcome of F does not affect the outcome of E and vice versa. If the outcome of F does not affect the outcome of E, then neither would the outcome of F'.

Example: Let E = event you get a six on a die and F = event you get a H on a coin toss. Now, E and F are independent. F' is the event you get a T on a coin toss. The events E and F' are independent, so whether you get a H or T on the coin toss does not affect the probability of the six on a die.

56. If you know $\Pr(E \cup F)$, then you can compute $\Pr(E \cap F) = \Pr(E) + \Pr(F) - \Pr(E \cup F)$. Alternatively, if you know that E and F are independent events, then you can compute $\Pr(E \cap F) = \Pr(E) \cdot \Pr(F)$.

Chapter 6 Chapter Test

1. a. $S = \{$PN, PD, PQ, PH, ND, NQ, NH, DQ, DH, QH$\}$

b. $E = \{$PN, PQ, NQ, DH$\}$

2. a. $\dfrac{3}{3+5} = \dfrac{3}{8}$

b. $\dfrac{1,000,000}{1,000,001}$

3. $\dfrac{.4}{1-.4} = \dfrac{2}{3}$; The odds of beating father are 2 to 3.

$\dfrac{1-.3}{.3} = \dfrac{7}{3}$; The odds of losing to mother are 7 to 3.

4. a. A male junior is elected.

b. A female junior is not elected.

c. A male or a junior is elected.

5. Pr(at least one head and at least one tail)
$= 1 - \left[\Pr\left(\text{all heads or all tails}\right) \right]$
$= 1 - \left[\left(\dfrac{1}{2}\right)^5 + \left(\dfrac{1}{2}\right)^5 \right]$
$= \dfrac{15}{16}$

6. Pr(at least one has had a measles vaccine)
$= 1 - \Pr(\text{all have had vaccine})$
$= 1 - (1-.15)^{20}$
$\approx .9612$

7. a. $\Pr(F) = 1 - \Pr(F') = 1 - \dfrac{3}{8} = \dfrac{5}{8}$

b. $\Pr(E \cup F)$
$= \Pr(E) + \Pr(E' \cap F)$
$= \dfrac{1}{4} + \dfrac{1}{2}$
$= \dfrac{3}{4}$

c. $\Pr\left(F \mid E'\right) = \dfrac{\Pr(F \cap E')}{\Pr(E')}$
$= \dfrac{\Pr(F \cap E')}{1 - \Pr(E)}$
$= \dfrac{\dfrac{1}{2}}{1 - \dfrac{1}{4}}$
$= \dfrac{2}{3}$

d. No; $\Pr(E \cup F) \neq \Pr(E) + \Pr(F)$

e. No; $\Pr(F \mid E') \neq \Pr(F)$

8. Pr(at least one defective)
$= 1 - \Pr(\text{none defective})$
$= 1 - (.9)^{10}$
$\approx .6513$
Pr(exactly one defective) $= (.1)(.9)^9 \times 10$
$\approx .3874$

9. a. Pr(red marble|face card) $= \dfrac{5}{15} = \dfrac{1}{3}$

b. Pr(red marble) $= \dfrac{3}{13} \times \dfrac{1}{3} + \dfrac{10}{13} \times \dfrac{12}{20}$
$= \dfrac{7}{13}$

c. No, they are not independent because $P(F) \neq P(F|E)$.
No, they are not mutually exclusive because $P(F|E) \neq 0$.

d. Pr(face card|green marble)
$= \dfrac{\dfrac{3}{13} \times \dfrac{10}{15}}{\dfrac{3}{13} \times \dfrac{10}{15} + \dfrac{10}{13} \times \dfrac{8}{20}} = \dfrac{1}{3}$

10. a. $\text{Pr(odd)} = .15 + .10 + .25 = .50$

   b. $\text{Pr(greater than 4|odd)} = \dfrac{.25}{.50} = .50$

11. a. Pr(female or sophomore)
$$= \frac{12+8+16}{50} = \frac{36}{50} = \frac{18}{25}$$

   b. Pr(male|freshman)
$$= \frac{14}{12+14} = \frac{14}{26} = \frac{7}{13}$$

Chapter 7

Exercises 7.1

1.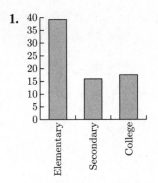

3.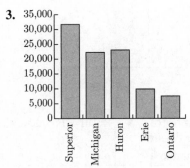

5.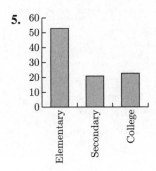

7. To find the central angle, multiply the percentage .089 by 360 to obtain 32°.

9.

Lake	Percent	$360° \times Percent$
Superior	33.5	120.6°
Michigan	23.6	85.0°
Huron	24.4	87.8°
Erie	10.5	37.8°
Ontario	8.0	28.8°

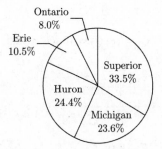

11. Pr(Master's or Doctorate) = .427 + .175 = .602

13.

0	⫿⫿⫿⫿ ⫿⫿	7
1	⫿⫿	2
2	⫿⫿	2
3	⫿⫿	2
4	⫿⫿⫿⫿	4
5		0
6		0
7	⫿⫿	2
8	⫿	1
9		0
10	⫿	1

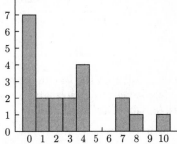

15. median of $\{2, 3, 5, 20\} = \dfrac{3+5}{2} = 4$

The median of $[n, 1, 3, 5, 15]$ equals 4 only if $n = 4$. Answer (c) is correct.

17.

2	5	7	10	15

19. min = 10, $Q_1 = 13$, $Q_2 = 17$, $Q_3 = 21$, max = 24;
IQR = $Q_3 - Q_1 = 21 - 13 = 8$;

10	13	17	21	24

21. min = 20, $Q_1 = 28$, $Q_2 = 42$, $Q_3 = 52.5$, max = 56;
IQR = $Q_3 - Q_1 = 52.5 - 28 = 24.5$;

20	28	42	52.5	56

23. (A) – (c), (B) – (d), (C) – (a), (D) – (b)

25. a. min = 200, $Q_1 = 400$, $Q_2 = 600$,
$Q_3 = 700$, max = 800

b. 25%

c. 25%

d. 50%

e. 75%

27. Braves: .200, .227, .229, .243, .281, .286, .296, .317, .345, .350
Brewers: .091, .150, .200, .250, .270, .280, .317, .333, .359

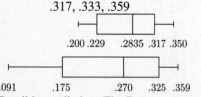

.200 .229   .2835 .317 .350

.091   .175   .270   .325 .359

Possible prediction: The Braves seem more likely to win because their hitting is more consistent with a higher median.

Exercises 7.2

1.

Grade	Relative Frequency
0	$\frac{2}{25} = .08$
1	$\frac{3}{25} = .12$
2	$\frac{10}{25} = .40$
3	$\frac{6}{25} = .24$
4	$\frac{4}{25} = .16$

3.

Number of calls during minute	Relative Frequency
20	$\frac{3}{60} = .05$
21	$\frac{3}{60} = .05$
22	$\frac{0}{60} = 0$
23	$\frac{6}{60} = .10$
24	$\frac{18}{60} = .30$
25	$\frac{12}{60} = .20$
26	$\frac{0}{60} = 0$
27	$\frac{9}{60} = .15$
28	$\frac{6}{60} = .10$
29	$\frac{3}{60} = .05$

5. HHH, HHT, HTH, THH,
HTT, THT, TTH, TTT

Number of Heads	Probability
0	$\dfrac{\binom{3}{0}}{2^3} = \dfrac{1}{8}$
1	$\dfrac{\binom{3}{1}}{2^3} = \dfrac{3}{8}$
2	$\dfrac{\binom{3}{2}}{2^3} = \dfrac{3}{8}$
3	$\dfrac{\binom{3}{3}}{2^3} = \dfrac{1}{8}$

7.

Number of Red Balls	Probability
0	$\dfrac{\binom{3}{0}\binom{4}{3}}{\binom{7}{3}} = \dfrac{4}{35}$
1	$\dfrac{\binom{3}{1}\binom{4}{2}}{\binom{7}{3}} = \dfrac{18}{35}$
2	$\dfrac{\binom{3}{2}\binom{4}{1}}{\binom{7}{3}} = \dfrac{12}{35}$
3	$\dfrac{\binom{3}{3}\binom{4}{0}}{\binom{7}{3}} = \dfrac{1}{35}$

9.

No. Red Balls	Player's Earnings	Probability
2	$5	$\dfrac{\binom{2}{2}\binom{4}{0}}{\binom{6}{2}} = \dfrac{1}{15}$
1	$1	$\dfrac{\binom{2}{1}\binom{4}{1}}{\binom{6}{2}} = \dfrac{8}{15}$
0	−1$	$\dfrac{\binom{2}{0}\binom{4}{2}}{\binom{6}{2}} = \dfrac{6}{15}$

11. $\Pr(5 \le X \le 7)$
$= \Pr(X = 5) + \Pr(X = 6) + \Pr(X = 7)$
$= .2 + .1 + .3$
$= .6$

13.

k	$\Pr(X^2 = k)$
0	.1
1	.2
4	.3
9	.2
16	.2

15.

k	$\Pr(X - 1 = k)$
−1	.1
0	.2
1	.3
2	.2
3	.2

17.

k	$\Pr\left(\frac{1}{5}Y = k\right)$
1	.3
2	.4
3	.1
4	.1
5	.1

19.

$(X+1)^2 = k$	$\Pr((X+1)^2 = k)$
$(0+1)^2 = 1$	.1
$(1+1)^2 = 4$	.2
$(2+1)^2 = 9$	.3
$(3+1)^2 = 16$	.2
$(4+1)^2 = 25$	.2

21.

	Relative Frequency	
Grade	9 A.M. class	10 A.M. class
F	$\frac{10}{60} \approx .17$	$\frac{16}{100} = .16$
D	$\frac{15}{60} = .25$	$\frac{23}{100} = .23$
C	$\frac{20}{60} \approx .33$	$\frac{15}{100} = .15$
B	$\frac{10}{60} \approx .17$	$\frac{21}{100} = .21$
A	$\frac{5}{60} \approx .08$	$\frac{25}{100} = .25$

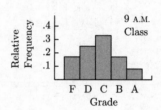

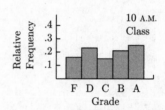

The 9 A.M. class has the distribution centered on the C grade with relatively few A's. The 10A.M. class has a large percentage of A's and D's with fewer C's.

23. percentage with C or higher:
$$\left(\frac{10 + 6 + 4}{25}\right) \times 100\% = 80\%$$

25. a. less than 22: $3 + 3 = 6$
more than 27: $6 + 3 = 9$

combined: $\left(\frac{6 + 9}{60}\right) \times 100\% = 25\%$

b. between 23 and 25:
$$\left(\frac{6 + 18 + 12}{60}\right) \times 100\% = 60\%$$

c.

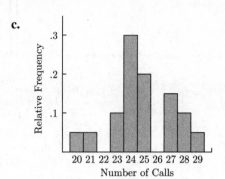

d. Estimated average number of calls would be 24 since that number has the highest frequency of occurrence.
It is actually ≈ 25.

27. a. 59

b. $\left(\dfrac{2}{40}\right) \times 100\% = 5\%$ of the time

c. 54

d. $\left(\dfrac{8+4+2}{40}\right) \times 100\% = 35\%$

e. Estimate an average of 54 items produced.

29. a. $Pr(U = 4) = 1 - \left(\dfrac{3}{15} + \dfrac{2}{15} + \dfrac{4}{15} + \dfrac{5}{15}\right)$

$= 1 - \dfrac{14}{15}$

$= \dfrac{1}{15}$

$\approx .07$

b. $Pr(U \geq 2)$
$= Pr(U = 2) + Pr(U = 3) + Pr(U = 4)$
$= \dfrac{4}{15} + \dfrac{5}{15} + \dfrac{1}{15}$
$= \dfrac{10}{15}$
$= \dfrac{2}{3}$
$\approx .67$

c. $Pr(U \leq 3) = 1 - Pr(U = 4)$
$= 1 - \dfrac{1}{15}$
$= \dfrac{14}{15}$
$\approx .93$

d.

$U + 2 = K$	$Pr(U + 2 = K)$
$0 + 2 = 2$	$\frac{3}{15}$
$1 + 2 = 3$	$\frac{2}{15}$
$2 + 2 = 4$	$\frac{4}{15}$
$3 + 2 = 5$	$\frac{5}{15}$
$4 + 2 = 6$	$\frac{1}{15}$

$Pr(U + 2 < 4)$
$= Pr(U + 2 = 2) + Pr(U + 2 = 3)$
$= \dfrac{3}{15} + \dfrac{2}{15}$
$= \dfrac{5}{15}$
$= \dfrac{1}{3}$
$\approx .33$

e.

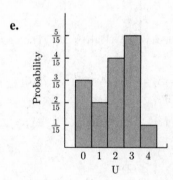

Exercises 7.3

1. Let "success" be the outcome "one." Then
$p = \dfrac{1}{6}$, $q = \dfrac{5}{6}$, $n = 4$.

$Pr(X = 2) = \dbinom{4}{2}\left(\dfrac{1}{6}\right)^2\left(\dfrac{5}{6}\right)^2 = \dfrac{25}{216} \approx .1157$

3. Let "success" be a "sale." Then $p = \dfrac{1}{4}$, $q = \dfrac{3}{4}$,
$n = 4$.

$Pr(X = 3) = \dbinom{4}{3}\left(\dfrac{1}{4}\right)^3\left(\dfrac{3}{4}\right) = \dfrac{3}{64} \approx .0469$

5. Let "success" be "a vote for" the candidate. Then $p = .6, q = .4, n = 5$.

$\Pr(X \le 2)$

$= \Pr(X = 0) + \Pr(X = 1) + \Pr(X = 2)$

$= \binom{5}{0}(.6)^0(.4)^5 + \binom{5}{1}(.6)^1(.4)^4 + \binom{5}{2}(.6)^2(.4)^3$

$= .01024 + .0768 + .2304$

$\approx .3174$

7. Let "success" be a "chemistry major." Then $p = .1, q = .9, n = 8$

$\Pr(X = 2) = \binom{8}{2}(.1)^2(.9)^6 \approx .1488$

9. Let "success" be "a car with a commuter sticker." Then $p = .3, q = .7, n = 10$.

$\Pr(X \ge 2)$

$= 1 - \Pr(X < 2)$

$= 1 - [\Pr(X = 0) + \Pr(X = 1)]$

$= 1 - \left[\binom{10}{0}(.3)^0(.7)^{10} + \binom{10}{1}(.3)^1(.7)^9 \right]$

$\approx 1 - (.0282 + .1211)$

$= .8507$

11. Let "success" be a "silver car." Then $p = .4$, $q = .6, n = 10$.

$\Pr(X = 5) = \binom{10}{5}(.4)^5(.6)^5 \approx .2007$

13. Let "success" be "brand X." Then $p = .2, q = .8$, $n = 9$.

$\Pr(X > 2)$

$= 1 - \Pr(X \le 2)$

$= 1 - [\Pr(X = 0) + \Pr(X = 1) + \Pr(X = 2)]$

$= 1 - \left[\binom{9}{0}(.2)^0(.8)^9 + \binom{9}{1}(.2)^1(.8)^8 + \binom{9}{2}(.2)^2(.8)^7 \right]$

$\approx 1 - (.1342 + .3020 + .3020)$

$= .2618$

15.

k	$\Pr(X = k)$
0	$\binom{8}{0}(.4)^0(.6)^8 \approx .0168$
1	$\binom{8}{1}(.4)^1(.6)^7 \approx .0896$
2	$\binom{8}{2}(.4)^2(.6)^6 \approx .2090$
3	$\binom{8}{3}(.4)^3(.6)^5 \approx .2787$
4	$\binom{8}{4}(.4)^4(.6)^4 \approx .2322$
5	$\binom{8}{5}(.4)^5(.6)^3 \approx .1239$
6	$\binom{8}{6}(.4)^6(.6)^2 \approx .0413$
7	$\binom{8}{7}(.4)^7(.6)^1 \approx .0079$
8	$\binom{8}{8}(.4)^8(.6)^0 \approx .0007$

17. Let "success" be "head". Then $p = \dfrac{1}{2}$, $q = \dfrac{1}{2}$, $n = 4$.

$\Pr(X > 0) = 1 - \Pr(X = 0)$

$= 1 - \binom{4}{0}\left(\dfrac{1}{2}\right)^0\left(\dfrac{1}{2}\right)^4$

$= 1 - \dfrac{1}{16} = \dfrac{15}{16}$

Answer (d) is correct.

19. Let "success" be "a guilty vote." Then $p = .80, q = .20, n = 12$.

$\Pr(X \ge 10)$

$= \Pr(X = 10) + \Pr(X = 11) + \Pr(X = 12)$

$= \binom{12}{10}(.8)^{10}(.2)^2 + \binom{12}{11}(.8)^{11}(.2)^1 + \binom{12}{12}(.8)^{12}(.2)^0$

$\approx .5583$

21. The fourth head occurs on the tenth toss if exactly three heads occur in the first nine tosses and the tenth toss is

heads. The probability of this is $\binom{9}{3}\left(\frac{1}{2}\right)^3\left(\frac{1}{2}\right)^6 \times \frac{1}{2} = \binom{9}{3}\left(\frac{1}{2}\right)^{10} \approx 0.0820$.

23. Let "success" be "team C is chosen." Then $p = \frac{2}{3}$, $q = \frac{1}{3}$, $n = 3$.

$$\begin{aligned}\Pr(X \geq 2) &= \Pr(X = 2) + \Pr(X = 3) \\ &= \binom{3}{2}\left(\frac{2}{3}\right)^2\left(\frac{1}{3}\right)^1 + \binom{3}{3}\left(\frac{2}{3}\right)^3\left(\frac{1}{3}\right)^0 \\ &= \frac{12}{27} + \frac{8}{27} \\ &= \frac{20}{27}\end{aligned}$$

25. Here $p = 0.82$. The expected value is $\mu = np = 10(0.82) = 8.2$

$\Pr(X = 8) = C(10,8)\, 0.82^8 (0.18)^2 = 0.2980$

$\Pr(X = 9) = C(10,9)\left(0.82\right)^9 (0.18)^1 = 0.3017$

So 9 is the most likely number.

27. For the server to win on the sixth point, the server must win three of the first five points and win the sixth point. Since the server has probability $p = .6$ of winning any given point, the probability that the server wins the game on

the sixth point is $\binom{5}{3}(.6)^3(.4)^2(.6) \approx .2074$.

29. Let "success" be "child is a boy." Then $p = .5$, $q = .5$, $n = 4$.

a. $\Pr(X = 2) = \binom{4}{2}(.5)^2(.5)^2 = .375$

b. $\Pr(X = 3) + \Pr(X = 1) = 2\Pr(X = 3)$ since $\Pr(X = 3) = \Pr(X = 1)$.

$2\Pr(X = 3) = 2 \cdot \binom{4}{3}(.5)^3(.5)^1 = .5$

c. $\Pr(X = 4) = \binom{4}{4}(.5)^4(.5)^0 = .0625$;

$\Pr(X = 4) + \Pr(X = 0) = 2\Pr(X = 4) = .125$.

31. a. $\Pr(\text{at least one three in six rolls}) = 1 - \left(\frac{5}{6}\right)^6$
$$\approx .6651$$

b. $\Pr(\text{at least two threes in twelve rolls}) = 1 - \Pr(\text{no threes}) - \Pr(\text{one three})$
$$= 1 - \left(\frac{5}{6}\right)^{12} - 12 \cdot \left(\frac{5}{6}\right)^{11}\left(\frac{1}{6}\right)^1$$
$$\approx .6187$$

(a) is more likely.

Chapter 7: Probability and Statistics

SSM: Finite Math

33. Let "success" = "lives to 100." Then $p = .0217$, $q = .9783$, $n = 77$.
$$\Pr(X \geq 2) = 1 - \Pr(X = 0) - \Pr(X = 1)$$
$$= 1 - (.9783)^{77} - 77(.0217)^1(.9783)^{76}$$
$$\approx .500$$

35. Let "success" = "adverse reaction." Then $p = .02$, $q = .98$, $n = 56$.
$$\Pr(X \geq 3) = 1 - \Pr(X = 0) - \Pr(X = 1) - \Pr(X = 2)$$
$$= 1 - (.98)^{56} - \binom{56}{1}(.02)^1(.98)^{55} - \binom{56}{2}(.02)^2(.98)^{54}$$
$$\approx .102$$

37. a. Substituting $p = .6$ in this formula gives a probability $\approx .2898$.

    b. Suppose as in Problem 32(d) above that all $2n - 1$ games are played. Then the probability that the favorite wins at least n games is $\binom{2n-1}{n}p^n(1-p)^{n-1} + \binom{2n-1}{n+1}p^{n+1}(1-p)^{n-2} + \cdots + \binom{2n-1}{2n-1}p^{2n-1}$.

    Some experimenting shows that for $p = .6$, $n = 21$ is the smallest value of n for which this probability exceeds .9.

39. Let "success" be "recovery." Then $p = .25$, $q = .75$, $n = 40$.

    a. $\Pr(X = 20) = \binom{40}{20}(.25)^{20}(.75)^{20} \approx .0003976$

    b. $\Pr(X \geq 16) \approx .02624$

41. Let "success" be "a seven." Then $p = \dfrac{1}{6}$, $q = \dfrac{5}{6}$, $n = 25$.

    a. $\Pr(X = 8) = \binom{25}{8}\left(\dfrac{1}{6}\right)^8\left(\dfrac{5}{6}\right)^{17} \approx .0290$

    b.

K	$\Pr(X = K)$
0	.0105
1	.0524
2	.1258
3	.1929
4	.2122
5	.1782
6	.1188

    c. $\Pr(X \geq 10) \approx .0047$

7-8

43. Let "success" = "becomes centenarian." Then $p = .0217$, $q = .9783$, n is unknown.

$\Pr(X \geq 2)$

$= 1 - \Pr(X = 0) - \Pr(X = 1)$

$= 1 - (.9783)^n - \binom{n}{1}(.0217)^1(.9783)^{n-1}$

$= 1 - (.9783)^n - n(.0217)(.9783)^{n-1}$

$\Pr(X \geq 2) > .9$

$\Leftrightarrow (.9783)^n + n(.0217)(.9783)^{n-1} < .1$

$\Leftrightarrow (.9783)^{n-1}(.9783 + n(.0217)) < .1$

The smallest value of n for which this occurs is $n = 178$.

Exercises 7.4

1. $E(X) = 0(.15) + 1(.2) + 2(.1) + 3(.25) + 4(.3)$
$= 2.35$

3. a. $GPA = \dfrac{4+4+4+3+3+3+3+2+2+1}{10}$

$= \dfrac{29}{10}$

$= 2.9$

b.

Grade	Relative Frequency
4	$\frac{3}{10} = .3$
3	$\frac{4}{10} = .4$
2	$\frac{2}{10} = .2$
1	$\frac{1}{10} = .1$

c. $E(X) = 4(.3) + 3(.4) + 2(.2) + 1(.1)$
$= 2.9$

5. $\bar{x}_A = 0(.3) + 1(.3) + 2(.2) + 3(.1) + 4(0) + 5(.1)$
$= 1.5$
$\bar{x}_B = 0(.2) + 1(.3) + 2(.3) + 3(.1) + 4(.1) + 5(0)$
$= 1.6$
Group A had fewer cavities.

7.

Earnings	Probability
–$1	$\frac{37}{38}$
$35	$\frac{1}{38}$

$E(X) = -1\left(\dfrac{37}{38}\right) + 35\left(\dfrac{1}{38}\right) \approx -.0526 \approx -\$.05$

9.

Earnings	Probability
–50¢	$\frac{2}{6} = \frac{1}{3}$
0¢	$\left(\frac{4}{6}\right)\left(\frac{2}{5}\right) = \frac{4}{15}$
50¢	$\left(\frac{4}{6}\right)\left(\frac{3}{5}\right)\left(\frac{2}{4}\right) = \frac{1}{5}$
$1	$\left(\frac{4}{6}\right)\left(\frac{3}{5}\right)\left(\frac{2}{4}\right)\left(\frac{2}{3}\right) = \frac{2}{15}$
$1.50	$\left(\frac{4}{6}\right)\left(\frac{3}{5}\right)\left(\frac{2}{4}\right)\left(\frac{1}{3}\right)\left(\frac{2}{2}\right) = \frac{1}{15}$

$E(X) = -.5\left(\dfrac{1}{3}\right) + 0\left(\dfrac{4}{15}\right) + .5\left(\dfrac{1}{5}\right) + 1\left(\dfrac{2}{15}\right) + 1.5\left(\dfrac{1}{15}\right)$

$\approx .1667$

$\approx \$.17$

11. Let x be the cost of the policy.
$\mu = (-x)(.9) + (10,000 - x)(.1) = -x + 1000$
The expected value is zero if $x = 1000$.
He should be willing to pay up to $1000.

13.

Recorded Value	Probability
1	$\frac{1}{36}$
2	$\frac{3}{36}$
3	$\frac{5}{36}$
4	$\frac{7}{36}$
5	$\frac{9}{36}$
6	$\frac{11}{36}$

$E(X) = 1\left(\dfrac{1}{36}\right) + 2\left(\dfrac{3}{36}\right) + 3\left(\dfrac{5}{36}\right) + 4\left(\dfrac{7}{36}\right)$

$+ 5\left(\dfrac{9}{36}\right) + 6\left(\dfrac{11}{36}\right) \approx 4.47$

15. The expected number of times that a 5 or a 6 will appear is $\mu = np = 30\left(\dfrac{1}{3}\right) = 10$

17. The expected value of points when taking one three point shot is $\mu = 1(0.40) = 0.4$ points. The expected value of taking three free-throws when the success probability is 60% is $\mu = 3(0.60) = 1.8$, which is the greater probability.

19. There are many possible answers, for example

    a. 1, 2, 3, 4, 5

    b. 1, 2, 3, 4, 6

    c. 0, 2, 3, 4, 5

21. Solve
$$\frac{16 \cdot 54 + 14x}{30} = 56 \cdot 1$$
$$14x + 864 = 1683$$
$$14x = 819$$
$$x = 58.5°$$

23.
$$\frac{5 + 6 + x}{3} = \frac{2 + 7 + 9}{3}$$
$$11 + x = 18$$
$$x = 7$$
Answer (d) is correct.

25.
$$\frac{7x + 4y}{x + y}$$
Answer (b) is correct.

27.
$$\frac{120 + 121 + x}{3} = 200$$
$$241 + x = 600$$
$$x = 359$$
Answer (c) is correct.

29. Let d = size of Dick's card collection.
$$\frac{\frac{3}{2}d + d + \frac{1}{2}d}{3} = 120$$
$$d = 120$$
Tom has $\dfrac{3}{2}d = \dfrac{3}{2}(120) = 180$
Answer (d) is correct.

31.
$$\frac{5 \times .300 + 4 \times .350}{9} \approx .322$$
Answer (b) is correct.

33. Let x = number of cases.

$$20\left(\frac{1}{2}x\right)+30\left(\frac{1}{2}x\right)=75,000$$
$$25x = 75,000$$
$$x = 3000 \text{ cases}$$

Answer (a) is correct.

35. Let x = chance of rain.

$$-8000+40,000x = 0$$
$$x = .20 \to 20\%$$

Answer (a) is correct. (Attendance revenue is irrelevant, since it is not affected by the decision whether or not to purchase insurance.)

Exercises 7.5

1. $m = 70(.5) + 71(.2) + 72(.1) + 73(.2) = 71$

$$\sigma^2 = (70-71)^2(.5)+(71-71)^2(.2)+(72-71)^2(.1)+(73-71)^2(.2)$$
$$= .5+0+.1+.8$$
$$= 1.4$$

3. B

5. a. $\mu_A = -10\left(\frac{1}{5}\right)+20\left(\frac{3}{5}\right)+25\left(\frac{1}{5}\right)=15$

$\mu_B = 0(.3)+10(.4)+30(.3)=13$

$\sigma_A^2 = (-10-15)^2\left(\frac{1}{5}\right)+(20-15)^2\left(\frac{3}{5}\right)+(25-15)^2\left(\frac{1}{5}\right)=125+15+20=160$

$\sigma_B^2 = (0-13)^2(.3)+(10-13)^2(.4)+(30-13)^2(.3)=50.7+3.6+86.7=141$

   b. Investment A

   c. Investment B

7. a. $\mu_A = 100(.1)+101(.2)+102(.3)+103(0)+104(0)+105(.2)+106(.2)=103$

$\sigma_A^2 = (100-103)^2(.1)+(101-103)^2(.2)+\cdots+(106-103)^2(.2)=4.6$

$\mu_B = 100(0)+101(.2)+102(0)+103(.2)+104(.1)+105(.2)+106(.3)=104$

$\sigma_A^2 = (100-104)^2(0)+(101-104)^2(.2)+\cdots+(106-104)^2(.3)=3.4$

   b. Business B

   c. Business B

9. The number of heads is a binomial random variable X with $n = 12, p = .5$ so $\mu_X = 12\times.5 = 6$, $\sigma_X = \sqrt{12\times.5\times.5} \approx 1.732$.

11. The number of defective widgets is a binomial random variable X with $n = 200, p = .015$ so $\mu_X = 200\times.015 = 3$, $\sigma_X = \sqrt{200\times.015\times.985} \approx 1.719$.

13. a. $35 - c = 25$ and $35 + c = 45 \Rightarrow c = 10$.

$$\text{Probability} \geq 1 - \frac{5^2}{10^2} = 1 - \frac{25}{100} = .75$$

b. $35 - c = 20$ and $35 + c = 50 \Rightarrow c = 15$

$$\text{Probability} \geq 1 - \frac{5^2}{15^2} \approx .89$$

c. $35 - c = 29$ and $35 + c = 41 \Rightarrow c = 6$

$$\text{Probability} \geq 1 - \frac{5^2}{6^2} \approx .31$$

15. $\mu = 3000$, $\sigma = 250$

$3000 - c = 2000$ and $3000 + c = 4000 \Rightarrow c = 1000$

$$\text{Probability} \geq 1 - \frac{250^2}{1000^2} = .9375$$

Number of bulbs to replace: $\geq 5000(.9375) \approx 4688$

17. $\text{Probability} = 1 - \dfrac{6^2}{c^2} = \dfrac{7}{16}$

$$16c^2 - 576 = 7c^2$$
$$9c^2 = 576$$
$$c = \sqrt{64} = 8$$

19. a. $\mu = 2\left(\dfrac{1}{36}\right) + 3\left(\dfrac{2}{36}\right) + \cdots + 12\left(\dfrac{1}{36}\right) = 7$

$$\sigma^2 = (2-7)^2\left(\frac{1}{36}\right) + \cdots + (12-7)^2\left(\frac{1}{36}\right)$$
$$= \frac{210}{36}$$
$$= \frac{35}{6}$$

b. $\Pr(4 \leq X \leq 10) = \dfrac{3}{36} + \dfrac{4}{36} + \dfrac{5}{36} + \dfrac{6}{36} + \dfrac{5}{36} + \dfrac{4}{36} + \dfrac{3}{36}$

$$= \frac{30}{36}$$
$$= \frac{5}{6}$$

c. $7 - c = 4$ and $7 + c = 10 \Rightarrow c = 3$

$$\text{Probability} \geq 1 - \frac{\frac{35}{6}}{3^2} = \frac{19}{54} \approx .35$$

21. $E(X) = (-2 - 1 + 0 + 1 + 2)(.2) = 0$

$E(X^2) = (4 + 1 + 0 + 1 + 4)(.2) = 2$

$\text{Var}(X) = E(X^2) - E(X)^2 = 2 - 0 = 2$

23.

$2X$	Probability
-2	$\frac{1}{8}$
-1	$\frac{3}{8}$
0	$\frac{1}{8}$
1	$\frac{1}{8}$
2	$\frac{2}{8}$

$$\mu = -2\left(\frac{1}{8}\right) - 1\left(\frac{3}{8}\right) + \cdots + 2\left(\frac{2}{8}\right) = 0$$

$$\sigma_{2X}^2 = (-2-0)^2\left(\frac{1}{8}\right) + \cdots + (2-0)^2\left(\frac{2}{8}\right) = 2$$

$$\sigma_{2X}^2 = 4\sigma_X^2 = 4\left(\frac{1}{2}\right) = 2$$

25. $\mu = \dfrac{54,169 + 51,612 + \cdots + 45,166}{8} = 50,334.75$

$$\sigma^2 = \frac{(54,169 - 50,334.75)^2 + \cdots + (45,166 - 50.334.75)^2}{8}$$

$\sigma \approx 2364.08$

27. $\mu = 4.72$, $\sigma \approx 1.40$

Exercises 7.6

1. $A(1.25) = .8944$

3. $1 - A(.25) = 1 - .5987 = .4013$

5. $A(1.5) - A(.5) = .9332 - .6915 = .2417$

7. $A(-.5) + (1 - A(.5)) = .3085 + (1 - .6915)$
$\qquad\qquad\qquad\qquad\quad = .6170$

9. $\Pr(Z \geq z) = .0401$
$A(z) = 1 - .0401 = .9599$
$z = 1.75$

11. $\Pr(-z \leq Z \leq z) = .5468$
$\quad A(-z) = \dfrac{1 - .5468}{2} = .2266$
$-z = -.75$
$z = .75$

13. The 90th percentile of the standard normal distribution is 1.28 (use table or InvNorm(0.90) on TI 83)

15. $\mu = 6$, $\sigma = 2$

17. $\mu = 9$, $\sigma = 1$

19. $\dfrac{6-8}{\frac{3}{4}} = -\dfrac{2}{1} \cdot \dfrac{4}{3} = -\dfrac{8}{3}$

21. $\dfrac{x-8}{\frac{3}{4}} = 10$

$x = \dfrac{30}{4} + 8 = \dfrac{62}{4} = 15\dfrac{1}{2}$

Answer (a) is correct.

23. $\Pr(X \geq 9) = \Pr\left(Z \geq \dfrac{9-10}{\frac{1}{2}} \right)$

$= \Pr(Z \geq -2)$
$= 1 - \Pr(Z \leq -2)$
$= 1 - .0228$
$= .9772$

25. $\Pr(6 \leq X \leq 10) = \Pr\left(\dfrac{6-7}{2} \leq Z \leq \dfrac{10-7}{2} \right)$

$= \Pr(-.50 \leq Z \leq 1.50)$
$= .9332 - .3085$
$= .6247$

27. $\Pr(-2 \leq Z \leq 2) = A(2) - A(-2)$
$= .9772 - .0228$
$= .9544$

29. From Table 2 we see that $\Pr(Z \leq 2) = .9772$ for standard normal Z. Solve $\dfrac{6-5}{\sigma} = 2$: $2\sigma = 1$, $\sigma = .5$.

31. $\mu = 3.3$, $\sigma = .2$

$\Pr(X \geq 4) = \Pr\left(Z \geq \dfrac{4-3.3}{.2} \right)$

$= \Pr(Z \geq 3.5)$
$= 1 - \Pr(Z \leq 3.5)$
$= 1 - .9998$
$= .0002$

33. $\mu = 6$, $\sigma = .02$

$\Pr(5.95 \leq X \leq 6.05)$

$= \Pr\left(\dfrac{5.95-6}{.02} x \leq Z \leq \dfrac{6.05-6}{.02} \right)$

$= \Pr(-2.5 \leq Z \leq 2.5)$
$= .9938 - .0062$
$= .9876$

35. $\mu = 30{,}000$, $\sigma = 4000$

$\Pr(X > 39{,}000)$

$= \Pr\left(Z > \dfrac{39{,}000 - 30{,}000}{4000} \right)$

$= \Pr(Z > 2.25)$
$= 1 - \Pr(Z \leq 2.25)$
$= 1 - .9878$
$= .0122$

37. $\mu = 520$, $\sigma = 75$

a. $x_{90} = 520 + 75 z_{90}$
$= 520 + 75 \times 1.28$
$= 616$

b. $\Pr(-z \leq Z \leq z) = .90$
$\Pr(Z \leq -z) = .05 \Rightarrow z_{05} \approx -1.65$

$\dfrac{x - \mu}{\sigma} = \dfrac{x - 520}{75} = -1.65$

$\Rightarrow x_{05} = 396.25 \approx 396$

$\dfrac{x - \mu}{\sigma} = \dfrac{x - 520}{75} = 1.65$

$\Rightarrow x_{95} = 643.75 \approx 644$

Between 396 and 644

c. $x_{98} = 520 + 75 z_{98} = 520 + 75 \times 2.05 = 674$

39. $\mu = 30{,}000$, $\sigma = 5000$

$\Pr(Z \leq z) = .02 \Rightarrow z_{02} \approx -2.05$

$\dfrac{x - \mu}{\sigma} = \dfrac{x - 30{,}000}{5000} = -2.05 \Rightarrow x_{02} = 19{,}750$

19,750 miles

41. $\mu = ?$, $\sigma = .25$

a. $\Pr(Z > z) = .005 \Rightarrow z_{99.5} \approx 2.60$

$\dfrac{x - \mu}{\sigma} = \dfrac{6 - \mu}{.25} = 2.60$

$\Rightarrow \mu \approx 5.35$ ounces

b. $\Pr(Z > z) = .99 \Rightarrow z_{01} \approx -2.35$

$\dfrac{x - \mu}{\sigma} = \dfrac{x - 5.35}{.25} = -2.35$

$\Rightarrow x_{01} \approx 4.76$ ounces

43.

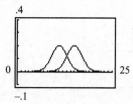

The curve is translated to the right.

45. $\mu = 5.4$, $\sigma = .6$

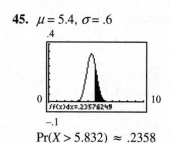

$Pr(X > 5.832) \approx .2358$

Exercises 7.7

1. $n = 25$, $p = \dfrac{1}{5}$

$$\mu = np = 25\left(\frac{1}{5}\right) = 5,$$

$$\sigma = \sqrt{npq} = \sqrt{25\left(\frac{1}{5}\right)\left(\frac{4}{5}\right)} = 2$$

a. $Pr(X = 5) \approx Pr\left(\dfrac{4.5-5}{2} \le Z \le \dfrac{5.5-5}{2}\right)$

$\qquad = Pr(-.25 \le Z \le .25)$

$\qquad = .5987 - .4013$

$\qquad = .1974$

b. $Pr(3 \le X \le 7)$

$\qquad \approx Pr\left(\dfrac{2.5-5}{2} \le Z \le \dfrac{7.5-5}{2}\right)$

$\qquad = Pr(-1.25 \le Z \le 1.25)$

$\qquad = .8944 - .1056$

$\qquad = .7888$

c. $Pr(X < 10) \approx Pr\left(Z \le \dfrac{9.5-5}{2}\right)$

$\qquad = Pr(Z \le 2.25)$

$\qquad = .9878$

3. $n = 20$, $p = \dfrac{1}{6}$

$$\mu = 20\left(\frac{1}{6}\right) = \frac{10}{3}, \ \sigma = \sqrt{20\left(\frac{1}{6}\right)\left(\frac{5}{6}\right)} = \frac{5}{3}$$

$$Pr(X \ge 8) \approx Pr\left(Z \ge \frac{7.5 - \frac{10}{3}}{\frac{5}{3}}\right)$$

$\qquad = Pr(Z \ge 2.5)$

$\qquad = 1 - .9938$

$\qquad = .0062$

5. $n = 90$, $p = \dfrac{9}{19}$

$$\mu = 90\left(\frac{9}{19}\right) = \frac{810}{19}, \ \sigma = \sqrt{90\left(\frac{9}{19}\right)\left(\frac{10}{19}\right)} = \frac{90}{19}$$

$$Pr(X > 45) \approx Pr\left(Z \ge \frac{45.5 - \frac{810}{19}}{\frac{90}{19}}\right)$$

$\qquad \approx Pr(Z \ge .61)$

$\qquad \approx Pr(Z \ge .60)$

$\qquad = 1 - .7257$

$\qquad = .2743$

7. $n = 75$, $p = \dfrac{3}{4}$

$$\mu = 75\left(\frac{3}{4}\right) = 56.25, \quad \sigma = \sqrt{75\left(\frac{3}{4}\right)\left(\frac{1}{4}\right)} = 3.75$$

$$Pr(X \ge 68) \approx Pr\left(Z \ge \frac{67.5 - 56.25}{3.75}\right)$$

$\qquad = Pr(Z \ge 3)$

$\qquad = 1 - .9987$

$\qquad = .0013$

9. $n = 20$, $p = .310$

$\qquad \mu = 20(.310) = 6.2$, $\sigma = \sqrt{20(.31)(.69)} \approx 2.068$

$$Pr(X \ge 6) \approx Pr\left(Z \ge \frac{5.5 - 6.2}{2.068}\right)$$

$\qquad \approx Pr(Z \ge -.34)$

$\qquad \approx Pr(Z \ge -.35)$

$\qquad = 1 - .3632$

$\qquad = .6368$

11. $n = 1000$, $p = .02$

$\qquad \mu = 1000(.02) = 20$,

$\qquad \sigma = \sqrt{1000(.02)(.98)} \approx 4.427$

$$Pr(X < 15) \approx Pr\left(Z \le \frac{14.5 - 20}{4.427}\right)$$

$\qquad \approx Pr(Z \le -1.24)$

$\qquad \approx Pr(Z \le -1.25)$

$\qquad = .1056$

13. probability of failure $= (.01)(.02)(.01) = .000002$
$n = 1,000,000,$
$E(X) = \mu = 1,000,000(.000002) = 2$
$\sigma = \sqrt{1,000,000(.000002)(.999998)} \approx 1.414$

$$\begin{aligned}
\Pr(X > 3) &\approx \Pr\left(Z \geq \frac{3.5-2}{1.414}\right)\\
&\approx \Pr(Z \geq 1.06)\\
&\approx \Pr(Z \geq 1.05)\\
&= 1 - .8531\\
&= .1469
\end{aligned}$$

15. $n = 100, p = .35$
$\mu = 100(.35) = 35,\ \sigma = \sqrt{100(.35)(.65)} \approx 4.770$
$\Pr(30 \leq X \leq 40)$

$$\begin{aligned}
&= \Pr\left(\frac{29.5-35}{4.77} \leq Z \leq \frac{40.5-35}{4.77}\right)\\
&\approx \Pr(-1.15 \leq Z \leq 1.15)\\
&= .8749 - .1251\\
&= .7498
\end{aligned}$$

17. $n = 1000, p = .03$
$\mu = 1000(.03) = 30,$
$\sigma = \sqrt{1000(.03)(.97)} \approx 5.394$

$$\begin{aligned}
\Pr(X \geq 29) &\approx \Pr\left(Z \geq \frac{28.5-30}{5.394}\right)\\
&\approx \Pr(Z \geq -0.278)\\
&\approx \Pr(Z \geq -0.28)\\
&= 1 - .3897\\
&= .6103
\end{aligned}$$

19. $n = 100,\ p = \frac{1}{2}$
Exact: $\Pr(49 \leq X \leq 51) \approx .2356$
Normal Approximation:

$$\mu = 100\left(\frac{1}{2}\right) = 50,\ \sigma = \sqrt{100\left(\frac{1}{2}\right)\left(\frac{1}{2}\right)} = 5$$

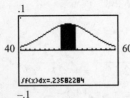

$\Pr(48.5 \leq X \leq 51.5) \approx .2358$

21. $n = 150, p = .2$
Exact: $\Pr(X = 30) \approx .0812$
Normal Approximation: $\mu = 150(.2) = 30,$
$\sigma = \sqrt{150(.2)(.8)} \approx 4.899$

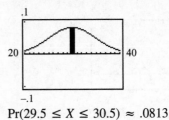

$\Pr(29.5 \leq X \leq 30.5) \approx .0813$

Chapter 7 Supplementary Exercises

1.

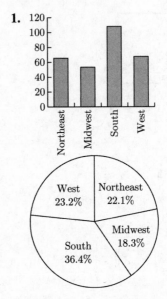

2. min $= 1,$
$Q_1 = 2.5,\ Q_2 = 4.5, Q_3 = 11.5, \text{max} = 23;$
IQR $= Q_3 - Q_1 = 11.5 - 2.5 = 9;$

3. $n = 3,\ p = \frac{1}{3}$

a.

k	$\Pr(X = k)$
0	$\binom{3}{0}\left(\frac{1}{3}\right)^0\left(\frac{2}{3}\right)^3 = \frac{8}{27}$
1	$\binom{3}{1}\left(\frac{1}{3}\right)^1\left(\frac{2}{3}\right)^2 = \frac{12}{27}$
2	$\binom{3}{2}\left(\frac{1}{3}\right)^2\left(\frac{2}{3}\right)^1 = \frac{6}{27}$
3	$\binom{3}{3}\left(\frac{1}{3}\right)^3\left(\frac{2}{3}\right)^0 = \frac{1}{27}$

b. $\mu = 0\left(\dfrac{8}{27}\right) + 1\left(\dfrac{12}{27}\right) + 2\left(\dfrac{6}{27}\right) + 3\left(\dfrac{1}{27}\right) = 1$

$\sigma^2 = (0-1)^2\left(\dfrac{8}{27}\right) + (1-1)^2\left(\dfrac{12}{27}\right) + (2-1)^2\left(\dfrac{6}{27}\right) + (3-1)^2\left(\dfrac{1}{27}\right)$

$= \dfrac{2}{3}$

4. $\Pr(Z \geq .75) = 1 - .7734 = .2266$

5. $\Pr(6.5 \leq X \leq 11) = \Pr\left(\dfrac{6.5-5}{3} \leq Z \leq \dfrac{11-5}{3}\right)$

$= A(2) - A(.5)$

$= .9772 - .6915$

$= .2857$

6. $n = 4, p = .3$

$\Pr(X = 2) = \dbinom{4}{2}(.3)^2(.7)^2 = .2646$

7. $\mu = 10, \ \sigma = \dfrac{1}{3}$

$10 - c = 9$ and $10 + c = 11 \Rightarrow c = 1$

Probability: $\geq 1 - \dfrac{\left(\frac{1}{3}\right)^2}{1^2} = \dfrac{8}{9} \approx .89$

8. $\mu = 0(.2) + 1(.3) + 5(.1) + 10(.4) = 4.8$

$\sigma^2 = (0-4.8)^2(.2) + (1-4.8)^2(.3) + (5-4.8)^2(.1) + (10-4.8)^2(.4)$

$= 19.76$

9. $\mu = 5.75, \ \sigma = .2$

$\Pr(X \geq 6) = \Pr\left(Z \geq \dfrac{6-5.75}{.2}\right)$

$= \Pr(Z \geq 1.25)$

$= 1 - .8944$

$= .1056$

10.56%

10. Let x be the number of red balls.

k	$\Pr(X = k)$
0	$\dfrac{\binom{4}{0}\binom{4}{4}}{\binom{8}{4}} = \dfrac{1}{70}$
1	$\dfrac{\binom{4}{1}\binom{4}{3}}{\binom{8}{4}} = \dfrac{16}{70}$
2	$\dfrac{\binom{4}{2}\binom{4}{2}}{\binom{8}{4}} = \dfrac{36}{70}$
3	$\dfrac{\binom{4}{3}\binom{4}{1}}{\binom{8}{4}} = \dfrac{16}{70}$
4	$\dfrac{\binom{4}{4}\binom{4}{0}}{\binom{8}{4}} = \dfrac{1}{70}$

$$\mu = 0\left(\frac{1}{70}\right) + 1\left(\frac{16}{70}\right) + 2\left(\frac{36}{70}\right) + 3\left(\frac{16}{70}\right) + 4\left(\frac{1}{70}\right) = 2$$

$$\sigma^2 = (0-2)^2\left(\frac{1}{70}\right) + (1-2)^2\left(\frac{16}{70}\right) + (2-2)^2\left(\frac{36}{70}\right) + (3-2)^2\left(\frac{16}{70}\right) + (4-2)^2\left(\frac{1}{70}\right)$$

$$= \frac{4}{7}$$

11. $n = 54$, $p = \dfrac{2}{5}$

$$\mu = 54\left(\frac{2}{5}\right) = 21.6$$

$$\sigma = \sqrt{54\left(\frac{2}{5}\right)\left(\frac{3}{5}\right)} = 3.6$$

$$\Pr(X \leq 13) \approx \Pr\left(Z \leq \frac{13.5 - 21.6}{3.6}\right) = \Pr(Z \leq -2.25) = .0122$$

12. $n = 75, \; p = \dfrac{1}{4}$

$\mu = 75\left(\dfrac{1}{4}\right) = 18.75$

$\sigma = \sqrt{75\left(\dfrac{1}{4}\right)\left(\dfrac{3}{4}\right)} = 3.75$

$\Pr(8 \le X \le 22) \approx \Pr\left(\dfrac{7.5 - 18.75}{3.75} \le Z \le \dfrac{22.5 - 18.75}{3.75}\right) = \Pr(-3 \le Z \le 1) = .8413 - .0013 = .84$

13. $\mu = 80, \; \sigma = 15$

$\Pr(80 - n \le X \le 80 + h) = .8664$

$\dfrac{1 - .8664}{2} = .0668 \Rightarrow$ (area left of $80 - h$)

$\Pr(Z \le z) = .0668$ when $z = -1.5$

$\Pr(-1.5 \le Z \le 1.5) = .8664$

Therefore, $\dfrac{x - \mu}{\sigma} = -1.5$ and $\dfrac{x + \mu}{\sigma} = 1.5.$

$\dfrac{(80 - h) - 80}{15} = -1.5$ and $\dfrac{(80 + h) - 80}{15} = 1.5$

$h = 22.5$

14. $\Pr(Z \ge z) = .7734$

$\Pr(Z < z) = 1 - .7734 = .2266$

$\quad z = -.75$

15. a. $\Pr(133 \le X) \approx \Pr\left(\dfrac{132.5 - 100}{15} \le Z\right)$

$\approx \Pr(2.167 \le Z)$

$\approx \Pr(2.20 \le Z)$

$= 1 - .9861$

$= .0139$

$= 1.39\%$

b. $x_{95} = 100 + 15 z_{95}$

$\quad\quad = 100 + 15 \cdot 1.65$

$\quad\quad = 124.75$

16. The student has a .6 probability of guessing correctly on the six questions with answer *true* and a .4 probability of guessing correctly on the four questions with answer *false*. Therefore the student's expected score is $6(.6) + 4(.4) = 5.2$ correct answers which gives 52 points or 52%.

A better strategy is to choose true for all the questions which guarantees a score of 60%.

Conceptual Exercises

17. a. scoring in the third quartile is not very good: 100, 40, 40, 40,

b. scoring in the third quartile corresponds to a perfect grade: 100, 100, 90, 80, 70

18. a. The mean and median are equal: 1, 2, 3, 4, 5, 6, 7, 8, 9, 10 : The mean is 5.5; the median is 5.5

b. the mean is less than the median: 1, 1, 1, 1, 4, 5, 6, 7, 8, 9 : The mean is 4.3; the median is 4.5

c. the median is less than the mean 1, 2, 3, 4, 5, 6, 10, 12, 14, 100. The median is 5.5; the mean is 15.7

19. A population mean is the average of all the data in the entire population. When a sample is taken from a population, the sample mean is the average of all the data in that particular sample. Sample means vary whereas the population mean is fixed.

20. Expected value is a concept similar to the mean. It is the long range number you would expect to occur if the experiment was repeated a great many times. An example: the expected value might be the long range profit or loss of an insurance company due to the probability that an individual lives for an additional year.

21. Yes; in general, if we add a constant to each number in a set, then the mean will increase by that constant.

22. Yes; in general, if we multiply each number in a set by some constant, then the standard deviation will be multiplied by that constant.

23. The binomial probability distribution applies when there is a fixed number of independent trials when the probability of success is constant. The outcome of each trial is classified as either a "success" or a "failure".

24. Repeated trials that do not produce a binomial distribution: 1) tossing a coin until a head appears. 2) Having children until a girl is born.

Chapter 7 Chapter Test

1.

Number Waiting in Line	Relative Frequency
0	.04
1	.10
2	.18
3	.26
4	.22
5	.14
6	.06

Pr(at most 3 customers in line)
$= .04 + .10 + .18 + .26$
$= .58$

2.

20   26   30   37   42
Interquartile range = 11

3. a. Possible outcomes are HH, HT, TH, TT

Number of Heads, k	$Pr(X = k)$
0	.25
1	.50
2	.25

b.

k	$Pr(2X + 5 = k)$
5	.25
7	.50
9	.25

4. a. $Pr(\text{get 7 twice}) = \binom{12}{2}\left(\frac{1}{6}\right)^2\left(\frac{5}{6}\right)^{10} \approx .296$

b. Pr (get 7 at least twice)
$= 1 - Pr(\text{get 7 zero or one time})$
$= 1 - \binom{12}{0}\left(\frac{1}{6}\right)^0\left(\frac{5}{6}\right)^{12} - \binom{12}{1}\left(\frac{1}{6}\right)^1\left(\frac{5}{6}\right)^{11}$
$\approx .619$

   c.   The expected number of 7's is $12 \cdot \dfrac{1}{6} = 2$.

5. X has mean

$\mu = (-2)(.3) + 0(.1) + 1(.4) + 3(.2) = .4,$

variance

$\sigma^2 = (-2 - .4)^2(.3) + (0 - .4)^2(.1) + (1 - .4)^2(.4) + (3 - .4)^2(.2)$
$\quad = 3.24,$

and standard deviation

$\sigma = \sqrt{3.24} = 1.8.$

6. When a pair of fair dice is rolled, the probabilities that the result is 7 or 11 are $\dfrac{1}{6}$ and $\dfrac{1}{18}$ respectively. Hence

Lucy's expected winnings are $(-10)\dfrac{2}{9} + 3 \cdot \dfrac{7}{9} = \dfrac{1}{9} \approx .11,$ or 11 cents per roll.

7. a. $\Pr(Z \le 1) = .8413$

   b. $\Pr(Z \ge -2.25) = 1 - \Pr(Z \le -2.25)$
$\qquad\qquad\qquad\quad = 1 - .0122$
$\qquad\qquad\qquad\quad = .9878$

   c. $\Pr(-1 \le Z \le 1.15) = \Pr(Z \le 1.15) - \Pr(Z \le -1)$
$\qquad\qquad\qquad\qquad\quad = .8749 - .1587$
$\qquad\qquad\qquad\qquad\quad = .7162$

8. Let X be a normally distributed random variable with mean 106 and standard deviation 10.

   a. $\Pr(X \ge 116) = \Pr\left(\dfrac{116 - 106}{10} \le Z\right)$
$\qquad\qquad\qquad = \Pr(1 \le Z)$
$\qquad\qquad\qquad = 1 - \Pr(Z \le 1)$
$\qquad\qquad\qquad = 1 - .8413$
$\qquad\qquad\qquad = .1587$

   b. $\Pr(96 \le X \le 121) = \Pr\left(\dfrac{96 - 106}{10} \le Z \le \dfrac{121 - 106}{10}\right)$
$\qquad\qquad\qquad\qquad = \Pr(-1 \le Z \le 1.5)$
$\qquad\qquad\qquad\qquad = \Pr(Z \le 1.5) - \Pr(Z \le -1)$
$\qquad\qquad\qquad\qquad = .9332 - .1587$
$\qquad\qquad\qquad\qquad = .7745$

   c. $\Pr(Z \le -.7) \approx .242$ so $z_{24.2} = -.7.$
$\qquad x_{24.2} = 106 + (-.7)10 = 99; \ 1 - .242 = .758$
$\qquad \Pr(Z \le .7) \approx .758$ so $z_{75.8} = .7.$
$\qquad x_{75.8} = 106 + (.7)10 = 113$

9. Let X be a binomial random variable with $p = 0.2$, $n = 25$. X has mean $\mu = 25(.2) = 5$ and standard deviation $\sigma = \sqrt{25(.2)(.8)} = 2$.

a. $\Pr(6 \le X \le 10) \approx \Pr\left(\dfrac{5.5 - 5}{2} \le Z \le \dfrac{10.5 - 5}{2} \right)$

$= \Pr(.25 \le Z \le 2.75)$

$= \Pr(Z \le 2.75) - \Pr(Z \le .25)$

$= .9970 - .5987$

$= .3983$

b. $\Pr(X = 10) = \dbinom{25}{10}(.2)^{10}(.8)^{15} = .0118$;

$\Pr(X = 10) \approx \Pr\left(\dfrac{9.5 - 5}{2} \le Z \le \dfrac{10.5 - 5}{2} \right)$

$= \Pr(2.25 \le Z \le 2.75)$

$= \Pr(Z \le 2.75) - \Pr(Z \le 2.25)$

$= .9970 - .9878$

$= .0092$;

difference is .0026.

c. $\Pr(X \ge 1) = 1 - \Pr(X = 0)$

$= 1 - (.8)^{25}$

$= .9962$;

$\Pr(X \ge 1) \approx \Pr\left(\dfrac{.5 - 5}{2} \le Z \right)$

$= \Pr(-2.25 \le Z)$

$= 1 - \Pr(Z \le -2.25)$

$= 1 - .0122$

$= .9878$;

difference is .0084.

Chapter 8

1. Yes; the matrix is square, all entries are ≥ 0, and the sum of the entries in each column is 1.

3. No; the matrix is not square.

5. Yes; the matrix is square, all entries are ≥ 0, and the sum of the entries in each column is 1.

7. $\begin{array}{c} \\ A \\ B \end{array} \begin{array}{c} A \quad B \\ \begin{bmatrix} .3 & .5 \\ .7 & .5 \end{bmatrix} \end{array}$

9. $\begin{array}{c} \\ A \\ B \\ C \end{array} \begin{array}{c} A \quad B \quad C \\ \begin{bmatrix} \frac{1}{3} & \frac{2}{9} & \frac{1}{3} \\ \frac{1}{3} & \frac{4}{9} & \frac{1}{6} \\ \frac{1}{3} & \frac{1}{3} & \frac{1}{2} \end{bmatrix} \end{array}$

11. $\begin{array}{c} \\ A \\ B \\ C \end{array} \begin{array}{c} A \quad B \quad C \\ \begin{bmatrix} .4 & .2 & 0 \\ .5 & 0 & 0 \\ .1 & .8 & 1 \end{bmatrix} \end{array}$

13.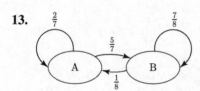

15.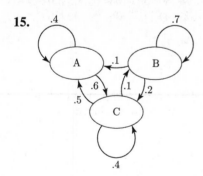

17.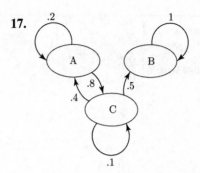

19. initial state matrix $= \begin{bmatrix} .47 \\ .53 \end{bmatrix}_0$

next state $= A \begin{bmatrix} .47 \\ .53 \end{bmatrix}_0 = \begin{bmatrix} .8 & .3 \\ .2 & .7 \end{bmatrix} \begin{bmatrix} .47 \\ .53 \end{bmatrix}_0$

$= \begin{bmatrix} .535 \\ .465 \end{bmatrix}_1$

next state $= A^2 \begin{bmatrix} .47 \\ .53 \end{bmatrix}_0 = \begin{bmatrix} .70 & .45 \\ .30 & .55 \end{bmatrix} \begin{bmatrix} .47 \\ .53 \end{bmatrix}_0$

$= \begin{bmatrix} .5675 \\ .4325 \end{bmatrix}_2$

After one generation, about 53.5% of French women will work. After two generations, about 56.75% of French women will work.

21. initial state matrix $= \begin{bmatrix} .40 \\ .40 \\ .20 \end{bmatrix}_0$

next state $= \begin{bmatrix} .5 & .4 & .2 \\ .4 & .3 & .6 \\ .1 & .3 & .2 \end{bmatrix} \begin{bmatrix} .4 \\ .4 \\ .2 \end{bmatrix}$

$= \begin{bmatrix} .4 \\ .4 \\ .2 \end{bmatrix}_1$

40% will be Zone I, 40% will be in Zone II, 20% will be in Zone III.

23. a. $\begin{array}{c} \\ S \\ O \end{array} \begin{array}{c} S \qquad O \\ \begin{bmatrix} .992 & .007 \\ .008 & .993 \end{bmatrix} \end{array}$

b. initial state $= \begin{bmatrix} .36 \\ .64 \end{bmatrix}_0$

next state $= \begin{bmatrix} .992 & .007 \\ .008 & .993 \end{bmatrix}\begin{bmatrix} .36 \\ .64 \end{bmatrix}_0 = \begin{bmatrix} .3616 \\ .6384 \end{bmatrix}_1$

next state $= \begin{bmatrix} .992 & .007 \\ .008 & .993 \end{bmatrix}\begin{bmatrix} .3616 \\ .6384 \end{bmatrix}_1$

$= \begin{bmatrix} .3632 \\ .6368 \end{bmatrix}_2$

25. a.
$$\begin{array}{c} \quad L \ \ R \\ \begin{matrix} L \\ R \end{matrix}\begin{bmatrix} .9 & .7 \\ .1 & .3 \end{bmatrix} \end{array}$$

b. $A^2 = \begin{bmatrix} .9 & .7 \\ .1 & .3 \end{bmatrix}\begin{bmatrix} .9 & .7 \\ .1 & .3 \end{bmatrix} = \begin{bmatrix} .88 & .84 \\ .12 & .16 \end{bmatrix}$

c. initial state $= \begin{bmatrix} .5 \\ .5 \end{bmatrix}_0$

$\begin{bmatrix} \ \\ \ \end{bmatrix}_1 = \begin{bmatrix} .9 & .7 \\ .1 & .3 \end{bmatrix}\begin{bmatrix} .5 \\ .5 \end{bmatrix}_0$

$= \begin{bmatrix} .8 \\ .2 \end{bmatrix}_1$

$\begin{bmatrix} \ \\ \ \end{bmatrix}_2 = A^2 \begin{bmatrix} .5 \\ .5 \end{bmatrix}_0$

$= \begin{bmatrix} .88 & .84 \\ .12 & .16 \end{bmatrix}\begin{bmatrix} .5 \\ .5 \end{bmatrix}_0$

$= \begin{bmatrix} .86 \\ .14 \end{bmatrix}_2$

d. Answers will vary. Correct answer is 87.5%.

27. a. 4%

b. 96% of the freshmen who held middle-of-the-road political views continued to hold these views as sophomores.

c.

d. initial state $= \begin{bmatrix} .32 \\ .43 \\ .25 \end{bmatrix}_0$

next state $= \begin{bmatrix} .94 & .02 & .01 \\ .05 & .96 & .04 \\ .01 & .02 & .95 \end{bmatrix}\begin{bmatrix} .32 \\ .43 \\ .25 \end{bmatrix}_0 = \begin{bmatrix} .3119 \\ .4388 \\ .2493 \end{bmatrix}_1$

next state $= \begin{bmatrix} .94 & .02 & .01 \\ .05 & .96 & .04 \\ .01 & .02 & .95 \end{bmatrix}\begin{bmatrix} .3119 \\ .4388 \\ .2493 \end{bmatrix}_1$

$\approx \begin{bmatrix} .3045 \\ .4468 \\ .2487 \end{bmatrix}_2$

43.88% of the students held middle-of-the-road political views as sophomores, 44.68% as juniors.

29. a.
$$\begin{array}{c} \quad U \ \ S \ \ R \\ \begin{matrix} U \\ S \\ R \end{matrix}\begin{bmatrix} .86 & .05 & .03 \\ .08 & .86 & .05 \\ .06 & .09 & .92 \end{bmatrix} \end{array}$$

b. Since we are concerned with only the people who live in urban areas of 2000, we use

$\begin{bmatrix} \ \\ \ \\ \ \end{bmatrix}_0 = \begin{bmatrix} 1 \\ 0 \\ 0 \end{bmatrix}_0.$

$\begin{bmatrix} \ \\ \ \\ \ \end{bmatrix}_2 = A^2 \begin{bmatrix} \ \\ \ \\ \ \end{bmatrix}_0$

$= \begin{bmatrix} .86 & .05 & .03 \\ .08 & .86 & .05 \\ .06 & .09 & .92 \end{bmatrix}\begin{bmatrix} .86 & .05 & .03 \\ .08 & .86 & .05 \\ .06 & .09 & .92 \end{bmatrix}\begin{bmatrix} 1 \\ 0 \\ 0 \end{bmatrix}_0$

$= \begin{bmatrix} .7454 \\ .1406 \\ .114 \end{bmatrix}_2$

11.4% of people who live in urban areas in 2000 will live in rural areas in 2002.

31. a.
$$A = \begin{array}{c} \quad V \quad\ \ C \\ \begin{matrix} V \\ C \end{matrix}\begin{bmatrix} .128 & .663 \\ .872 & .337 \end{bmatrix} \end{array}$$

b. $A^2 = \begin{bmatrix} .59452 & .308295 \\ .40548 & .691705 \end{bmatrix}$

The probability that the second letter following a vowel is also a vowel is .59452.

33. $A^2 = \begin{bmatrix} .45 & .44 \\ .55 & .56 \end{bmatrix}$, $A^3 = \begin{bmatrix} .445 & .444 \\ .555 & .556 \end{bmatrix}$,

$A^4 = \begin{bmatrix} .4445 & .4444 \\ .5555 & .5556 \end{bmatrix}$

$\begin{bmatrix} \quad \\ \quad \end{bmatrix}_3 = \begin{bmatrix} .445 & .444 \\ .555 & .556 \end{bmatrix}\begin{bmatrix} .3 \\ .7 \end{bmatrix}_0 = \begin{bmatrix} .4443 \\ .5557 \end{bmatrix}_3 \approx \begin{bmatrix} .44 \\ .56 \end{bmatrix}_3$

$\begin{bmatrix} \quad \\ \quad \end{bmatrix}_4 = \begin{bmatrix} .4445 & .4444 \\ .5555 & .5556 \end{bmatrix}\begin{bmatrix} .3 \\ .7 \end{bmatrix}_0$

$= \begin{bmatrix} .44443 \\ .55557 \end{bmatrix}_4$

$\approx \begin{bmatrix} .44 \\ .56 \end{bmatrix}_4$

35. $A^1 = A = \begin{bmatrix} \frac{1}{3} & \frac{1}{3} \\ \frac{2}{3} & \frac{2}{3} \end{bmatrix} \approx \begin{bmatrix} .33 & .33 \\ .67 & .67 \end{bmatrix}$

$A^2 = A \cdot A = \begin{bmatrix} \frac{1}{3} & \frac{1}{3} \\ \frac{2}{3} & \frac{2}{3} \end{bmatrix}\begin{bmatrix} \frac{1}{3} & \frac{1}{3} \\ \frac{2}{3} & \frac{2}{3} \end{bmatrix} = \begin{bmatrix} \frac{1}{3} & \frac{1}{3} \\ \frac{2}{3} & \frac{2}{3} \end{bmatrix} \approx \begin{bmatrix} .33 & .33 \\ .67 & .67 \end{bmatrix}$

The pattern continues. All powers are $\begin{bmatrix} \frac{1}{3} & \frac{1}{3} \\ \frac{2}{3} & \frac{2}{3} \end{bmatrix}$.

37. $A^1 = A = \begin{bmatrix} .1 & .3 \\ .9 & .7 \end{bmatrix}$

$A^2 = A \cdot A = \begin{bmatrix} .1 & .3 \\ .9 & .7 \end{bmatrix}\begin{bmatrix} .1 & .3 \\ .9 & .7 \end{bmatrix} = \begin{bmatrix} .28 & .24 \\ .72 & .76 \end{bmatrix}$

$A^3 = A^2 \cdot A = \begin{bmatrix} .28 & .24 \\ .72 & .76 \end{bmatrix}\begin{bmatrix} .1 & .3 \\ .9 & .7 \end{bmatrix}$

$= \begin{bmatrix} .244 & .252 \\ .756 & .748 \end{bmatrix} \approx \begin{bmatrix} .24 & .25 \\ .76 & .75 \end{bmatrix}$

$A^4 = A^3 \cdot A = \begin{bmatrix} .244 & .252 \\ .756 & .748 \end{bmatrix}\begin{bmatrix} .1 & .3 \\ .9 & .7 \end{bmatrix}$

$= \begin{bmatrix} .2512 & .2496 \\ .7488 & .7504 \end{bmatrix} \approx \begin{bmatrix} .25 & .25 \\ .75 & .75 \end{bmatrix}$

$A^5 = A^4 \cdot A = \begin{bmatrix} .2512 & .2496 \\ .7488 & .7504 \end{bmatrix}\begin{bmatrix} .1 & .3 \\ .9 & .7 \end{bmatrix}$

$= \begin{bmatrix} .24976 & .25008 \\ .75024 & .74992 \end{bmatrix}$

$\approx \begin{bmatrix} .25 & .25 \\ .75 & .75 \end{bmatrix}$

39. No; all powers have a zero entry in the upper right corner, so the matrix is not regular.

41. a. Use the method described in the text to generate the next four distribution matrices.

$\begin{bmatrix} .35 \\ .65 \end{bmatrix}$, $\begin{bmatrix} .425 \\ .575 \end{bmatrix}$, $\begin{bmatrix} .3875 \\ .6125 \end{bmatrix}$, $\begin{bmatrix} .40625 \\ .59375 \end{bmatrix}$

b. $A^4 B = \begin{bmatrix} .4375 & .375 \\ .5625 & .625 \end{bmatrix}\begin{bmatrix} .5 \\ .5 \end{bmatrix} = \begin{bmatrix} .40625 \\ .59375 \end{bmatrix}$

43. The distribution matrix gets closer and closer to $\begin{bmatrix} .4 \\ .6 \end{bmatrix}$.

45. The matrices get closer and closer to $\begin{bmatrix} .4 & .4 \\ .6 & .6 \end{bmatrix}$.

Each column of this matrix is the same as the 2 by 1 matrix found in Exercise 43.

Exercises 8.2

1. Yes; the matrix is stochastic, since all entries are positive.

3. Yes; the matrix is stochastic, since the second power is $\begin{bmatrix} .79 & .3 \\ .21 & .7 \end{bmatrix}$, which has all positive entries.

5. Yes; the matrix is stochastic, since the second power is $\begin{bmatrix} .8 & .08 & .4 \\ .1 & .86 & .3 \\ .1 & .06 & .3 \end{bmatrix}$, which has all positive entries.

7. $\begin{cases} x + y = 1 \\ \begin{bmatrix} .5 & .1 \\ .5 & .9 \end{bmatrix}\begin{bmatrix} x \\ y \end{bmatrix} = \begin{bmatrix} x \\ y \end{bmatrix} \end{cases}$

$\begin{cases} x + y = 1 \\ .5x + .1y = x \\ .5x + .9y = y \end{cases}$

$\begin{cases} x + y = 1 \\ -.5x + .1y = 0 \\ .5x - .1y = 0 \end{cases}$

The second and third equations in this system are equivalent.

$\begin{bmatrix} 1 & 1 & | & 1 \\ -.5 & .1 & | & 0 \end{bmatrix} \xrightarrow{[2]+.5[1]} \begin{bmatrix} 1 & 1 & | & 1 \\ 0 & .6 & | & .5 \end{bmatrix}$

$\xrightarrow{\frac{5}{3}[2]} \begin{bmatrix} 1 & 1 & | & 1 \\ 0 & 1 & | & \frac{5}{6} \end{bmatrix}$

$$\xrightarrow{[1]+(-1)[2]} \begin{bmatrix} 1 & 0 & \frac{1}{6} \\ 0 & 1 & \frac{5}{6} \end{bmatrix}$$

$$x = \frac{1}{6}, \ y = \frac{5}{6}$$

The stable distribution is $\begin{bmatrix} x \\ y \end{bmatrix} = \begin{bmatrix} \frac{1}{6} \\ \frac{5}{6} \end{bmatrix}$.

9. $\begin{cases} x + y = 1 \\ \begin{bmatrix} .8 & .3 \\ .2 & .7 \end{bmatrix}\begin{bmatrix} x \\ y \end{bmatrix} = \begin{bmatrix} x \\ y \end{bmatrix} \end{cases}$

$\begin{cases} x + y = 1 \\ .8x + .3y = x \\ .2x + .7y = y \end{cases}$

$\begin{cases} x + y = 1 \\ -.2x + .3y = 0 \\ .2x - .3y = 0 \end{cases}$

The second and third equations in this system are equivalent.

$$\begin{bmatrix} 1 & 1 & 1 \\ -.2 & .3 & 0 \end{bmatrix} \xrightarrow{[2]+.2[1]} \begin{bmatrix} 1 & 1 & 1 \\ 0 & .5 & .2 \end{bmatrix}$$

$$\xrightarrow{2[2]} \begin{bmatrix} 1 & 1 & 1 \\ 0 & 1 & .4 \end{bmatrix}$$

$$\xrightarrow{[1]+(-1)[2]} \begin{bmatrix} 1 & 0 & .6 \\ 0 & 1 & .4 \end{bmatrix}$$

$x = .6, \ y = .4$

The stable distribution is $\begin{bmatrix} x \\ y \end{bmatrix} = \begin{bmatrix} .6 \\ .4 \end{bmatrix}$.

11. $\begin{cases} x + y + z = 1 \\ \begin{bmatrix} .1 & .4 & .7 \\ .6 & .4 & .2 \\ .3 & .2 & .1 \end{bmatrix}\begin{bmatrix} x \\ y \\ z \end{bmatrix} = \begin{bmatrix} x \\ y \\ z \end{bmatrix} \end{cases}$

$\begin{cases} x + y + z = 1 \\ .1x + .4y + .7z = x \\ .6x + .4y + .2z = y \\ .3x + .2y + .1z = z \end{cases}$

$\begin{cases} x + y + z = 1 \\ -.9x + .4y + .7z = 0 \\ .6x - .6y + .2z = 0 \\ .3x + .2y - .9z = 0 \end{cases}$

$$\begin{bmatrix} 1 & 1 & 1 & 1 \\ -.9 & .4 & .7 & 0 \\ .6 & -.6 & .2 & 0 \\ .3 & .2 & -.9 & 0 \end{bmatrix}$$

$$\begin{matrix} 10[2] \\ 10[3] \\ 10[4] \end{matrix} \longrightarrow \begin{bmatrix} 1 & 1 & 1 & 1 \\ -9 & 4 & 7 & 0 \\ 6 & -6 & 2 & 0 \\ 3 & 2 & -9 & 0 \end{bmatrix}$$

$$\begin{matrix} [2]+9[1] \\ [3]+(-6)[1] \\ [4]+(-3)[1] \end{matrix} \longrightarrow \begin{bmatrix} 1 & 1 & 1 & 1 \\ 0 & 13 & 16 & 9 \\ 0 & -12 & -4 & -6 \\ 0 & -1 & -12 & -3 \end{bmatrix}$$

$$\begin{matrix} \frac{1}{13}[2] \\ [1]+(-1)[2] \\ [3]+12[2] \\ [4]+1[2] \end{matrix} \longrightarrow \begin{bmatrix} 1 & 0 & -\frac{3}{13} & \frac{4}{13} \\ 0 & 1 & \frac{16}{13} & \frac{9}{13} \\ 0 & 0 & \frac{140}{13} & \frac{30}{13} \\ 0 & 0 & -\frac{140}{13} & -\frac{30}{13} \end{bmatrix}$$

$$\begin{matrix} \frac{13}{140}[3] \\ [1]+\frac{3}{13}[3] \\ [2]-\frac{16}{13}[3] \\ [4]+\frac{140}{13}[3] \end{matrix} \longrightarrow \begin{bmatrix} 1 & 0 & 0 & \frac{5}{14} \\ 0 & 1 & 0 & \frac{3}{7} \\ 0 & 0 & 1 & \frac{3}{14} \\ 0 & 0 & 0 & 0 \end{bmatrix}$$

$$x = \frac{5}{14}, \ y = \frac{3}{7}, \ z = \frac{3}{14}$$

The stable distribution is $\begin{bmatrix} x \\ y \\ z \end{bmatrix} = \begin{bmatrix} \frac{5}{14} \\ \frac{3}{7} \\ \frac{3}{14} \end{bmatrix}$.

13. The stochastic matrix is $\begin{matrix} & \ \ L & R \\ \begin{matrix} L \\ R \end{matrix} & \begin{bmatrix} .9 & .7 \\ .1 & .3 \end{bmatrix} \end{matrix}$.

Find the stable distribution.

$\begin{cases} x + y = 1 \\ \begin{bmatrix} .9 & .7 \\ .1 & .3 \end{bmatrix}\begin{bmatrix} x \\ y \end{bmatrix} = \begin{bmatrix} x \\ y \end{bmatrix} \end{cases}$

$\begin{cases} x + y = 1 \\ .9x + .7y = x \\ .1x + .3y = y \end{cases}$

$\begin{cases} x + y = 1 \\ -.1x + .7y = 0 \\ .1x - .7y = 0 \end{cases}$

The second and third equations in this system are equivalent.

$$\begin{bmatrix} 1 & 1 & | & 1 \\ -.1 & .7 & | & 0 \end{bmatrix} \xrightarrow{[2]+.1[1]} \begin{bmatrix} 1 & 1 & | & 1 \\ 0 & .8 & | & .1 \end{bmatrix}$$

$$\xrightarrow{1.25[2]} \begin{bmatrix} 1 & 1 & | & 1 \\ 0 & 1 & | & .125 \end{bmatrix}$$

$$\xrightarrow{[1]+(-1)[2]} \begin{bmatrix} 1 & 0 & | & .875 \\ 0 & 1 & | & .125 \end{bmatrix}$$

$x = .785, y = .125$

The stable distribution is $\begin{bmatrix} x \\ y \end{bmatrix} = \begin{bmatrix} .875 \\ .125 \end{bmatrix}$.

After many days, 87.5% of the mice will be going to the left.

15. The stochastic matrix is $\begin{array}{c} \\ \text{GM} \\ \text{non-GM} \end{array} \overset{\text{GM non-GM}}{\begin{bmatrix} .6 & .1 \\ .4 & .9 \end{bmatrix}}$.

Find the stable distribution.

$$\begin{cases} x+y=1 \\ \begin{bmatrix} .6 & .1 \\ .4 & .9 \end{bmatrix}\begin{bmatrix} x \\ y \end{bmatrix} = \begin{bmatrix} x \\ y \end{bmatrix} \end{cases}$$

$$\begin{cases} x+y=1 \\ .6x+.1y=x \\ .4x+.9y=y \end{cases}$$

$$\begin{cases} x+y=1 \\ -.4x+.1y=0 \\ .4x-.1y=0 \end{cases}$$

The second and third equations in the system are equivalent.

$$\begin{bmatrix} 1 & 1 & | & 1 \\ -.4 & .1 & | & 0 \end{bmatrix} \xrightarrow{[2]+.4[1]} \begin{bmatrix} 1 & 1 & | & 1 \\ 0 & .5 & | & .4 \end{bmatrix}$$

$$\xrightarrow{2[2]} \begin{bmatrix} 1 & 1 & | & 1 \\ 0 & 1 & | & .8 \end{bmatrix} \xrightarrow{[1]+(-1)[2]} \begin{bmatrix} 1 & 0 & | & .2 \\ 0 & 1 & | & .8 \end{bmatrix}$$

The stable distribution is $\begin{bmatrix} .2 \\ .8 \end{bmatrix}$.

In the long run General Motors market share is 20%.

17. The stochastic matrix is $\begin{array}{c} R \\ S \end{array} \overset{R \quad S}{\begin{bmatrix} .1 & .6 \\ .9 & .4 \end{bmatrix}}$.

Find the stable distribution.

$$\begin{cases} x+y=1 \\ \begin{bmatrix} .1 & .6 \\ .9 & .4 \end{bmatrix}\begin{bmatrix} x \\ y \end{bmatrix} = \begin{bmatrix} x \\ y \end{bmatrix} \end{cases}$$

$$\begin{cases} x+y=1 \\ .1x+.6y=x \\ .9x+.4y=y \end{cases}$$

$$\begin{cases} x+y=1 \\ -.9x+.6y=0 \\ .9x+.6y=0 \end{cases}$$

The second and third equations in this system are equivalent.

$$\begin{bmatrix} 1 & 1 & | & 1 \\ -.9 & .6 & | & 0 \end{bmatrix} \xrightarrow{[2]+.9[1]} \begin{bmatrix} 1 & 1 & | & 1 \\ 0 & 1.5 & | & .9 \end{bmatrix}$$

$$\xrightarrow{\frac{2}{3}[2]} \begin{bmatrix} 1 & 1 & | & 1 \\ 0 & 1 & | & .6 \end{bmatrix}$$

$$\xrightarrow{[1]+(-1)[2]} \begin{bmatrix} 1 & 0 & | & .4 \\ 0 & 1 & | & .6 \end{bmatrix}$$

$x = .4, y = .6$

The stable distribution is $\begin{bmatrix} x \\ y \end{bmatrix} = \begin{bmatrix} .4 \\ .6 \end{bmatrix}$.

In the log run, the daily likelihood of rain is 40% or $\dfrac{2}{5}$.

19. a. $\begin{array}{c} A \\ B \\ C \end{array} \overset{A \quad B \quad C}{\begin{bmatrix} .7 & .1 & .1 \\ .2 & .8 & .3 \\ .1 & .1 & .6 \end{bmatrix}}$

b. $\begin{bmatrix} \\ \\ \end{bmatrix}_1 = \begin{bmatrix} .7 & .1 & .1 \\ .2 & .8 & .3 \\ .1 & .1 & .6 \end{bmatrix}\begin{bmatrix} .4 \\ .3 \\ .3 \end{bmatrix}_0 = \begin{bmatrix} .34 \\ .41 \\ .25 \end{bmatrix}_1$

$\begin{bmatrix} \\ \\ \end{bmatrix}_2 = \begin{bmatrix} .7 & .1 & .1 \\ .2 & .8 & .3 \\ .1 & .1 & .6 \end{bmatrix}\begin{bmatrix} .34 \\ .41 \\ .25 \end{bmatrix}_1 = \begin{bmatrix} .304 \\ .471 \\ .225 \end{bmatrix}_2$

34% of the cars are at location *A* after one day, and 30.4% after two days.

c. Find the stable distribution.

$$\begin{cases} x+y+z=1 \\ .7x+.1y+.1z=x \\ .2x+.8y+.3z=y \\ .1x+.1y+.6z=z \end{cases}$$

$$\begin{cases} x+y+z=1 \\ -.3x+.1y+.1z=0 \\ .2x-.2y+.3z=0 \\ .1x+.1y-.4z=0 \end{cases}$$

The fourth equation is equivalent to the second plus the third and so is redundant.

$$\begin{bmatrix} 1 & 1 & 1 & | & 1 \\ -.3 & .1 & .1 & | & 0 \\ .2 & -.2 & .3 & | & 0 \end{bmatrix} \xrightarrow[\substack{[2]+.3[1] \\ [3]-.2[1]}]{} \begin{bmatrix} 1 & 1 & 1 & | & 1 \\ 0 & .4 & .4 & | & .3 \\ 0 & -.4 & .1 & | & -.2 \end{bmatrix}$$

$$\xrightarrow[\substack{\frac{10}{4}[2] \\ \frac{-10}{4}[3]}]{} \begin{bmatrix} 1 & 1 & 1 & | & 1 \\ 0 & 1 & 1 & | & .75 \\ 0 & 1 & -.25 & | & .5 \end{bmatrix}$$

$$\xrightarrow[\substack{[1]-[2] \\ [3]-[2]}]{} \begin{bmatrix} 1 & 0 & 0 & | & .25 \\ 0 & 1 & 1 & | & .75 \\ 0 & 0 & -1.25 & | & -.25 \end{bmatrix}$$

$$\xrightarrow[\substack{[2]+\frac{8}{10}[3] \\ -\frac{8}{10}[3]}]{} \begin{bmatrix} 1 & 0 & 0 & | & .25 \\ 0 & 1 & 0 & | & .55 \\ 0 & 0 & 1 & | & .2 \end{bmatrix}$$

In the long run there are $\dfrac{1}{4}$ at A, $\dfrac{11}{20}$ at B

and $\dfrac{1}{5}$ at C.

21. The stochastic matrix is $\begin{array}{c} \\ D \\ R \\ H \end{array}\begin{array}{c} \begin{array}{ccc} D & R & H \end{array} \\ \begin{bmatrix} .5 & 0 & .25 \\ 0 & .5 & 2.5 \\ .5 & .5 & .5 \end{bmatrix} \end{array}.$

Find the stable distribution.

$$\begin{cases} x+y+z=1 \\ \begin{bmatrix} .5 & 0 & .25 \\ 0 & .5 & .25 \\ .5 & .5 & .5 \end{bmatrix}\begin{bmatrix} x \\ y \\ z \end{bmatrix} = \begin{bmatrix} x \\ y \\ z \end{bmatrix} \end{cases}$$

$$\begin{cases} x & + & y & + & z & = & 1 \\ .5x & & & + & .25z & = & x \\ & & .5y & + & .25z & = & y \\ .5x & + & .5y & + & .5z & = & z \end{cases}$$

$$\begin{cases} x & + & y & + & z & = & 1 \\ -.5x & & & + & .25z & = & 0 \\ & & -.5y & + & .25z & = & 0 \\ .5x & + & .5y & - & .5z & = & 0 \end{cases}$$

$$\begin{bmatrix} 1 & 1 & 1 & | & 1 \\ -.5 & 0 & .25 & | & 0 \\ 0 & -.5 & .25 & | & 0 \\ .5 & .5 & -.5 & | & 0 \end{bmatrix}$$

$$\xrightarrow[\substack{4[2] \\ 4[3] \\ 2[4]}]{} \begin{bmatrix} 1 & 1 & 1 & | & 1 \\ -2 & 0 & 1 & | & 0 \\ 0 & -2 & 1 & | & 0 \\ 1 & 1 & -1 & | & 0 \end{bmatrix}$$

$$\xrightarrow[\substack{[4]+(-1)[1]}]{[2]+2[1]} \begin{bmatrix} 1 & 1 & 1 & | & 1 \\ 0 & 2 & 3 & | & 2 \\ 0 & -2 & 1 & | & 0 \\ 0 & 0 & -2 & | & -1 \end{bmatrix}$$

$$\xrightarrow[\substack{.5[2] \\ [1]+(-1)[2] \\ [3]+2[2]}]{} \begin{bmatrix} 1 & 0 & -.5 & | & 0 \\ 0 & 1 & 1.5 & | & 1 \\ 0 & 0 & 4 & | & 2 \\ 0 & 0 & -2 & | & -1 \end{bmatrix}$$

$$\xrightarrow[\substack{.25[3] \\ [1]+(.5)[3] \\ [2]+(-1.5)[3] \\ [4]+2[3]}]{} \begin{bmatrix} 1 & 0 & 0 & | & .25 \\ 0 & 1 & 0 & | & .25 \\ 0 & 0 & 1 & | & .5 \\ 0 & 0 & 0 & | & 0 \end{bmatrix}$$

$x = .25$, $y = .25$, $z = .5$

The stable distribution is $\begin{bmatrix} x \\ y \\ z \end{bmatrix} = \begin{bmatrix} .25 \\ .25 \\ .5 \end{bmatrix}$.

In the long run, 25% will be dominant.

23. $\begin{bmatrix} .5 \\ .5 \end{bmatrix}$ is a stable distribution for the matrix

$A = \begin{bmatrix} 0 & 1 \\ 1 & 0 \end{bmatrix}$ because $.5 + .5 = 1$ and

$\begin{bmatrix} 0 & 1 \\ 1 & 0 \end{bmatrix}\begin{bmatrix} .5 \\ .5 \end{bmatrix} = \begin{bmatrix} .5 \\ .5 \end{bmatrix}$. However, given an arbitrary

initial distribution $\begin{bmatrix} \\ \end{bmatrix}_0 \neq \begin{bmatrix} .5 \\ .5 \end{bmatrix}$, $A^n \begin{bmatrix} \\ \end{bmatrix}_0$ will

not approach $\begin{bmatrix} .5 \\ .5 \end{bmatrix}$ as n gets large, so the

existence of a stable distribution for A does not contradict the main premise of this section.

25. Calculating $[A]^{255}$ gives $\begin{bmatrix} .7 & .7 \\ .3 & .3 \end{bmatrix}$.

This suggests that the stable distribution is $\begin{bmatrix} .7 \\ .3 \end{bmatrix}$.

Check: $x = .7$, $y = .3$ is indeed a solution to the

system $\begin{cases} x+y=1 \\ \begin{bmatrix} .85 & .35 \\ .15 & .65 \end{bmatrix}\begin{bmatrix} x \\ y \end{bmatrix} = \begin{bmatrix} x \\ y \end{bmatrix} \end{cases}$.

Also, $\begin{bmatrix} .85 & .35 \\ .15 & .65 \end{bmatrix}\begin{bmatrix} .7 \\ .3 \end{bmatrix} = \begin{bmatrix} .7 \\ .3 \end{bmatrix}$.

27. Calculating $[A]$ ^ 255 → Frac gives $\begin{bmatrix} \frac{8}{35} & \frac{8}{35} & \frac{8}{35} \\ \frac{3}{7} & \frac{3}{7} & \frac{3}{7} \\ \frac{12}{35} & \frac{12}{35} & \frac{12}{35} \end{bmatrix}$.

This suggests that the stable distribution is $\begin{bmatrix} \frac{8}{35} \\ \frac{3}{7} \\ \frac{12}{35} \end{bmatrix}$.

Check: $x = \dfrac{8}{35}$, $y = \dfrac{3}{7}$, $z = \dfrac{12}{35}$ is indeed a

solution to the system $\begin{cases} x + y + z = 1 \\ \begin{bmatrix} .1 & .4 & .1 \\ .3 & .2 & .8 \\ .6 & .4 & .1 \end{bmatrix} \begin{bmatrix} x \\ y \\ z \end{bmatrix} = \begin{bmatrix} x \\ y \\ z \end{bmatrix} \end{cases}$.

Also, $\begin{bmatrix} .1 & .4 & .1 \\ .3 & .2 & .8 \\ .6 & .4 & .1 \end{bmatrix} \begin{bmatrix} \frac{8}{35} \\ \frac{3}{7} \\ \frac{12}{35} \end{bmatrix} = \begin{bmatrix} \frac{8}{35} \\ \frac{3}{7} \\ \frac{12}{35} \end{bmatrix}$.

Exercises 8.3

1. Yes; the states A and B are absorbing, and it's possible to get A and B from states C and D.

3. No; the state A is absorbing, but it's not possible to get to A from C or D.

5. No; states 1 and 2 are absorbing states, but states 3 and 4 do not lead to absorbing states.

7. Yes; state 1 is an absorbing state, and state 3 leads to state 1. Furthermore, it is possible to go from state 2 to state 1 through an intermediate step (state 2 to state 3 to state 1).

9. $\begin{array}{c} \\ B \\ A \\ C \end{array} \begin{array}{c} B \;\; A \;\; C \\ \begin{bmatrix} 1 & .3 & .4 \\ 0 & .2 & .5 \\ 0 & .5 & .1 \end{bmatrix} \end{array}$

11. $\begin{array}{c} \\ D \\ A \\ B \\ C \end{array} \begin{array}{c} D \;\; A \;\; B \;\; C \\ \begin{bmatrix} 1 & .4 & 0 & .1 \\ 0 & .1 & 1 & .6 \\ 0 & .2 & 0 & .1 \\ 0 & .3 & 0 & .2 \end{bmatrix} \end{array}$

13. $\begin{bmatrix} 1 & 0 & .3 \\ 0 & 1 & .2 \\ 0 & 0 & .5 \end{bmatrix} = \left[\begin{array}{c|c} I & S \\ \hline 0 & R \end{array} \right]$

$R = [.5]; \; S = \begin{bmatrix} .3 \\ .2 \end{bmatrix};$

$F = (I - R)^{-1} = [1 - .5]^{-1} = [2]$

Find the stable matrix:

$S(I - R)^{-1} = \begin{bmatrix} .3 \\ .2 \end{bmatrix} [2] = \begin{bmatrix} .6 \\ .4 \end{bmatrix}$

$\left[\begin{array}{c|c} I & S(I-R)^{-1} \\ \hline 0 & 0 \end{array} \right] = \begin{bmatrix} 1 & 0 & .6 \\ 0 & 1 & .4 \\ 0 & 0 & 0 \end{bmatrix}$

15. $\begin{bmatrix} 1 & 0 & .1 & 0 \\ 0 & 1 & .5 & .2 \\ 0 & 0 & .3 & .6 \\ 0 & 0 & .1 & .2 \end{bmatrix} = \left[\begin{array}{c|c} I & S \\ \hline 0 & R \end{array} \right]$

$R = \begin{bmatrix} .3 & .6 \\ .1 & .2 \end{bmatrix}; \; S = \begin{bmatrix} .1 & 0 \\ .5 & .2 \end{bmatrix}$

$I - R = \begin{bmatrix} 1 & 0 \\ 0 & 1 \end{bmatrix} - \begin{bmatrix} .3 & .6 \\ .1 & .2 \end{bmatrix} = \begin{bmatrix} .7 & -.6 \\ -.1 & .8 \end{bmatrix}$

$= \begin{bmatrix} a & b \\ c & d \end{bmatrix}$

$\Delta = ad - bc = (.7)(.8) - (-.6)(-.1) = .5$

$F = (I - R)^{-1} = \begin{bmatrix} \frac{d}{\Delta} & -\frac{b}{\Delta} \\ -\frac{c}{\Delta} & \frac{a}{\Delta} \end{bmatrix} = \begin{bmatrix} \frac{.8}{.5} & \frac{-.6}{.5} \\ \frac{-.1}{.5} & \frac{.7}{.5} \end{bmatrix}$

$= \begin{bmatrix} 1.6 & 1.2 \\ .2 & 1.4 \end{bmatrix}$

Find the stable matrix:

$S(I - R)^{-1} = \begin{bmatrix} .1 & 0 \\ .5 & .2 \end{bmatrix} \begin{bmatrix} 1.6 & 1.2 \\ .2 & 1.4 \end{bmatrix}$

$= \begin{bmatrix} .16 & .12 \\ .84 & .88 \end{bmatrix}$

$\left[\begin{array}{c|c} I & S(I-R)^{-1} \\ \hline 0 & 0 \end{array} \right] = \begin{bmatrix} 1 & 0 & .16 & .12 \\ 0 & 1 & .84 & .88 \\ 0 & 0 & 0 & 0 \\ 0 & 0 & 0 & 0 \end{bmatrix}$

17.
$$\begin{bmatrix} 1 & 0 & 0 & | & .1 & .2 \\ 0 & 1 & 0 & | & .3 & 0 \\ 0 & 0 & 1 & | & 0 & .2 \\ \hline 0 & 0 & 0 & | & .5 & 0 \\ 0 & 0 & 0 & | & .1 & .6 \end{bmatrix} = \begin{bmatrix} I & | & S \\ \hline 0 & | & R \end{bmatrix}$$

$$R = \begin{bmatrix} .5 & 0 \\ .1 & .6 \end{bmatrix}; \ S = \begin{bmatrix} .1 & .2 \\ .3 & 0 \\ 0 & .2 \end{bmatrix}$$

Find the fundamental matrix:
$$I - R = \begin{bmatrix} 1 & 0 \\ 0 & 1 \end{bmatrix} - \begin{bmatrix} .5 & 0 \\ .1 & .6 \end{bmatrix} = \begin{bmatrix} .5 & 0 \\ -.1 & .4 \end{bmatrix}$$
$$= \begin{bmatrix} a & b \\ c & d \end{bmatrix}$$
$$\Delta = ad - bc = (.5)(.4) - (0)(-.1) = .2$$
$$F = (I - R)^{-1} = \begin{bmatrix} \frac{d}{\Delta} & -\frac{b}{\Delta} \\ -\frac{c}{\Delta} & \frac{d}{\Delta} \end{bmatrix} = \begin{bmatrix} \frac{.4}{.2} & -\frac{0}{.2} \\ -\frac{-.1}{.2} & \frac{.5}{.2} \end{bmatrix}$$
$$= \begin{bmatrix} 2 & 0 \\ .5 & 2.5 \end{bmatrix}$$

Find the stable matrix:
$$S(I - R)^{-1} = \begin{bmatrix} .1 & .2 \\ .3 & 0 \\ 0 & .2 \end{bmatrix} \begin{bmatrix} 2 & 0 \\ .5 & 2.5 \end{bmatrix}$$
$$= \begin{bmatrix} .3 & .5 \\ .6 & 0 \\ .1 & .5 \end{bmatrix}$$
$$\begin{bmatrix} I & | & S(I-R)^{-1} \\ \hline 0 & | & 0 \end{bmatrix} = \begin{bmatrix} 1 & 0 & 0 & | & .3 & .5 \\ 0 & 1 & 0 & | & .6 & 0 \\ 0 & 0 & 1 & | & .1 & .5 \\ \hline 0 & 0 & 0 & | & 0 & 0 \\ 0 & 0 & 0 & | & 0 & 0 \end{bmatrix}$$

19. If the gambler begins with $2, he should have $1 for an expected number of .79 plays.

21. a.

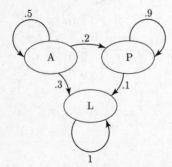

b.
$$\begin{array}{c} \quad\ L\ \ \ A\ \ \ P \\ \begin{array}{c} L \\ A \\ P \end{array} \begin{bmatrix} 1 & | & .3 & .1 \\ 0 & | & .5 & 0 \\ 0 & | & .2 & .9 \end{bmatrix} \end{array}$$

c. $R = \begin{bmatrix} .5 & 0 \\ .2 & .9 \end{bmatrix}; \ S = [.3 \ \ .1]$

$$I - R = \begin{bmatrix} 1 & 0 \\ 0 & 1 \end{bmatrix} - \begin{bmatrix} .5 & 0 \\ .2 & .9 \end{bmatrix} = \begin{bmatrix} .5 & 0 \\ -.2 & .1 \end{bmatrix}$$
$$\Delta = (.5)(.1) - (0)(-.2) = .05$$
$$(I - R)^{-1} = \begin{bmatrix} \frac{.1}{.05} & 0 \\ \frac{.2}{.05} & \frac{.5}{.05} \end{bmatrix} = \begin{bmatrix} 2 & 0 \\ 4 & 10 \end{bmatrix}$$
$$S(I - R)^{-1} = [.3 \ .1]\begin{bmatrix} 2 & 0 \\ 4 & 10 \end{bmatrix} = [1 \ \ 1]$$

The stable matrix is
$$\begin{array}{c} \quad\ L\ \ A\ \ P \\ \begin{array}{c} L \\ A \\ P \end{array} \begin{bmatrix} 1 & | & 1 & 1 \\ 0 & | & 0 & 0 \\ 0 & | & 0 & 0 \end{bmatrix} \end{array} .$$

d. Add the numbers in the *A* column of the fundamental matrix: $2 + 4 = 6$ yrs.

23. a.

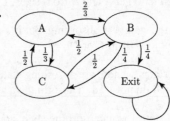

b.
$$\begin{array}{c} \quad\ E\ \ A\ \ B\ \ C \\ \begin{array}{c} E \\ A \\ B \\ C \end{array} \begin{bmatrix} 1 & | & 0 & \frac{1}{4} & 0 \\ 0 & | & 0 & \frac{1}{2} & \frac{1}{2} \\ 0 & | & \frac{2}{3} & 0 & \frac{1}{2} \\ 0 & | & \frac{1}{3} & \frac{1}{4} & 0 \end{bmatrix} \end{array}$$

c. $R = \begin{bmatrix} 0 & \frac{1}{2} & \frac{1}{2} \\ \frac{2}{3} & 0 & \frac{1}{2} \\ \frac{1}{3} & \frac{1}{4} & 0 \end{bmatrix}; \ S = \begin{bmatrix} 0 & \frac{1}{4} & 0 \end{bmatrix}$

$$I - R = \begin{bmatrix} 1 & 0 & 0 \\ 0 & 1 & 0 \\ 0 & 0 & 1 \end{bmatrix} - \begin{bmatrix} 0 & \frac{1}{2} & \frac{1}{2} \\ \frac{2}{3} & 0 & \frac{1}{2} \\ \frac{1}{3} & \frac{1}{4} & 0 \end{bmatrix} = \begin{bmatrix} 1 & -\frac{1}{2} & -\frac{1}{2} \\ -\frac{2}{3} & 1 & -\frac{1}{2} \\ -\frac{1}{3} & -\frac{1}{4} & 1 \end{bmatrix}$$

$$(I-R)^{-1} = \begin{bmatrix} 4.2 & 3 & 3.6 \\ 4 & 4 & 4 \\ 2.4 & 2 & 3.2 \end{bmatrix}$$

$$S(I-R)^{-1} = \begin{bmatrix} 0 & \frac{1}{4} & 0 \end{bmatrix} \begin{bmatrix} 4.2 & 3 & 3.6 \\ 4 & 4 & 4 \\ 2.4 & 2 & 3.2 \end{bmatrix} = [1 \ 1 \ 1]$$

The stable matrix is
$$\begin{array}{c} \\ E \\ A \\ B \\ C \end{array} \begin{array}{cccc} E & A & B & C \\ \end{array} \\ \begin{bmatrix} 1 & 1 & 1 & 1 \\ 0 & 0 & 0 & 0 \\ 0 & 0 & 0 & 0 \\ 0 & 0 & 0 & 0 \end{bmatrix}.$$

d. Add the numbers in the A column of the fundamental matrix: 10.6 minutes.

25. a.
$$\begin{array}{c} D \\ G \\ F \\ S \end{array} \begin{array}{cccc} D & G & F & S \\ \end{array} \\ \begin{bmatrix} 1 & 0 & .2 & .1 \\ 0 & 1 & 0 & .9 \\ 0 & 0 & 0 & 0 \\ 0 & 0 & .8 & 0 \end{bmatrix}$$

b. $R = \begin{bmatrix} 0 & 0 \\ .8 & 0 \end{bmatrix}; \ S = \begin{bmatrix} .2 & .1 \\ 0 & .9 \end{bmatrix}$

$$I-R = \begin{bmatrix} 1 & 0 \\ 0 & 1 \end{bmatrix} - \begin{bmatrix} 0 & 0 \\ .8 & 0 \end{bmatrix} = \begin{bmatrix} 1 & 0 \\ -.8 & 1 \end{bmatrix}$$

$$= \begin{bmatrix} a & b \\ c & d \end{bmatrix}$$

$\Delta = ad - bc = (1)(1) - (0)(-.8) = 1$

$$(I-R)^{-1} = \begin{bmatrix} \frac{d}{\Delta} & -\frac{b}{\Delta} \\ -\frac{c}{\Delta} & \frac{a}{\Delta} \end{bmatrix} = \begin{bmatrix} \frac{1}{1} & -\frac{0}{1} \\ -\frac{-.8}{1} & \frac{1}{1} \end{bmatrix}$$

$$= \begin{bmatrix} 1 & 0 \\ .8 & 1 \end{bmatrix}$$

$$S(I-R)^{-1} = \begin{bmatrix} .2 & .1 \\ 0 & .9 \end{bmatrix} \begin{bmatrix} 1 & 0 \\ .8 & 1 \end{bmatrix}$$

$$= \begin{bmatrix} .28 & .1 \\ .72 & .9 \end{bmatrix}$$

$$\begin{bmatrix} I & S(I-R)^{-1} \\ \hline 0 & 0 \end{bmatrix} = \begin{array}{c} D \\ G \\ F \\ S \end{array} \begin{array}{cccc} D & G & F & S \\ \end{array} \\ \begin{bmatrix} 1 & 0 & .28 & .1 \\ 0 & 1 & .72 & .9 \\ 0 & 0 & 0 & 0 \\ 0 & 0 & 0 & 0 \end{bmatrix}$$

c. From the F column of the stable matrix, the probability that a freshman will eventually graduate is .72.

d. The fundamental matrix is
$$F = (I-R)^{-1} = \begin{array}{c} F \\ S \end{array} \begin{array}{cc} F & S \\ \end{array} \\ \begin{bmatrix} 1 & 0 \\ .8 & 1 \end{bmatrix}.$$

Add the numbers in the F column:
$1 + .8 = 1.8$.
The student will attend an expected number of 1.8 years.

27. First, arrange the matrix so that the absorbing states come first and calculate the fundamental and stable matrices.

$$\begin{array}{c} \\ Paid \\ Bad \\ \leq 30 \\ < 60 \end{array} \begin{array}{cccc} Paid & Bad & \leq 30 & < 60 \\ \end{array} \\ \begin{bmatrix} 1 & 0 & .4 & .1 \\ 0 & 1 & 0 & .1 \\ 0 & 0 & .4 & .4 \\ 0 & 0 & .2 & .4 \end{bmatrix} = \begin{bmatrix} I & S \\ \hline 0 & R \end{bmatrix}$$

$$R = \begin{bmatrix} .4 & .4 \\ .2 & .4 \end{bmatrix}; \ S = \begin{bmatrix} .4 & .1 \\ 0 & .1 \end{bmatrix}$$

$$I-R = \begin{bmatrix} 1 & 0 \\ 0 & 1 \end{bmatrix} - \begin{bmatrix} .4 & .4 \\ .2 & .4 \end{bmatrix} = \begin{bmatrix} .6 & -.4 \\ -.2 & .6 \end{bmatrix}$$

$$= \begin{bmatrix} a & b \\ c & d \end{bmatrix}$$

$\Delta = ad - bc = (.6)(.6) - (-.4)(-.2) = .28$

$$F = (I-R)^{-1} = \begin{bmatrix} \frac{d}{\Delta} & -\frac{b}{\Delta} \\ -\frac{c}{\Delta} & \frac{a}{\Delta} \end{bmatrix}$$

$$= \begin{bmatrix} \frac{.6}{.28} & \frac{-.4}{.28} \\ \frac{-.2}{.28} & \frac{.6}{.28} \end{bmatrix} = \begin{bmatrix} \frac{15}{7} & \frac{10}{7} \\ \frac{5}{7} & \frac{15}{7} \end{bmatrix}$$

$$S(I-R)^{-1} = \begin{bmatrix} .4 & .1 \\ 0 & .1 \end{bmatrix} \begin{bmatrix} \frac{15}{7} & \frac{10}{7} \\ \frac{5}{7} & \frac{15}{7} \end{bmatrix}$$

$$= \begin{bmatrix} \frac{13}{14} & \frac{11}{14} \\ \frac{1}{14} & \frac{3}{14} \end{bmatrix}$$

The stable matrix is
$$\begin{array}{c} \\ Paid \\ Bad \\ \leq 30 \\ < 60 \end{array} \begin{array}{cccc} Paid & Bad & \leq 30 & < 60 \\ \end{array} \\ \begin{bmatrix} 1 & 0 & \frac{13}{14} & \frac{11}{14} \\ 0 & 1 & \frac{1}{14} & \frac{3}{14} \\ 0 & 0 & 0 & 0 \\ 0 & 0 & 0 & 0 \end{bmatrix}.$$

a. From the stable matrix, the probability of an account eventually being paid off is $\dfrac{13}{14}$ if it is currently at most 30 days overdue and $\dfrac{11}{14}$ if it is less than 60 days overdue (but more than 30 days overdue).

b. The fundamental matrix is

$$\begin{array}{c} \\ \leq 30 \\ < 60 \end{array} \begin{array}{cc} \leq 30 & < 60 \\ \left[\begin{array}{cc} \frac{15}{7} & \frac{10}{7} \\ \frac{5}{7} & \frac{15}{7} \end{array}\right] \end{array}$$. Add the numbers in the

first column: $\dfrac{15}{7} + \dfrac{5}{7} = \dfrac{20}{7}$. An account that is overdue at most 30 days is expected to reach an absorbing state (paid or bad) after $\dfrac{20}{7}$ months.

c. From the stable matrix, about $\dfrac{13}{14}$ of the "≤ 30 day" debt will be paid, and about $\dfrac{11}{14}$ of the "<60 day" debt will be paid.
$\dfrac{13}{14}(\$2000) + \dfrac{11}{14}(\$5000) \approx \$5786$. About $\$5786$ will be paid and about $\$1214$ will become bad debt.

29. First, find the absorbing stochastic matrix and calculate the fundamental and stable matrices.

$$\begin{array}{c} \\ \$0 \\ \$4 \\ \$1 \\ \$2 \\ \$3 \end{array} \begin{array}{c} \$0\ \$4\ \$1\ \$2\ \$3 \\ \left[\begin{array}{ccc|cc} 1 & 0 & \frac{1}{2} & 0 & 0 \\ 0 & 1 & 0 & 0 & \frac{1}{2} \\ \hline 0 & 0 & 0 & \frac{1}{2} & 0 \\ 0 & 0 & \frac{1}{2} & 0 & \frac{1}{2} \\ 0 & 0 & 0 & \frac{1}{2} & 0 \end{array}\right] \end{array} = \left[\begin{array}{c|c} I & S \\ \hline 0 & R \end{array}\right]$$

$$R = \left[\begin{array}{ccc} 0 & \frac{1}{2} & 0 \\ \frac{1}{2} & 0 & \frac{1}{2} \\ 0 & \frac{1}{2} & 0 \end{array}\right]; S = \left[\begin{array}{ccc} \frac{1}{2} & 0 & 0 \\ 0 & 0 & \frac{1}{2} \end{array}\right]$$

$$I - R = \left[\begin{array}{ccc} 1 & 0 & 0 \\ 0 & 1 & 0 \\ 0 & 0 & 1 \end{array}\right] - \left[\begin{array}{ccc} 0 & \frac{1}{2} & 0 \\ \frac{1}{2} & 0 & \frac{1}{2} \\ 0 & \frac{1}{2} & 0 \end{array}\right]$$

$$= \left[\begin{array}{ccc} 1 & -\frac{1}{2} & 0 \\ -\frac{1}{2} & 1 & -\frac{1}{2} \\ 0 & -\frac{1}{2} & 1 \end{array}\right]$$

$$F = (I - R)^{-1} = \left[\begin{array}{ccc} 1 & -\frac{1}{2} & 0 \\ -\frac{1}{2} & 1 & -\frac{1}{2} \\ 0 & -\frac{1}{2} & 1 \end{array}\right]^{-1}$$

$$= \left[\begin{array}{ccc} \frac{3}{2} & 1 & \frac{1}{2} \\ 1 & 2 & 1 \\ \frac{1}{2} & 1 & \frac{3}{2} \end{array}\right]$$

$$S(I - R)^{-1} = \left[\begin{array}{ccc} \frac{1}{2} & 0 & 0 \\ 0 & 0 & \frac{1}{2} \end{array}\right] \left[\begin{array}{ccc} \frac{3}{2} & 1 & \frac{1}{2} \\ 1 & 2 & 1 \\ \frac{1}{2} & 1 & \frac{3}{2} \end{array}\right]$$

$$= \left[\begin{array}{ccc} \frac{3}{4} & \frac{1}{2} & \frac{1}{4} \\ \frac{1}{4} & \frac{1}{2} & \frac{3}{4} \end{array}\right]$$

The stable matrix is

$$\begin{array}{c} \\ \$0 \\ \$4 \\ \$1 \\ \$2 \\ \$3 \end{array} \begin{array}{c} \$0\ \$4\ \ \$1\ \ \$2\ \ \$3 \\ \left[\begin{array}{cc|ccc} 0 & 1 & \frac{3}{4} & \frac{1}{2} & \frac{1}{4} \\ 1 & 0 & \frac{1}{4} & \frac{1}{2} & \frac{3}{4} \\ \hline 0 & 0 & 0 & 0 & 0 \\ 0 & 0 & 0 & 0 & 0 \\ 0 & 0 & 0 & 0 & 0 \end{array}\right] \end{array}$$.

a. From the top row of the stable matrix, the probability of eventually going broke is $\dfrac{3}{4}$ if he starts with $1, $\dfrac{1}{2}$ if he starts with $2, and $\dfrac{1}{4}$ if he starts with $3.

b. The fundamental matrix is

$$\begin{array}{c} \\ \$1 \\ \$2 \\ \$3 \end{array} \begin{array}{c} \$1\ \ \$2\ \ \$3 \\ \left[\begin{array}{ccc} \frac{3}{2} & 1 & \frac{1}{2} \\ 1 & 2 & 1 \\ \frac{1}{2} & 1 & \frac{3}{2} \end{array}\right] \end{array}$$.

Add the entries in the middle column:
$1 + 2 + 1 = 4$
Starting with $2, he will play for an expected number of 4 times.

31. Consider the Markov process with states 1, 2 and 3 corresponding to the number of different quotations received so far. This process has absorbing stochastic matrix

$$A = \begin{matrix} 3 \\ 2 \\ 1 \end{matrix} \begin{bmatrix} 1 & \frac{1}{3} & 0 \\ 0 & \frac{2}{3} & \frac{2}{3} \\ 0 & 0 & \frac{1}{3} \end{bmatrix}.$$

The distribution matrix after purchasing one bottle is $B = \begin{bmatrix} 0 \\ 0 \\ 1 \end{bmatrix}$, so the distribution matrix after purchasing four more bottles (for a total of five) is $A^4 B = \begin{pmatrix} 1 & \frac{65}{81} & \frac{50}{81} \\ 0 & \frac{16}{81} & \frac{10}{27} \\ 0 & 0 & \frac{1}{81} \end{pmatrix} \begin{pmatrix} 0 \\ 0 \\ 1 \end{pmatrix} = \begin{pmatrix} \frac{50}{81} \\ \frac{10}{27} \\ \frac{1}{81} \end{pmatrix}.$

Therefore, the probability of receiving all three quotations after purchasing five bottles is $\frac{50}{81}$.

We have $R = \begin{bmatrix} \frac{2}{3} & \frac{2}{3} \\ 0 & \frac{1}{3} \end{bmatrix}$ so the fundamental matrix

$$(I - R)^{-1} = \begin{bmatrix} \frac{1}{3} & -\frac{2}{3} \\ 0 & \frac{2}{3} \end{bmatrix}^{-1} = \begin{bmatrix} 3 & 3 \\ 0 & \frac{3}{2} \end{bmatrix}.$$

The expected number of soft drinks you have to purchase to get all three quotations is the sum of the numbers in the "1" column of the fundamental matrix plus one for the first bottle, so it's $3 + \frac{3}{2} + 1 = 5\frac{1}{2}$.

Chapter 8 Supplementary Exercises

1. Stochastic, neither; the matrix is stochastic because it is square, the entries are all ≥ 0, and the sum of the entries in each column is 1. It is not regular because all powers of the matrix include zero entries. The first state is an absorbing state, but the matrix is not an absorbing matrix because an object that begins in the third or fourth state will always remain within these two (nonabsorbing) states.

2. Stochastic, regular; the matrix is stochastic because it is square, the entries are all ≥ 0, and the sum of the entries in each column is 1. It is regular because it has no zero entries.

3. Stochastic, regular; the matrix is stochastic because it is square, the entries are all ≥ 0, and the sum of the entries in each column is 1. It is regular because the second power, $\begin{bmatrix} .3 & .21 \\ .7 & .79 \end{bmatrix}$, contains no zero entries.

4. Stochastic, absorbing; the matrix is stochastic because it is square, the entries are all ≥ 0, and the sum of the entries in each column is 1. It is absorbing because the first and second states are absorbing states, and it is possible for an object to get from the third state to an absorbing state.

5. Not stochastic because the sum of the entries in the middle column is not 1.

6. Stochastic, absorbing; the matrix is stochastic because it is square, the entries are all ≥ 0, and the sum of the entries in each column is 1. It is absorbing because the first and second states are absorbing states and it is possible to get from the third state (indirectly) or the fourth state (directly) to an absorbing state.

7. $\begin{cases} x + y = 1 \\ \begin{bmatrix} .6 & .5 \\ .4 & .5 \end{bmatrix} \begin{bmatrix} x \\ y \end{bmatrix} = \begin{bmatrix} x \\ y \end{bmatrix} \end{cases}$

$\begin{cases} x + y = 1 \\ .6x + .5y = x \\ .4x + .5y = y \end{cases}$

$\begin{cases} x + y = 1 \\ -.4x + .5y = 0 \\ .4x - .5y = 0 \end{cases}$

The second and third equations in this system are equivalent.

$$\begin{bmatrix} 1 & 1 & | & 1 \\ -.4 & .5 & | & 0 \end{bmatrix} \xrightarrow{[2]+.4[1]} \begin{bmatrix} 1 & 1 & | & 1 \\ 0 & .9 & | & .4 \end{bmatrix}$$

$$\xrightarrow{\frac{10}{9}[2]} \begin{bmatrix} 1 & 1 & | & 1 \\ 0 & 1 & | & \frac{4}{9} \end{bmatrix} \xrightarrow{[1]+(-1)[2]} \begin{bmatrix} 1 & 0 & | & \frac{5}{9} \\ 0 & 1 & | & \frac{4}{9} \end{bmatrix}$$

$$x = \frac{5}{9}, \ y = \frac{4}{9}$$

The stable distribution is $\begin{bmatrix} \frac{5}{9} \\ \frac{4}{9} \end{bmatrix}$.

8.
$$\begin{bmatrix} 1 & 0 & 0 & \frac{1}{8} & \frac{1}{4} \\ 0 & 1 & 0 & \frac{1}{8} & 0 \\ 0 & 0 & 1 & 0 & \frac{1}{4} \\ \hline 0 & 0 & 0 & \frac{1}{4} & \frac{1}{2} \\ 0 & 0 & 0 & \frac{1}{2} & 0 \end{bmatrix} = \begin{bmatrix} I & S \\ \hline 0 & R \end{bmatrix}$$

$$R = \begin{bmatrix} \frac{1}{4} & \frac{1}{2} \\ \frac{1}{2} & 0 \end{bmatrix}; S = \begin{bmatrix} \frac{1}{8} & \frac{1}{4} \\ \frac{1}{8} & 0 \\ 0 & \frac{1}{4} \end{bmatrix}$$

$$I - R = \begin{bmatrix} 1 & 0 \\ 0 & 1 \end{bmatrix} - \begin{bmatrix} \frac{1}{4} & \frac{1}{2} \\ \frac{1}{2} & 0 \end{bmatrix}$$

$$= \begin{bmatrix} \frac{3}{4} & -\frac{1}{2} \\ -\frac{1}{2} & 1 \end{bmatrix}$$

$$= \begin{bmatrix} a & b \\ c & d \end{bmatrix}$$

$$\Delta = ad - bc = \left(\frac{3}{4}\right)(1) - \left(-\frac{1}{2}\right)\left(-\frac{1}{2}\right)$$

$$= \frac{1}{2}$$

$$(I - R)^{-1} = \begin{bmatrix} \frac{d}{\Delta} & -\frac{b}{\Delta} \\ -\frac{c}{\Delta} & \frac{a}{\Delta} \end{bmatrix}$$

$$= \begin{bmatrix} \frac{1}{1/2} & \frac{-1/2}{1/2} \\ -\frac{-1/2}{1/2} & \frac{3/4}{1/2} \end{bmatrix}$$

$$= \begin{bmatrix} 2 & 1 \\ 1 & \frac{3}{2} \end{bmatrix}$$

$$S(I - R)^{-1} = \begin{bmatrix} \frac{1}{8} & \frac{1}{4} \\ \frac{1}{8} & 0 \\ 0 & \frac{1}{4} \end{bmatrix} \begin{bmatrix} 2 & 1 \\ 1 & \frac{3}{2} \end{bmatrix}$$

$$= \begin{bmatrix} \frac{1}{2} & \frac{1}{2} \\ \frac{1}{4} & \frac{1}{8} \\ \frac{1}{4} & \frac{3}{8} \end{bmatrix}$$

$$\begin{bmatrix} I & S(I-R)^{-1} \\ \hline 0 & 0 \end{bmatrix} = \begin{bmatrix} 1 & 0 & 0 & \frac{1}{2} & \frac{1}{2} \\ 0 & 1 & 0 & \frac{1}{4} & \frac{1}{8} \\ 0 & 0 & 1 & \frac{1}{4} & \frac{3}{8} \\ \hline 0 & 0 & 0 & 0 & 0 \\ 0 & 0 & 0 & 0 & 0 \end{bmatrix}$$

9. a.
$$\begin{array}{c} \quad H \quad M \quad L \\ \begin{array}{c} H \\ M \\ L \end{array} \begin{bmatrix} .5 & .4 & .3 \\ .4 & .3 & .5 \\ .1 & .3 & .2 \end{bmatrix} \end{array}$$

b.
$$\begin{bmatrix} .5 & .4 & .3 \\ .4 & .3 & .5 \\ .1 & .3 & .2 \end{bmatrix} \begin{bmatrix} .1 \\ .6 \\ .3 \end{bmatrix} = \begin{bmatrix} .38 \\ .37 \\ .25 \end{bmatrix}$$

38% of the children of the current generation will have high incomes.

c. Find the stable distribution.
$$\begin{cases} x + y + z = 1 \\ \begin{bmatrix} .5 & .4 & .3 \\ .4 & .3 & .5 \\ .1 & .3 & .2 \end{bmatrix} \begin{bmatrix} x \\ y \\ z \end{bmatrix} = \begin{bmatrix} x \\ y \\ z \end{bmatrix} \end{cases}$$

$$\begin{cases} x + y + z = 1 \\ .5x + .4y + .3z = x \\ .4x + .3y + .5z = y \\ .1x + .3y + .2z = z \end{cases}$$

$$\begin{cases} x + y + z = 1 \\ -.5x + .4y + .3z = 0 \\ .4x - .7y + .5z = 0 \\ .1x + .3y + .8z = 0 \end{cases}$$

The fourth equation is equivalent to the second plus third and so is redundant.

$$\begin{bmatrix} 1 & 1 & 1 & 1 \\ -.5 & .4 & .3 & 0 \\ .4 & -.7 & .5 & 0 \end{bmatrix} \xrightarrow[10[3]]{10[2]} \begin{bmatrix} 1 & 1 & 1 & 1 \\ -5 & 4 & 3 & 0 \\ 4 & -7 & 5 & 0 \end{bmatrix}$$

$$\xrightarrow[{[3]+(-4)[1]}]{[2]+5[1]} \begin{bmatrix} 1 & 1 & 1 & 1 \\ 0 & 9 & 8 & 5 \\ 0 & -11 & 1 & -4 \end{bmatrix}$$

$$\xrightarrow[{[3+(11)[2]}]{\frac{1}{9}[2] \\ [1]+(-1)[2]} \begin{bmatrix} 1 & 0 & \frac{1}{9} & \frac{4}{9} \\ 0 & 1 & \frac{8}{9} & \frac{5}{9} \\ 0 & 0 & \frac{97}{9} & \frac{19}{9} \end{bmatrix}$$

$$\xrightarrow[{[2]+\left(-\frac{8}{9}\right)[3]}]{\frac{9}{97}[3] \\ [1]+\left(-\frac{1}{9}\right)[3]} \begin{bmatrix} 1 & 0 & 0 & \frac{41}{97} \\ 0 & 1 & 0 & \frac{37}{97} \\ 0 & 0 & 1 & \frac{19}{97} \end{bmatrix}$$

$$x = \frac{41}{97}, y = \frac{37}{97}, z = \frac{19}{97}.$$

The stable distribution is $\begin{bmatrix} \frac{41}{97} \\ \frac{37}{97} \\ \frac{19}{97} \end{bmatrix}$. In the long

run, $\frac{19}{97}$ of the population will have low

incomes.

10. a. $\begin{array}{cc} & P \quad N \\ \begin{array}{c} P \\ N \end{array} & \begin{bmatrix} .8 & .3 \\ .2 & .7 \end{bmatrix} \end{array}$

b. $A^2 \begin{bmatrix} \\ \end{bmatrix}_0 = \begin{bmatrix} .8 & .3 \\ .2 & .7 \end{bmatrix} \begin{bmatrix} .8 & .3 \\ .2 & .7 \end{bmatrix} \begin{bmatrix} 1 \\ 0 \end{bmatrix}_0 = \begin{bmatrix} .7 \\ .3 \end{bmatrix}_2$

30% will need adjusting after 2 days.

c. Find the stable distribution.

$\begin{cases} x + y = 1 \\ \begin{bmatrix} .8 & .3 \\ .2 & .7 \end{bmatrix} \begin{bmatrix} x \\ y \end{bmatrix} = \begin{bmatrix} x \\ y \end{bmatrix} \end{cases}$

$\begin{cases} x + y = 1 \\ .8x + .3y = x \\ .2x + .7y = y \end{cases}$

$\begin{cases} x + y = 1 \\ -.2x + .3y = 0 \\ .2x - .3y = 0 \end{cases}$

The second and third equations in this system are equivalent.

$\begin{bmatrix} 1 & 1 & | & 1 \\ -.2 & .3 & | & 0 \end{bmatrix} \xrightarrow{[2]+.2[1]} \begin{bmatrix} 1 & 1 & | & 1 \\ 0 & .5 & | & .2 \end{bmatrix}$

$\xrightarrow{2[2]} \begin{bmatrix} 1 & 1 & | & 1 \\ 0 & 1 & | & .4 \end{bmatrix}$

$\xrightarrow{[1]+(-1)[2]} \begin{bmatrix} 1 & 0 & | & .6 \\ 0 & 1 & | & .4 \end{bmatrix}$

$x = .6$, $y = .4$

The stable distribution is $\begin{bmatrix} .6 \\ .4 \end{bmatrix}$. In the long

run, 60% will be properly adjusted.

11. $\begin{bmatrix} 1 & 0 & | & \frac{1}{6} & \frac{1}{2} & \frac{2}{5} \\ 0 & 1 & | & 0 & 0 & \frac{2}{5} \\ \hline 0 & 0 & | & 0 & 0 & 0 \\ 0 & 0 & | & \frac{2}{3} & \frac{1}{2} & 0 \\ 0 & 0 & | & \frac{1}{6} & 0 & \frac{1}{5} \end{bmatrix} = \begin{bmatrix} I & | & S \\ \hline 0 & | & R \end{bmatrix}$

$R = \begin{bmatrix} 0 & 0 & 0 \\ \frac{2}{3} & \frac{1}{2} & 0 \\ \frac{1}{6} & 0 & \frac{1}{5} \end{bmatrix}$; $S = \begin{bmatrix} \frac{1}{6} & \frac{1}{2} & \frac{2}{5} \\ 0 & 0 & \frac{2}{5} \end{bmatrix}$

$I - R = \begin{bmatrix} 1 & 0 & 0 \\ 0 & 1 & 0 \\ 0 & 0 & 1 \end{bmatrix} - \begin{bmatrix} 0 & 0 & 0 \\ \frac{2}{3} & \frac{1}{2} & 0 \\ \frac{1}{6} & 0 & \frac{1}{5} \end{bmatrix}$

$= \begin{bmatrix} 1 & 0 & 0 \\ -\frac{2}{3} & \frac{1}{2} & 0 \\ -\frac{1}{6} & 0 & \frac{4}{5} \end{bmatrix}$

Use the Gauss-Jordan method to find $(I - R)^{-1}$.

$\begin{bmatrix} 1 & 0 & 0 & | & 1 & 0 & 0 \\ -\frac{2}{3} & \frac{1}{2} & 0 & | & 0 & 1 & 0 \\ -\frac{1}{6} & 0 & \frac{4}{5} & | & 0 & 0 & 1 \end{bmatrix}$

$\begin{array}{c} [2]+\frac{2}{3}[1] \\ \xrightarrow{\hspace{1cm}} \\ [3]+\frac{1}{6}[1] \end{array} \begin{bmatrix} 1 & 0 & 0 & | & 1 & 0 & 0 \\ 0 & \frac{1}{2} & 0 & | & \frac{2}{3} & 1 & 0 \\ 0 & 0 & \frac{4}{5} & | & \frac{1}{6} & 0 & 1 \end{bmatrix}$

$\begin{array}{c} 2[2] \\ \xrightarrow{\hspace{1cm}} \\ \frac{5}{4}[3] \end{array} \begin{bmatrix} 1 & 0 & 0 & | & 1 & 0 & 0 \\ 0 & 1 & 0 & | & \frac{4}{3} & 2 & 0 \\ 0 & 0 & 1 & | & \frac{5}{24} & 0 & \frac{5}{4} \end{bmatrix}$

$S(I - R)^{-1} = \begin{bmatrix} \frac{1}{6} & \frac{1}{2} & \frac{2}{5} \\ 0 & 0 & \frac{2}{5} \end{bmatrix} \begin{bmatrix} 1 & 0 & 0 \\ \frac{4}{3} & 2 & 0 \\ \frac{5}{24} & 0 & \frac{5}{4} \end{bmatrix}$

$= \begin{bmatrix} \frac{11}{12} & 1 & \frac{1}{2} \\ \frac{1}{12} & 0 & \frac{1}{2} \end{bmatrix}$

$\begin{bmatrix} I & | & S(I-R)^{-1} \\ \hline 0 & | & 0 \end{bmatrix} = \begin{bmatrix} 1 & 0 & | & \frac{11}{12} & 1 & \frac{1}{2} \\ 0 & 1 & | & \frac{1}{12} & 0 & \frac{1}{2} \\ \hline 0 & 0 & | & 0 & 0 & 0 \\ 0 & 0 & | & 0 & 0 & 0 \\ 0 & 0 & | & 0 & 0 & 0 \end{bmatrix}$

12. a.

$$
\begin{array}{c}
\text{I} \\
\text{II} \\
\text{III} \\
\text{IV}
\end{array}
\begin{array}{cccc}
\text{I} & \text{II} & \text{III} & \text{IV}
\end{array}
\left[
\begin{array}{cc|cc}
1 & 0 & 0 & \frac{1}{4} \\
0 & 1 & \frac{1}{3} & \frac{1}{4} \\
0 & 0 & 0 & \frac{1}{2} \\
0 & 0 & \frac{2}{3} & 0
\end{array}
\right]
$$

b. The mouse will find the cheese after two minutes if he either goes to room II after one minute, or goes to room III after one minute and then room II after two minutes. Therefore the probability that the mouse finds the cheese after two minutes is

$$\frac{1}{4}+\frac{1}{2}\cdot\frac{1}{3}=\frac{5}{12}.$$

c. Find the stable matrix.

$$R=\begin{bmatrix}0 & \frac{1}{2} \\ \frac{2}{3} & 0\end{bmatrix}, \; S=\begin{bmatrix}0 & \frac{1}{4} \\ \frac{1}{3} & \frac{1}{4}\end{bmatrix}$$

$$I-R=\begin{bmatrix}1 & 0 \\ 0 & 1\end{bmatrix}-\begin{bmatrix}0 & \frac{1}{2} \\ \frac{2}{3} & 0\end{bmatrix}$$

$$=\begin{bmatrix}1 & -\frac{1}{2} \\ -\frac{2}{3} & 1\end{bmatrix}$$

$$=\begin{bmatrix}a & b \\ c & d\end{bmatrix}$$

$$\Delta = ad-bc = (1)(1)-\left(-\frac{2}{3}\right)\left(-\frac{1}{2}\right)=\frac{2}{3}$$

$$F=(I-R)^{-1}=\begin{bmatrix}\frac{d}{\Delta} & -\frac{b}{\Delta} \\ -\frac{c}{\Delta} & \frac{a}{\Delta}\end{bmatrix}$$

$$=\begin{bmatrix}\frac{1}{2/3} & \frac{-1/2}{2/3} \\ -\frac{2/3}{2/3} & \frac{1}{2/3}\end{bmatrix}$$

$$=\begin{bmatrix}\frac{3}{2} & \frac{3}{4} \\ 1 & \frac{3}{2}\end{bmatrix}$$

$$S(I-R)^{-1}=\begin{bmatrix}0 & \frac{1}{4} \\ \frac{1}{3} & \frac{1}{4}\end{bmatrix}\begin{bmatrix}\frac{3}{2} & \frac{3}{4} \\ 1 & \frac{3}{2}\end{bmatrix}=\begin{bmatrix}\frac{1}{4} & \frac{3}{8} \\ \frac{3}{4} & \frac{5}{8}\end{bmatrix}$$

The stable matrix is

$$
\begin{bmatrix}I & S(I-R)^{-1} \\ \hline 0 & 0\end{bmatrix}=
\begin{array}{c}
\text{I} \\ \text{II} \\ \text{III} \\ \text{IV}
\end{array}
\begin{array}{cccc}
\text{I} & \text{II} & \text{III} & \text{IV}
\end{array}
\left[
\begin{array}{cc|cc}
1 & 0 & \frac{1}{4} & \frac{3}{8} \\
0 & 1 & \frac{3}{4} & \frac{5}{8} \\
0 & 0 & 0 & 0 \\
0 & 0 & 0 & 0
\end{array}
\right]
$$

From the fourth column, if he starts in room IV the probability of finding cheese in the long run is $\frac{5}{8}=.625$.

d. The fundamental matrix is $\begin{array}{c}\text{III} \\ \text{IV}\end{array}\begin{array}{cc}\text{III} & \text{IV}\end{array}\begin{bmatrix}\frac{3}{2} & \frac{3}{4} \\ 1 & \frac{3}{2}\end{bmatrix}$. Add the entries in the column for room III:

$$\frac{3}{2}+1=\frac{5}{2}.$$

A mouse that starts in room III will spend an expected $2\frac{1}{2}$ minutes before finding the cheese or being trapped.

13. (c) is the correct choice because it satisfies the system $\begin{cases}x+y+z=1 \\ \begin{bmatrix}.4 & .4 & .2 \\ .1 & .1 & .3 \\ .5 & .5 & .5\end{bmatrix}\begin{bmatrix}x \\ y \\ z\end{bmatrix}=\begin{bmatrix}x \\ y \\ z\end{bmatrix}\end{cases}$

14. $A=\begin{array}{c}\text{A} \\ \text{B}\end{array}\begin{array}{cc}\text{A} & \text{B}\end{array}\begin{bmatrix}.9 & .2 \\ .1 & .8\end{bmatrix}$

$$A^2\begin{bmatrix} \\ \end{bmatrix}_0=\begin{bmatrix}.9 & .2 \\ .1 & .8\end{bmatrix}\begin{bmatrix}.9 & .2 \\ .1 & .8\end{bmatrix}\begin{bmatrix}.5 \\ .5\end{bmatrix}_0$$

$$=\begin{bmatrix}.585 \\ .415\end{bmatrix}_2$$

58.5% of the regular listeners will listen to station A two days from now.

15. a. If the traffic is moderate on a particular day then for the next day the probability of light traffic is .2, the probability of moderate traffic is .75, and the probability of heavy traffic is .05.

b. Find the stable distribution.

$$\begin{cases}x+y+z=1 \\ \begin{bmatrix}.70 & .20 & .10 \\ .20 & .75 & .30 \\ .10 & .05 & .60\end{bmatrix}\begin{bmatrix}x \\ y \\ z\end{bmatrix}=\begin{bmatrix}x \\ y \\ z\end{bmatrix}\end{cases}$$

$$\begin{cases}x + y + z = 1 \\ .7x + .2y + .1z = x \\ .2x + .75y + .3z = y \\ .1x + .05y + .6z = z\end{cases}$$

$$\begin{cases} x + y + z = 1 \\ -.3x + .2y + .1z = 0 \\ .2x - .25y + .3z = 0 \\ .1x + .05y - .4z = 0 \end{cases}$$

The fourth equation is equivalent to the second plus the third and so is redundant.

$$\begin{bmatrix} 1 & 1 & 1 & | & 1 \\ -.3 & .2 & .1 & | & 0 \\ .2 & -.25 & .3 & | & 0 \end{bmatrix} \xrightarrow[20[3]]{10[2]} \begin{bmatrix} 1 & 1 & 1 & | & 1 \\ -3 & 2 & 1 & | & 0 \\ 4 & -5 & 6 & | & 0 \end{bmatrix}$$

$$\xrightarrow[\substack{[2]+3[1] \\ [3]+(-4)[1]}]{} \begin{bmatrix} 1 & 1 & 1 & | & 1 \\ 0 & 5 & 4 & | & 3 \\ 0 & -9 & 2 & | & -4 \end{bmatrix}$$

$$\xrightarrow[\substack{\frac{1}{5}[2] \\ [1]+(-1)[2] \\ [3]+9[2]}]{} \begin{bmatrix} 1 & 0 & \frac{1}{5} & | & \frac{2}{5} \\ 0 & 1 & \frac{4}{5} & | & \frac{3}{5} \\ 0 & 0 & \frac{46}{5} & | & \frac{7}{5} \end{bmatrix}$$

$$\xrightarrow[\substack{\frac{5}{46}[3] \\ [1]+\left(-\frac{1}{5}\right)[3] \\ [2]+\left(-\frac{4}{5}\right)[3]}]{} \begin{bmatrix} 1 & 0 & 0 & | & \frac{17}{46} \\ 0 & 1 & 0 & | & \frac{11}{23} \\ 0 & 0 & 1 & | & \frac{7}{46} \end{bmatrix}$$

The stable distribution is $\begin{bmatrix} \frac{17}{46} \\ \frac{11}{23} \\ \frac{7}{46} \end{bmatrix}$ or about $\begin{bmatrix} .370 \\ .478 \\ .152 \end{bmatrix}$.

About 37.0% of workdays will have light traffic, 47.8% will have moderate traffic, and 15.2% will have heavy traffic.

c. $\dfrac{7}{46} \cdot 20 \approx 3.04$

About 3 workdays will have heavy traffic.

16.

$$\begin{matrix} & C & L & G & S \\ C & \begin{bmatrix} 1 & 0 & | & .60 & .05 \\ L & 0 & 1 & | & .10 & .40 \\ G & 0 & 0 & | & .20 & .50 \\ S & 0 & 0 & | & .10 & .05 \end{bmatrix} \end{matrix}$$

$$R = \begin{bmatrix} .20 & .50 \\ .10 & .05 \end{bmatrix}$$

$$I - R = \begin{bmatrix} 1 & 0 \\ 0 & 1 \end{bmatrix} - \begin{bmatrix} .20 & .50 \\ .10 & .05 \end{bmatrix}$$

$$= \begin{bmatrix} .80 & -.50 \\ -.10 & .95 \end{bmatrix}$$

$$= \begin{bmatrix} a & b \\ c & d \end{bmatrix}$$

$$\Delta = ad - bc = (.80)(.95) - (-.50)(-.10) = .71$$

$$F = (I - R)^{-1}$$

$$= \begin{bmatrix} \frac{d}{\Delta} & -\frac{b}{\Delta} \\ -\frac{c}{\Delta} & \frac{a}{\Delta} \end{bmatrix}$$

$$= \begin{bmatrix} \frac{.95}{.71} & -\frac{-.50}{.71} \\ -\frac{.10}{.71} & \frac{.80}{.71} \end{bmatrix}$$

$$= \begin{matrix} & G & S \\ G & \begin{bmatrix} \frac{95}{71} & \frac{50}{71} \\ S & \frac{10}{71} & \frac{80}{71} \end{bmatrix} \end{matrix}$$

Add the entries in each column:

$$\frac{95}{71} + \frac{10}{71} = \frac{105}{71} = 1\frac{34}{71} \approx 1.48$$

$$\frac{50}{71} + \frac{86}{71} = \frac{130}{71} = 1\frac{59}{71} \approx 1.83$$

A patient who begins in state G has an expected number of approximately 1.48 months; a patient who begins in state S has an expected number of approximately 1.83 months.

17. a.

$$\begin{cases} x + y + z + w = 1 \\ \begin{bmatrix} .20 & .10 & .05 & .05 \\ .30 & .20 & .20 & .30 \\ .40 & .40 & .50 & .40 \\ .10 & .30 & .25 & .25 \end{bmatrix} \begin{bmatrix} x \\ y \\ z \\ w \end{bmatrix} = \begin{bmatrix} x \\ y \\ z \\ w \end{bmatrix} \end{cases}$$

$$\begin{cases} x + y + z + w = l \\ .2x + .1y + .05z + .05w = x \\ .3x + .2y + .2z + .3w = y \\ .4x + .4y + .5z + .4w = z \\ .1x + .3y + .25z + .25w = w \end{cases}$$

$$\begin{cases} x + y + z + w = 1 \\ -.8x + .1y + .05z + .05w = 0 \\ .3x - .8y + .2z + .3w = 0 \\ .4x + .4y - .5z + .4w = 0 \\ .1x + .3y + .25z - .75w = 0 \end{cases}$$

The fifth equation is equivalent to the sum of the second, third and fourth and so is redundant.

$$\begin{bmatrix} 1 & 1 & 1 & 1 & 1 \\ -.8 & .1 & .05 & .05 & 0 \\ .3 & -.8 & .2 & .3 & 0 \\ .4 & .4 & -.5 & .4 & 0 \end{bmatrix} \begin{matrix} 20[2] \\ 10[3] \\ \overrightarrow{10[4]} \end{matrix} \begin{bmatrix} 1 & 1 & 1 & 1 & 1 \\ -16 & 2 & 1 & 1 & 0 \\ 3 & -8 & 2 & 3 & 0 \\ 4 & 4 & -5 & 4 & 0 \end{bmatrix} \begin{matrix} [2]+16[1] \\ [3]+(-3)[1] \\ \overrightarrow{[4]+(-4)[1]} \end{matrix} \begin{bmatrix} 1 & 1 & 1 & 1 & 1 \\ 0 & 18 & 17 & 17 & 16 \\ 0 & -11 & -1 & 0 & -3 \\ 0 & 0 & -9 & 0 & -4 \end{bmatrix}$$

$$\begin{matrix} \frac{1}{18}[2] \\ [1]+(-1)[2] \\ \overrightarrow{[3]+11[2]} \end{matrix} \begin{bmatrix} 1 & 0 & \frac{1}{18} & \frac{1}{18} & \frac{1}{9} \\ 0 & 1 & \frac{17}{18} & \frac{17}{18} & \frac{8}{9} \\ 0 & 0 & \frac{169}{18} & \frac{187}{18} & \frac{61}{9} \\ 0 & 0 & -9 & 0 & -4 \end{bmatrix} \begin{matrix} \frac{18}{169}[3] \\ [1]+(-\frac{1}{18})[3] \\ [2]+(-\frac{17}{18})[3] \\ \overrightarrow{[4]+9[3]} \end{matrix} \begin{bmatrix} 1 & 0 & 0 & -\frac{1}{169} & \frac{12}{169} \\ 0 & 1 & 0 & -\frac{17}{169} & \frac{35}{169} \\ 0 & 0 & 1 & \frac{187}{169} & \frac{122}{169} \\ 0 & 0 & 0 & \frac{1683}{169} & \frac{422}{169} \end{bmatrix}$$

$$\begin{matrix} \frac{169}{1683}[4] \\ [1]+\frac{1}{169}[4] \\ [2]+\frac{17}{169}[4] \\ \overrightarrow{[3]+(-\frac{187}{169})[4]} \end{matrix} \begin{bmatrix} 1 & 0 & 0 & 0 & \frac{122}{1683} \\ 0 & 1 & 0 & 0 & \frac{23}{99} \\ 0 & 0 & 1 & 0 & \frac{4}{9} \\ 0 & 0 & 0 & 1 & \frac{422}{1683} \end{bmatrix}$$

The stable distribution is $\begin{bmatrix} \frac{122}{1683} \\ \frac{23}{99} \\ \frac{4}{9} \\ \frac{422}{1683} \end{bmatrix}$.

In the long run, the probability of having 1, 2, 3, or 4 units of water in the reservoir at any given time will be $\frac{122}{1683}, \frac{23}{99}, \frac{4}{9},$ or $\frac{422}{1683}$, respectively.

b. $\frac{122}{1683}(\$4000) + \frac{23}{99}(\$6000) + \frac{4}{9}(\$10,000) + \frac{422}{1683}(\$3000) \approx \$6881$

The average weekly benefits will be about \$6881.

Conceptual Exercises

18. The entries in each column are the probabilities of the transitions to the various states from the state associated to the column. These must add up to 1, just as the probabilities of all branches emanating from any node in a tree diagram must add up to 1.

19. If we label the states I and II corresponding to the rows and columns, then .26 is the probability of reaching state II from state I after four time periods.

20. The stochastic matrix *A* represents a Markov process; A^2 represents this Markov process after two times periods, and so is also a stochastic matrix.

Chapter 8 Chapter Test

1. **a.** Stochastic

   b. Not stochastic; column sums are not equal to 1.

   c. Not stochastic; entries are not all greater than or equal to 0

   d. Stochastic

2. **a.** Not regular; all powers are identical to the original matrix and have 2 zeros.

   b. Regular; the matrix is stochastic, and all entries are positive.

   c. Regular; the matrix is stochastic, and the second power is $\begin{bmatrix} .76 & .4 \\ .24 & .6 \end{bmatrix}$, which has all positive entries.

3. (b) $\begin{bmatrix} .2 & .2 & .3 \\ .1 & .1 & .4 \\ .7 & .7 & .3 \end{bmatrix} \begin{bmatrix} .25 \\ .25 \\ .50 \end{bmatrix} = \begin{bmatrix} .25 \\ .25 \\ .50 \end{bmatrix}$

4. $\begin{cases} x + y = 1 \\ \begin{bmatrix} \frac{1}{5} & \frac{3}{5} \\ \frac{4}{5} & \frac{2}{5} \end{bmatrix} \begin{bmatrix} x \\ y \end{bmatrix} = \begin{bmatrix} x \\ y \end{bmatrix} \end{cases}$

$\begin{cases} x + y = 1 \\ \frac{1}{5}x + \frac{3}{5}y = x \\ \frac{4}{5}x + \frac{2}{5}y = y \end{cases}$

$\begin{cases} x + y = 1 \\ -\frac{4}{5}x + \frac{3}{5}y = 0 \\ \frac{4}{5}x - \frac{3}{5}y = 0 \end{cases}$

The second and third equations in this system are equivalent.

$\begin{bmatrix} 1 & 1 & | & 1 \\ -\frac{4}{5} & \frac{3}{5} & | & 0 \end{bmatrix} \xrightarrow{[2]+\frac{4}{5}[1]} \begin{bmatrix} 1 & 1 & | & 1 \\ 0 & \frac{7}{5} & | & \frac{4}{5} \end{bmatrix}$

$\xrightarrow{\frac{5}{7}[2]} \begin{bmatrix} 1 & 1 & | & 1 \\ 0 & 1 & | & \frac{4}{7} \end{bmatrix}$

$\xrightarrow{[1]+(-1)[2]} \begin{bmatrix} 1 & 0 & | & \frac{3}{7} \\ 0 & 1 & | & \frac{4}{7} \end{bmatrix}$

$x = \frac{3}{7}, y = \frac{4}{7}$

The stable distribution is $\begin{bmatrix} x \\ y \end{bmatrix} = \begin{bmatrix} \frac{3}{7} \\ \frac{4}{7} \end{bmatrix}$.

The stable matrix is $\begin{bmatrix} \frac{3}{7} & \frac{3}{7} \\ \frac{4}{7} & \frac{4}{7} \end{bmatrix}$.

5. a. $\begin{bmatrix} .70 & .60 \\ .30 & .40 \end{bmatrix}$

b. $\begin{bmatrix} .50 \\ .50 \end{bmatrix}$

c. $\begin{bmatrix} .70 & .60 \\ .30 & .40 \end{bmatrix} \begin{bmatrix} .50 \\ .50 \end{bmatrix} = \begin{bmatrix} .65 \\ .35 \end{bmatrix}$

$\begin{bmatrix} .70 & .60 \\ .30 & .40 \end{bmatrix} \begin{bmatrix} .65 \\ .35 \end{bmatrix} = \begin{bmatrix} .665 \\ .335 \end{bmatrix}$

66.5% will use Internet Explorer two days later.

d. $\begin{bmatrix} .70 & .60 \\ .30 & .40 \end{bmatrix} \begin{bmatrix} \frac{2}{3} \\ \frac{1}{3} \end{bmatrix} = \begin{bmatrix} \frac{2}{3} \\ \frac{1}{3} \end{bmatrix}$

e. Answers may vary. *Sample answer:* In the long run, $\frac{2}{3}$ of the students will use Internet Explorer on any given day.

6. a. Absorbing; states 1 and 2 are absorbing states, and it is possible to get to state 1 or state 2 from state 3.

b. Not absorbing; state 1 is absorbing, but it is not possible to get to state 1 from either state 2 or state 3, neither of which is an absorbing state.

c. Absorbing; state 1 is an absorbing state, state 2 leads to state 1, and state 3 can reach state 1 through state 2.

7. a.

	PC	Apple	None
PC	1	0	.3
Apple	0	1	.1
None	0	0	.6

b. $S = \begin{bmatrix} .3 \\ .1 \end{bmatrix}$ and $R = [.6]$

The stable matrix is

$\left[\begin{array}{cc|c} 1 & 0 & \\ 0 & 1 & S(I-R)^{-1} \\ \hline 0 & 0 & 0 \end{array} \right]$

$I - R = [1] - [.6] = [.4]$

$(I-R)^{-1} = \begin{bmatrix} \frac{1}{.4} \end{bmatrix}$

$S(I-R)^{-1} = \begin{bmatrix} .3 \\ .1 \end{bmatrix} \begin{bmatrix} \frac{1}{.4} \end{bmatrix} = \begin{bmatrix} \frac{.3}{.4} \\ \frac{.1}{.4} \end{bmatrix} = \begin{bmatrix} \frac{3}{4} \\ \frac{1}{4} \end{bmatrix}$

The stable matrix is given by

$\left[\begin{array}{cc|c} 1 & 0 & \frac{3}{4} \\ 0 & 1 & \frac{1}{4} \\ \hline 0 & 0 & 0 \end{array} \right]$.

c. 25%

    d. The fundamental matrix is

$$F = (I - R)^{-1} = \left[\frac{1}{.4}\right] = [2.5]$$

The expected number of years required for a math department to decide to set up its own computer lab is 2.5 years.

Chapter 9

Exercises 9.1

1. $R: \begin{bmatrix} -1 & -2 \\ \underline{0} & 3 \end{bmatrix}$, row 2;

$C: \begin{bmatrix} \underline{-1} & -2 \\ 0 & 3 \end{bmatrix}$, column 1

3. $R: \begin{bmatrix} \underline{-2} & 4 & 1 \\ \underline{-1} & 3 & 5 \\ \underline{-3} & 5 & 2 \end{bmatrix}$, row 2;

$C: \begin{bmatrix} -2 & 4 & 1 \\ \underline{-1} & 3 & \underline{5} \\ -3 & \underline{5} & 2 \end{bmatrix}$, column 1

5. $R: \begin{bmatrix} \underline{0} & 3 \\ \underline{-1} & 1 \\ -2 & \underline{-4} \end{bmatrix}$, row 1;

$C: \begin{bmatrix} \underline{0} & 3 \\ -1 & 1 \\ -2 & -4 \end{bmatrix}$, column 1

7. Row minima: $\begin{bmatrix} 1 & \underline{0} \\ 0 & \underline{-1} \end{bmatrix}$,

column maxima: $\begin{bmatrix} \underline{1} & 0 \\ 0 & -1 \end{bmatrix}$

a. Row 1, column 2

b. 0

9. $\begin{array}{c} \\ H \\ T \end{array} \begin{array}{c} H \quad\; T \\ \begin{bmatrix} 2 & -1 \\ -1 & -4 \end{bmatrix} \end{array}$

Row minima: $\begin{bmatrix} 2 & \underline{-1} \\ -1 & \underline{-4} \end{bmatrix}$,

column maxima: $\begin{bmatrix} \underline{2} & \underline{-1} \\ -1 & -4 \end{bmatrix}$

Row 1, column 2 is a saddle point, so the game is strictly determined. R should show heads, C should show tails.

11.
$$\begin{array}{c} \\ F \\ A \\ N \end{array} \begin{array}{c} \quad F \qquad\quad A \qquad\quad N \\ \begin{bmatrix} 8000 & -1000 & 1000 \\ -7000 & 4000 & -2000 \\ 3000 & 3000 & 2000 \end{bmatrix} \end{array}$$

Row minima:
$$\begin{bmatrix} 1000 & \underline{-1000} & 1000 \\ \underline{-7000} & 4000 & -2000 \\ 3000 & 3000 & \underline{2000} \end{bmatrix}$$

Column maxima:
$$\begin{bmatrix} \underline{8000} & -1000 & 1000 \\ -7000 & \underline{4000} & -2000 \\ 3000 & 3000 & \underline{2000} \end{bmatrix}$$

Row 3, column 3 is a saddle point, so the game is strictly determined. Both candidates should be neutral.

13.
$$\begin{array}{c} \\ 5 \\ 10 \end{array} \begin{array}{c} 6 \quad\;\; 7 \quad\;\; 8 \\ \begin{bmatrix} 1 & -5 & -5 \\ -6 & 3 & 2 \end{bmatrix} \end{array}$$

Row minima: $\begin{bmatrix} 1 & -5 & \underline{-5} \\ \underline{-6} & 3 & 2 \end{bmatrix}$

Column maxima: $\begin{bmatrix} \underline{1} & -5 & -5 \\ -6 & \underline{3} & \underline{2} \end{bmatrix}$

No saddle point, so the game is not strictly determined.

Exercises 9.2

1. a. $[.5\ .5]\begin{bmatrix} 3 & -1 \\ -7 & 5 \end{bmatrix}\begin{bmatrix} .5 \\ .5 \end{bmatrix} = [0]$

   b. $[1\ 0]\begin{bmatrix} 3 & -1 \\ -7 & 5 \end{bmatrix}\begin{bmatrix} .5 \\ .5 \end{bmatrix} = [1]$

   c. $[.3\ .7]\begin{bmatrix} 3 & -1 \\ -7 & 5 \end{bmatrix}\begin{bmatrix} .6 \\ .4 \end{bmatrix} = [-1.12]$

   d. $[.75\ .25]\begin{bmatrix} 3 & -1 \\ -7 & 5 \end{bmatrix}\begin{bmatrix} .2 \\ .8 \end{bmatrix} = [.5]$

   (b) is most advantageous to R.

3. $[.3\ .7]\begin{bmatrix} -20,000 & 0 \\ 0 & -50,000 \end{bmatrix}\begin{bmatrix} .2 \\ .8 \end{bmatrix} = [-29,200]$

   $29,200

5. The payoff matrix is $\begin{array}{c} \\ V \\ C \end{array}\begin{array}{cc} V & C \\ \begin{bmatrix} 0 & 2 \\ -1 & 0 \end{bmatrix} \end{array}$.

   $[.25\ .75]\begin{bmatrix} 0 & 2 \\ -1 & 0 \end{bmatrix}\begin{bmatrix} .4 \\ .6 \end{bmatrix} = [0]$

   Zero

7. $\begin{array}{c} C \\ R\begin{bmatrix} 3 & -1 \\ -2 & 2 \end{bmatrix} \end{array}$ max min of rows = -1

   $\begin{array}{c} C \\ R\begin{bmatrix} 3 & -1 \\ -2 & 2 \end{bmatrix} \end{array}$ min max of columns = 2

   $[0.3\ 0.7]\begin{bmatrix} 3 & -1 \\ -2 & 2 \end{bmatrix} \rightarrow \begin{bmatrix} -0.5 \\ 1.1 \end{bmatrix}[c_1\ \ c_2] \rightarrow -0.5c_1 + 1.1c_2 = 0 \rightarrow c_1 = \dfrac{1.1c_2}{0.5}$ (to be fair)

   $c_1 + c_2 = 1$

   Further, $\dfrac{11}{5}c_2 + c_2 = 1 \rightarrow \dfrac{16}{5}c_2 = 1 \rightarrow c_2 = \dfrac{5}{16} \rightarrow c_1 = \dfrac{11}{16}$

   For fair game, Carol's strategy is $\begin{bmatrix} \dfrac{11}{16} & \dfrac{5}{16} \end{bmatrix}$

9.
$$\begin{matrix} & C \\ R & \begin{bmatrix} 5 & -1 \\ -2 & 2 \end{bmatrix} \end{matrix} \quad \text{max min of rows} = -1$$

$$\begin{matrix} & C \\ R & \begin{bmatrix} 5 & -1 \\ -2 & 2 \end{bmatrix} \end{matrix} \quad \text{max min of columns} = 2 \qquad \text{Not strictly determined.}$$

$$\begin{bmatrix} 0.7 & 0.3 \end{bmatrix} \begin{bmatrix} 5 & -1 \\ -2 & 2 \end{bmatrix} = \begin{bmatrix} 2.9 \\ -0.1 \end{bmatrix}$$

$$\begin{bmatrix} 2.9 \\ -0.1 \end{bmatrix} \begin{bmatrix} c_1 & c_2 \end{bmatrix} \rightarrow 2.9c_1 - 0.1c_2 = 0 \rightarrow c_2 = 29c_1$$

Carol's strategy: $\begin{bmatrix} \dfrac{1}{30} & \dfrac{29}{30} \end{bmatrix}$

11.
$$\begin{matrix} & C \\ R & \begin{bmatrix} 1 & 2 & 4 \\ 1 & 0 & 5 \\ 0 & 1 & -1 \end{bmatrix} \end{matrix} \quad \text{max min of rows} = 1$$

$$\begin{matrix} & C \\ R & \begin{bmatrix} 1 & 2 & 4 \\ 1 & 0 & 5 \\ 0 & 1 & -1 \end{bmatrix} \end{matrix} \quad \text{min max of rows} = 1$$

Game strictly determined. Optimal strategy- play row 1, column 1.

$$R \rightarrow \begin{bmatrix} 1 & 0 & 0 \end{bmatrix}$$

$$C$$

$$\begin{bmatrix} 1 \\ 0 \\ 0 \end{bmatrix}$$

Value of game = \$1.00 in favor of Renee'. The game is not fair since the value of the game is different from zero.

$$[1 \quad 0 \quad 0]\begin{bmatrix} -3 & -2 & 6 \\ 2 & 0 & 2 \\ 5 & -2 & -4 \end{bmatrix} = [-3 \quad -2 \quad 6]$$

$$[-3 \quad -2 \quad 6]\begin{bmatrix} \frac{1}{3} \\ c_2 \\ c_3 \end{bmatrix} = 0 \rightarrow$$

$$-1 - 2c_2 + 6c_3 = 0$$

$$-2c_2 = -6c_3 + 1$$

$$c_2 = 3c_3 - \frac{1}{2}$$

$$c_2 + c_3 = \frac{2}{3}$$

$$3c_3 - \frac{1}{2} + c_3 = \frac{2}{3}$$

$$4c_3 = \frac{7}{6}$$

$$c_3 = \frac{7}{24} \rightarrow c_2 = \frac{9}{24}$$

Strategy: $\begin{bmatrix} \frac{1}{3} & \frac{3}{8} & \frac{7}{24} \end{bmatrix}$

Exercises 9.3

1. Objective function : Minimize $y_1 + y_2$

Constraints: $\begin{cases} y_1 + 4y_2 \geq 1 \\ 6y_1 + 3y_2 \geq 1 \\ y_1 \geq 0 \\ y_2 \geq 0 \end{cases}$

3. Maximize $M = z_1 + z_2$ subject to

$$\begin{cases} 2z_1 + 4z_2 \leq 1 \\ 5z_1 + 3z_2 \leq 1. \\ z_1 \geq 0,\ z_2 \geq 0 \end{cases}$$

	z_1	z_2	t	u	M	
t	2	4	1	0	0	1
u	5	3	0	1	0	1
M	−1	−1	0	0	1	0

	z_1	z_2	t	u	M	
t	0	$\frac{14}{5}$	1	$-\frac{2}{5}$	0	$\frac{3}{5}$
z_1	1	$\frac{3}{5}$	0	$\frac{1}{5}$	0	$\frac{1}{5}$
M	0	$-\frac{2}{5}$	0	$\frac{1}{5}$	1	$\frac{1}{5}$

	z_1	z_2	t	u	M	
z_2	0	1	$\frac{5}{14}$	$-\frac{1}{7}$	0	$\frac{3}{14}$
z_1	1	0	$-\frac{3}{14}$	$\frac{2}{7}$	0	$\frac{1}{14}$
M	0	0	$\frac{1}{7}$	$\frac{1}{7}$	1	$\frac{2}{7}$

$$z_1 = \frac{3}{14},\ z_2 = \frac{1}{14},\ M = \frac{2}{7},\ v = \frac{1}{M} = \frac{7}{2},$$

and the optimal strategy for C is

$$\begin{bmatrix} vz_1 \\ vz_2 \end{bmatrix} = \begin{bmatrix} \frac{1}{4} \\ \frac{3}{4} \end{bmatrix}.$$ The optimal strategy

for R is given by the bottom entries under t and u:

$$[vt \quad vu] = \begin{bmatrix} \frac{7}{2} \cdot \frac{1}{7} & \frac{7}{2} \cdot \frac{1}{7} \end{bmatrix} = \begin{bmatrix} \frac{1}{2} & \frac{1}{2} \end{bmatrix}.$$

5. Add 7 to each entry to make all the entries positive. We get $\begin{bmatrix} 10 & 1 \\ 2 & 11 \end{bmatrix}$. Then maximize $M = z_1 + z_2$ subject to

$$\begin{cases} 10z_1 + z_2 \le 1 \\ 2z_1 + 11z_2 \le 1. \\ z_1 \ge 0, \, z_2 \ge 0 \end{cases}$$

$$\begin{array}{c} \begin{array}{ccccc} z_1 & z_2 & t & u & M \end{array} \\ \begin{array}{c} t \\ u \\ M \end{array} \left[\begin{array}{ccccc|c} \underline{10} & 1 & 1 & 0 & 0 & 1 \\ 2 & 11 & 0 & 1 & 0 & 1 \\ -1 & -1 & 0 & 0 & 1 & 0 \end{array} \right] \end{array}$$

$$\begin{array}{c} \begin{array}{ccccc} z_1 & z_2 & t & u & M \end{array} \\ \begin{array}{c} z_1 \\ u \\ M \end{array} \left[\begin{array}{ccccc|c} 1 & \frac{1}{10} & \frac{1}{10} & 0 & 0 & \frac{1}{10} \\ 0 & \frac{54}{5} & -\frac{1}{5} & 1 & 0 & \frac{4}{5} \\ 0 & -\frac{9}{10} & \frac{1}{10} & 0 & 1 & \frac{1}{10} \end{array} \right] \end{array}$$

$$\begin{array}{c} \begin{array}{ccccc} z_1 & z_2 & t & u & M \end{array} \\ \begin{array}{c} z_1 \\ z_2 \\ M \end{array} \left[\begin{array}{ccccc|c} 1 & 0 & \frac{11}{108} & -\frac{1}{108} & 0 & \frac{5}{54} \\ 0 & 1 & -\frac{1}{54} & \frac{5}{54} & 0 & \frac{2}{27} \\ 0 & 0 & \frac{1}{12} & \frac{1}{12} & 1 & \frac{1}{6} \end{array} \right] \end{array}$$

$z_1 = \dfrac{5}{54}$, $z_2 = \dfrac{2}{27}$, $M = \dfrac{1}{6}$, $v = \dfrac{1}{M} = 6$, and the optimal strategy for C is

$\begin{bmatrix} vz_1 \\ vz_2 \end{bmatrix} = \begin{bmatrix} \frac{5}{9} \\ \frac{4}{9} \end{bmatrix}$. The optimal strategy for R is given by the bottom entries under t and u:

$[vt \quad vu] = \left[6 \cdot \frac{1}{12} \quad 6 \cdot \frac{1}{12} \right] = \left[\frac{1}{2} \quad \frac{1}{2} \right]$

7. Maximize $M = z_1 + z_2$ subject to

$$\begin{cases} 4z_1 + z_2 \le 1 \\ 2z_1 + 4z_2 \le 1 \, . \\ z_1 \ge 0, \, z_2 \ge 0 \end{cases}$$

$$\begin{array}{c} \begin{array}{ccccc} z_1 & z_2 & t & u & M \end{array} \\ \begin{array}{c} t \\ u \\ M \end{array} \left[\begin{array}{ccccc|c} 4 & 1 & 1 & 0 & 0 & 1 \\ 2 & \underline{4} & 0 & 1 & 0 & 1 \\ -1 & -1 & 0 & 0 & 1 & 0 \end{array} \right] \end{array}$$

$$\begin{array}{c} \begin{array}{ccccc} z_1 & z_2 & t & u & M \end{array} \\ \begin{array}{c} t \\ z_2 \\ M \end{array} \left[\begin{array}{ccccc|c} \frac{7}{2} & 0 & 1 & -\frac{1}{4} & 0 & \frac{3}{4} \\ \frac{1}{2} & 1 & 0 & \frac{1}{4} & 0 & \frac{1}{4} \\ -\frac{1}{2} & 0 & 0 & \frac{1}{4} & 1 & \frac{1}{4} \end{array} \right] \end{array}$$

$$\begin{array}{c} \begin{array}{ccccc} z_1 & z_2 & t & u & M \end{array} \\ \begin{array}{c} z_1 \\ z_2 \\ M \end{array} \left[\begin{array}{ccccc|c} 1 & 0 & \frac{2}{7} & -\frac{1}{14} & 0 & \frac{3}{14} \\ 0 & 1 & -\frac{1}{7} & \frac{2}{7} & 0 & \frac{1}{7} \\ 0 & 0 & \frac{1}{7} & \frac{3}{14} & 1 & \frac{5}{14} \end{array} \right] \end{array}$$

$z_1 = \dfrac{3}{14}$, $z_2 = \dfrac{1}{7}$, $M = \dfrac{5}{14}$, $v = \dfrac{1}{M} = \dfrac{14}{5}$, and the optimal strategy for C is

$\begin{bmatrix} vz_1 \\ vz_2 \end{bmatrix} = \begin{bmatrix} \frac{3}{5} \\ \frac{2}{5} \end{bmatrix}$.

The optimal strategy for R is given by the bottom entries under t and u:

$[vt \quad vu] = \left[\frac{14}{5} \cdot \frac{1}{7} \quad \frac{14}{5} \cdot \frac{3}{14} \right] = \left[\frac{2}{5} \quad \frac{3}{5} \right]$

9. Add 2 to each entry to make all entries of the payoff matrix.

$$\begin{bmatrix} 5 & 7 & 1 \\ 6 & 1 & 8 \end{bmatrix}$$

Set the tableaux up to find C's optimal strategy, then read the dual's solution off the bottom row:

	z_1	z_2	z_3	t	u	M	
t	5	$\underline{7}$	1	1	0	0	1
u	6	1	7	0	1	0	1
M	-1	-1	-1	0	0	1	0

	z_1	z_2	z_3	t	u	M	
z_2	$\frac{5}{7}$	1	$\frac{1}{7}$	$\frac{1}{7}$	0	0	$\frac{1}{7}$
u	$\frac{37}{7}$	0	$\underline{\frac{55}{7}}$	$-\frac{1}{7}$	1	0	$\frac{6}{7}$
M	$-\frac{2}{7}$	0	$-\frac{6}{7}$	$\frac{1}{7}$	0	1	$\frac{1}{7}$

	z_1	z_2	z_3	t	u	M	
z_2	$\frac{34}{55}$	1	0	$\frac{8}{55}$	$-\frac{1}{55}$	0	$\frac{7}{55}$
z_3	$\frac{37}{55}$	0	1	$-\frac{1}{55}$	$\frac{7}{55}$	0	$\frac{6}{55}$
M	$\frac{16}{55}$	0	0	$\frac{7}{55}$	$\frac{6}{55}$	1	$\frac{13}{55}$

$t = \dfrac{7}{55}, u = \dfrac{6}{55}, M = \dfrac{13}{55}, v = \dfrac{1}{M} = \dfrac{55}{13},$

and R's optimal strategy is

$[vt \quad vu] = \begin{bmatrix} \frac{7}{13} & \frac{6}{13} \end{bmatrix}.$

11. Add 4 to each entry to make them all positive. Maximize $M = z_1 + z_2$ subject

to $\begin{cases} z_1 + 5z_2 \le 1 \\ 8z_1 + 3z_2 \le 1 \\ 5z_1 + 4z_2 \le 1 \\ z_1 \ge 0, z_2 \ge 0 \end{cases}.$

	z_1	z_2	s	t	u	M	
s	1	$\underline{5}$	1	0	0	0	1
t	8	3	0	1	0	0	1
u	5	4	0	0	1	0	1
M	-1	-1	0	0	0	1	0

	z_1	z_2	s	t	u	M	
z_2	$\frac{1}{5}$	1	$\frac{1}{5}$	0	0	0	$\frac{1}{5}$
t	$\frac{37}{5}$	0	$-\frac{3}{5}$	1	0	0	$\frac{2}{5}$
u	$\frac{21}{5}$	0	$-\frac{4}{5}$	0	1	0	$\frac{1}{5}$
M	$-\frac{4}{5}$	0	$\frac{1}{5}$	0	0	1	$\frac{1}{5}$

	z_1	z_2	s	t	u	M	
z_2	0	1	$\frac{8}{37}$	$-\frac{1}{37}$	0	0	$\frac{7}{37}$
z_1	1	0	$-\frac{3}{37}$	$\frac{5}{37}$	0	0	$\frac{2}{37}$
u	0	0	$-\frac{17}{37}$	$-\frac{21}{37}$	1	0	$-\frac{1}{37}$
M	0	0	$\frac{5}{37}$	$\frac{4}{37}$	0	1	$\frac{9}{37}$

$z_1 = \dfrac{2}{37}, z_2 = \dfrac{7}{37}, M = \dfrac{9}{37}, v = \dfrac{1}{M} = \dfrac{37}{9},$

and C's optimal strategy is $\begin{bmatrix} vz_1 \\ vz_2 \end{bmatrix} = \begin{bmatrix} \frac{2}{9} \\ \frac{7}{9} \end{bmatrix}.$

13. The original matrix is not strictly determined.

The original matrix is $\begin{bmatrix} 5 & -3 \\ -3 & 1 \end{bmatrix}$. Add

4 to each entry to make all the entries positive. We get $\begin{bmatrix} 9 & 1 \\ 1 & 5 \end{bmatrix}$. Then maximize

$M = z_1 + z_2$ subject to $\begin{cases} 9z_1 + z_2 \leq 1 \\ z_1 + 5z_2 \leq 1 \\ z_1 \geq 0, \ z_2 \geq 0 \end{cases}$.

$$
\begin{array}{c}
\begin{array}{ccccc} z_1 & z_2 & t & u & M \end{array} \\
\begin{array}{c} t \\ u \\ M \end{array}
\left[\begin{array}{ccccc|c}
9 & 1 & 1 & 0 & 0 & 1 \\
1 & 5 & 0 & 1 & 0 & 1 \\
-1 & -1 & 0 & 0 & 1 & 0
\end{array}\right]
\end{array}
$$

$$
\begin{array}{c}
\begin{array}{ccccc} z_1 & z_2 & t & u & M \end{array} \\
\begin{array}{c} z_2 \\ u \\ M \end{array}
\left[\begin{array}{ccccc|c}
9 & 1 & 1 & 0 & 0 & 1 \\
-44 & 0 & -5 & 1 & 0 & -4 \\
8 & 0 & 1 & 0 & 1 & 1
\end{array}\right]
\end{array}
$$

$$
\begin{array}{c}
\begin{array}{ccccc} z_1 & z_2 & t & u & M \end{array} \\
\begin{array}{c} z_2 \\ z_1 \\ M \end{array}
\left[\begin{array}{ccccc|c}
0 & 1 & -\frac{1}{44} & -\frac{9}{44} & 0 & \frac{2}{11} \\
1 & 0 & \frac{5}{44} & -\frac{1}{44} & 0 & \frac{1}{11} \\
0 & 0 & \frac{1}{11} & \frac{2}{11} & 1 & \frac{3}{11}
\end{array}\right]
\end{array}
$$

$$v = \frac{1}{M} = \frac{11}{3}$$

Renee's optimal strategy is given by the bottom entries under t and u:

$$[vt \quad vu] = \begin{bmatrix} \frac{1}{3} & \frac{2}{3} \end{bmatrix}$$

Carlos' optimal strategy is $\begin{bmatrix} vz_1 \\ vz_2 \end{bmatrix} = \begin{bmatrix} \frac{1}{3} \\ \frac{2}{3} \end{bmatrix}$.

The value of the game is

$$v - 4 = \frac{1}{M} - 4 = \frac{11}{3} - 4 = -\frac{1}{3}.$$

15. The payoff matrix, with entries in

thousands of dollars, is $\begin{array}{c} \\ 1 \\ 2 \end{array}\begin{array}{c} 1 \quad 2 \\ \begin{bmatrix} -2 & 7 \\ 7 & -1 \end{bmatrix} \end{array}$. Add

3 to each entry to make all the entries positive. We get $\begin{bmatrix} 1 & 10 \\ 10 & 2 \end{bmatrix}$. Then

maximize $M = z_1 + z_2$ subject to

$$\begin{cases} z_1 + 10z_2 \leq 1 \\ 10z_1 + 2z_2 \leq 1. \\ z_1 \geq 0, \ z_2 \geq 0 \end{cases}$$

$$
\begin{array}{c}
\begin{array}{ccccc} z_1 & z_2 & t & u & M \end{array} \\
\begin{array}{c} t \\ u \\ M \end{array}
\left[\begin{array}{ccccc|c}
1 & 10 & 1 & 0 & 0 & 1 \\
10 & 2 & 0 & 1 & 0 & 1 \\
-1 & -1 & 0 & 0 & 1 & 0
\end{array}\right]
\end{array}
$$

$$
\begin{array}{c}
\begin{array}{ccccc} z_1 & z_2 & t & u & M \end{array} \\
\begin{array}{c} z_2 \\ u \\ M \end{array}
\left[\begin{array}{ccccc|c}
\frac{1}{10} & 1 & \frac{1}{10} & 0 & 0 & \frac{1}{10} \\
\frac{49}{5} & 0 & -\frac{1}{5} & 1 & 0 & \frac{4}{5} \\
-\frac{9}{10} & 0 & \frac{1}{10} & 0 & 1 & \frac{1}{10}
\end{array}\right]
\end{array}
$$

$$
\begin{array}{c}
\begin{array}{ccccc} z_1 & z_2 & t & u & M \end{array} \\
\begin{array}{c} z_2 \\ z_1 \\ M \end{array}
\left[\begin{array}{ccccc|c}
0 & 1 & \frac{5}{49} & -\frac{1}{98} & 0 & \frac{9}{98} \\
1 & 0 & -\frac{1}{49} & \frac{5}{49} & 0 & \frac{4}{49} \\
0 & 0 & \frac{4}{49} & \frac{9}{98} & 1 & \frac{17}{98}
\end{array}\right]
\end{array}
$$

$$v = \frac{1}{M} = \frac{98}{17}$$

a. R's optimal strategy is given by the bottom entries under t and u:

$$[vt \quad vu] = \begin{bmatrix} \frac{8}{17} & \frac{9}{17} \end{bmatrix}$$

b. C's optimal strategy is $\begin{bmatrix} vz_1 \\ vz_2 \end{bmatrix} = \begin{bmatrix} \frac{8}{17} \\ \frac{9}{17} \end{bmatrix}$.

c. $v - 3 = \frac{1}{M} - 3 = \frac{98}{17} - 3 \approx 2.765$

The value of the game is about $2765.

17. Add 4 to each entry to make them all positive. Maximize $M = z_1 + z_2$

$$\text{subject to } \begin{cases} 2z_1 + 5z_2 \le 1 \\ 6z_1 + z_2 \le 1 \\ 5z_1 + 2z_2 \le 1 \\ z_1 \ge 0,\ z_2 \ge 0 \end{cases}.$$

$$\begin{array}{c}\begin{array}{cccccc} z_1 & z_2 & s & t & u & M \end{array}\\ \begin{array}{c} s \\ t \\ u \\ M \end{array} \left[\begin{array}{cccccc|c} 2 & \underline{5} & 1 & 0 & 0 & 0 & 1 \\ 6 & 1 & 0 & 1 & 0 & 0 & 1 \\ 5 & 2 & 0 & 0 & 1 & 0 & 1 \\ \hline -1 & -1 & 0 & 0 & 0 & 1 & 0 \end{array}\right]\end{array}$$

$$\begin{array}{c}\begin{array}{cccccc} z_1 & z_2 & s & t & u & M \end{array}\\ \begin{array}{c} z_2 \\ t \\ u \\ M \end{array} \left[\begin{array}{cccccc|c} \frac{2}{5} & 1 & \frac{1}{5} & 0 & 0 & 0 & \frac{1}{5} \\ \frac{28}{5} & 0 & -\frac{1}{5} & 1 & 0 & 0 & \frac{4}{5} \\ \frac{21}{5} & 0 & -\frac{2}{5} & 0 & 1 & 0 & \frac{3}{5} \\ \hline -\frac{3}{5} & 0 & \frac{1}{5} & 0 & 0 & 1 & \frac{1}{5} \end{array}\right]\end{array}$$

$$\begin{array}{c}\begin{array}{cccccc} z_1 & z_2 & s & t & u & M \end{array}\\ \begin{array}{c} z_2 \\ z_1 \\ u \\ M \end{array} \left[\begin{array}{cccccc|c} 0 & 1 & \frac{3}{14} & -\frac{1}{14} & 0 & 0 & \frac{1}{7} \\ 1 & 0 & -\frac{1}{28} & \frac{5}{28} & 0 & 0 & \frac{1}{7} \\ 0 & 0 & -\frac{1}{4} & -\frac{3}{4} & 1 & 0 & 0 \\ \hline 0 & 0 & \frac{5}{28} & \frac{3}{28} & 0 & 1 & \frac{2}{7} \end{array}\right]\end{array}$$

$$v = \frac{1}{M} = \frac{7}{2}$$

Rosedale's optimal strategy is given by the bottom entries under s, t and u:

$$[vs \quad vt \quad vu] = \left[\begin{array}{ccc} \frac{5}{8} & \frac{3}{8} & 0 \end{array}\right]$$

Carter's optimal strategy is

$$\begin{bmatrix} vz_1 \\ vz_2 \end{bmatrix} = \begin{bmatrix} \frac{1}{2} \\ \frac{1}{2} \end{bmatrix}.$$

Chapter 9 Supplementary Exercises

1. Row minima: $\begin{bmatrix} 5 & \underline{-1} & 1 \\ \underline{-3} & 5 & 1 \\ 4 & 3 & \underline{2} \end{bmatrix}$,

column maxima: $\begin{bmatrix} \underline{5} & -1 & 1 \\ -3 & \underline{5} & 1 \\ 4 & 3 & \underline{2} \end{bmatrix}$

The game is strictly determined, with a saddle point at row 3, column 3 and a value of 2.

2. Row minima: $\begin{bmatrix} \underline{1} & 2 & 3 \\ 3 & 2 & \underline{1} \end{bmatrix}$,

column maxima: $\begin{bmatrix} 1 & \underline{2} & 3 \\ \underline{3} & 2 & 1 \end{bmatrix}$

The game has no saddle point and so is not strictly determined.

3. Row minima: $\begin{bmatrix} \underline{0} & 1 \\ 1 & \underline{0} \\ 2 & \underline{-1} \end{bmatrix}$,

column maxima: $\begin{bmatrix} 0 & \underline{1} \\ 1 & 0 \\ \underline{2} & -1 \end{bmatrix}$

The game has no saddle point and so is not strictly determined.

4. Row minima: $\begin{bmatrix} 2 & 1 & 2 \\ \underline{-1} & 0 & 3 \\ 4 & 1 & \underline{-4} \end{bmatrix}$,

column maxima: $\begin{bmatrix} 2 & \underline{1} & 2 \\ -1 & 0 & \underline{3} \\ \underline{4} & \underline{1} & -4 \end{bmatrix}$

The game is strictly determined, with a saddle point at row 1, column 2 and a value of 1.

5. $\begin{bmatrix} \frac{3}{4} & \frac{1}{4} \end{bmatrix}\begin{bmatrix} 0 & 24 \\ 12 & -36 \end{bmatrix}\begin{bmatrix} \frac{1}{3} \\ \frac{2}{3} \end{bmatrix} = [7]$

7

6. $\begin{bmatrix} \frac{1}{2} & \frac{1}{2} \end{bmatrix} \begin{bmatrix} -6 & 6 & 0 \\ 0 & -12 & 24 \end{bmatrix} \begin{bmatrix} \frac{1}{3} \\ \frac{1}{3} \\ \frac{1}{3} \end{bmatrix} = [2]$

2

7. $[.2 \ .3 \ .5] \begin{bmatrix} 1 & 0 \\ -3 & 1 \\ 0 & 5 \end{bmatrix} \begin{bmatrix} .4 \\ .6 \end{bmatrix} = [1.4]$

1.4

8. $[.1 \ .1 \ .8] \begin{bmatrix} 0 & 1 & 3 \\ -1 & 0 & 2 \\ -3 & -2 & 0 \end{bmatrix} \begin{bmatrix} .4 \\ .3 \\ .3 \end{bmatrix} = [-1.3]$

−1.3

9. Add 4 to each entry to get $\begin{bmatrix} 1 & 8 \\ 6 & 2 \end{bmatrix}$. Then use the simplex method.

$$\begin{array}{c} \\ t \\ u \\ M \end{array} \begin{array}{cccccc} z_1 & z_2 & t & u & M & \\ \left[\begin{array}{ccccc|c} 1 & 8 & 1 & 0 & 0 & 1 \\ \underline{6} & 2 & 0 & 1 & 0 & 1 \\ \hline -1 & -1 & 0 & 0 & 1 & 0 \end{array}\right] \end{array}$$

$$\begin{array}{c} \\ t \\ z_1 \\ M \end{array} \begin{array}{cccccc} z_1 & z_2 & t & u & M & \\ \left[\begin{array}{ccccc|c} 0 & \frac{23}{3} & 1 & -\frac{1}{6} & 0 & \frac{5}{7} \\ 1 & \frac{1}{3} & 0 & \frac{1}{6} & 0 & \frac{1}{6} \\ \hline 0 & -\frac{2}{3} & 0 & \frac{1}{6} & 1 & \frac{1}{6} \end{array}\right] \end{array}$$

$$\begin{array}{c} \\ z_2 \\ z_1 \\ M \end{array} \begin{array}{cccccc} z_1 & z_2 & t & u & M & \\ \left[\begin{array}{ccccc|c} 0 & 1 & \frac{3}{23} & -\frac{1}{46} & 0 & \frac{5}{46} \\ 1 & 0 & -\frac{1}{23} & \frac{4}{23} & 0 & \frac{3}{23} \\ \hline 0 & 0 & \frac{2}{23} & \frac{7}{46} & 1 & \frac{11}{46} \end{array}\right] \end{array}$$

$v = \dfrac{1}{M} = \dfrac{46}{11}$

R's optimal strategy:

$[vt \ \ vu] = \begin{bmatrix} \frac{46}{11} \cdot \frac{2}{23} & \frac{46}{11} \cdot \frac{7}{46} \end{bmatrix} = \begin{bmatrix} \frac{4}{11} & \frac{7}{11} \end{bmatrix}$

C's optimal strategy:

$\begin{bmatrix} vz_1 \\ vz_2 \end{bmatrix} = \begin{bmatrix} \frac{46}{11} \cdot \frac{3}{23} \\ \frac{46}{11} \cdot \frac{5}{46} \end{bmatrix} = \begin{bmatrix} \frac{6}{11} \\ \frac{5}{11} \end{bmatrix}$

10. Add 7 to each entry to get $\begin{bmatrix} 10 & 1 \\ 3 & 11 \end{bmatrix}$.

Then apply the simplex method.

$$\begin{array}{c} \\ \\ \\ M \end{array} \begin{array}{cccccc} z_1 & z_2 & t & u & M & \\ \left[\begin{array}{ccccc|c} \underline{10} & 1 & 1 & 0 & 0 & 1 \\ 3 & 11 & 0 & 1 & 0 & 1 \\ \hline -1 & -1 & 0 & 0 & 1 & 0 \end{array}\right] \end{array}$$

$$\begin{array}{c} \\ z_1 \\ u \\ M \end{array} \begin{array}{cccccc} z_1 & z_2 & t & u & M & \\ \left[\begin{array}{ccccc|c} 1 & \frac{1}{10} & \frac{1}{10} & 0 & 0 & \frac{1}{10} \\ 0 & \frac{107}{10} & -\frac{3}{10} & 1 & 0 & \frac{7}{10} \\ \hline 0 & -\frac{9}{10} & \frac{1}{10} & 0 & 1 & \frac{1}{10} \end{array}\right] \end{array}$$

$$\begin{array}{c} \\ z_1 \\ z_2 \\ M \end{array} \begin{array}{cccccc} z_1 & z_2 & t & u & M & \\ \left[\begin{array}{ccccc|c} 1 & 0 & \frac{11}{107} & -\frac{1}{107} & 0 & \frac{10}{107} \\ 0 & 1 & -\frac{3}{107} & \frac{10}{107} & 0 & \frac{7}{107} \\ \hline 0 & 0 & \frac{8}{107} & \frac{9}{107} & 1 & \frac{17}{107} \end{array}\right] \end{array}$$

$v = \dfrac{1}{M} = \dfrac{107}{17}$

R's optimal strategy:

$[vt \ \ vu] = \begin{bmatrix} \frac{107}{17} \cdot \frac{8}{107} & \frac{107}{17} \cdot \frac{9}{107} \end{bmatrix} = \begin{bmatrix} \frac{8}{17} & \frac{9}{17} \end{bmatrix}$

C's optimal strategy:

$\begin{bmatrix} vz_1 \\ vz_2 \end{bmatrix} = \begin{bmatrix} \frac{107}{17} \cdot \frac{10}{107} \\ \frac{107}{17} \cdot \frac{7}{107} \end{bmatrix} = \begin{bmatrix} \frac{10}{17} \\ \frac{7}{17} \end{bmatrix}$

11. Row 2, column 3 is a saddle point: [0 1]

12.

$$\begin{array}{c} \\ s \\ t \\ u \\ M \end{array} \begin{array}{ccccccc} z_1 & z_2 & s & t & u & M & \\ \left[\begin{array}{cccccc|c} 1 & 3 & 1 & 0 & 0 & 0 & 1 \\ 3 & 1 & 0 & 1 & 0 & 0 & 1 \\ \underline{4} & 2 & 0 & 0 & 1 & 0 & 1 \\ \hline -1 & -1 & 0 & 0 & 0 & 1 & 0 \end{array}\right] \end{array}$$

$$\begin{array}{c} \\ s \\ t \\ z_1 \\ M \end{array} \begin{array}{ccccccc} z_1 & z_2 & s & t & u & M & \\ \left[\begin{array}{cccccc|c} 0 & \frac{5}{2} & 1 & 0 & -\frac{1}{4} & 0 & \frac{3}{4} \\ 0 & -\frac{1}{2} & 0 & 1 & -\frac{3}{4} & 0 & \frac{1}{4} \\ 1 & \frac{1}{2} & 0 & 0 & \frac{1}{4} & 0 & \frac{1}{4} \\ \hline 0 & -\frac{1}{2} & 0 & 0 & \frac{1}{4} & 1 & \frac{1}{4} \end{array}\right] \end{array}$$

$$\begin{array}{c}\quad\quad z_1 \quad z_2 \quad s \quad t \quad u \quad M \\ \begin{array}{c}z_2\\t\\z_1\\M\end{array}\left[\begin{array}{cccccc|c} 0 & 1 & \frac{2}{5} & 0 & -\frac{1}{10} & 0 & \frac{3}{10} \\ 0 & 0 & \frac{1}{5} & 1 & -\frac{4}{5} & 0 & \frac{2}{5} \\ 1 & 0 & -\frac{1}{5} & 0 & \frac{3}{10} & 0 & \frac{1}{10} \\ \hline 0 & 0 & \frac{1}{5} & 0 & \frac{1}{5} & 1 & \frac{2}{5} \end{array}\right]\end{array}$$

$$v = \frac{1}{M} = \frac{5}{2}$$

$$\begin{bmatrix} vz_1 \\ vz_2 \end{bmatrix} = \begin{bmatrix} \frac{5}{2}\cdot\frac{1}{10} \\ \frac{5}{2}\cdot\frac{3}{10} \end{bmatrix} = \begin{bmatrix} \frac{1}{4} \\ \frac{3}{4} \end{bmatrix}$$

13. The payoff matrix is $\begin{array}{c}\\2\\6\end{array}\begin{array}{c}2\quad 6\\\left[\begin{array}{cc}-3 & 2 \\ 6 & -3\end{array}\right]\end{array}$. Add 4

to each entry to get $\begin{bmatrix} 1 & 6 \\ 10 & 1 \end{bmatrix}$. Then apply

the simplex method.

$$\begin{array}{c}\quad\quad z_1 \quad z_2 \quad t \quad u \quad M \\ \begin{array}{c}t\\u\\M\end{array}\left[\begin{array}{ccccc|c} 1 & \underline{6} & 1 & 0 & 0 & 1 \\ 10 & 1 & 0 & 1 & 0 & 1 \\ \hline -1 & -1 & 0 & 0 & 1 & 0 \end{array}\right]\end{array}$$

$$\begin{array}{c}\quad\quad z_1 \quad z_2 \quad t \quad u \quad M \\ \begin{array}{c}z_2\\u\\M\end{array}\left[\begin{array}{ccccc|c} \frac{1}{6} & 1 & \frac{1}{6} & 0 & 0 & \frac{1}{6} \\ \frac{59}{6} & 0 & -\frac{1}{6} & 1 & 0 & \frac{5}{6} \\ \hline -\frac{5}{6} & 0 & \frac{1}{6} & 0 & 1 & \frac{1}{6} \end{array}\right]\end{array}$$

$$\begin{array}{c}\quad\quad z_1 \quad z_2 \quad t \quad u \quad M \\ \begin{array}{c}z_2\\z_1\\M\end{array}\left[\begin{array}{ccccc|c} 0 & 1 & \frac{10}{59} & -\frac{1}{59} & 0 & \frac{9}{59} \\ 1 & 0 & -\frac{1}{59} & \frac{6}{59} & 0 & \frac{5}{59} \\ \hline 0 & 0 & \frac{9}{59} & \frac{5}{59} & 1 & \frac{14}{59} \end{array}\right]\end{array}$$

$$v = \frac{1}{M} = \frac{59}{14}$$

a. Carol's optimal strategy is

$$\begin{bmatrix} \frac{59}{14}\cdot\frac{5}{59} \\ \frac{59}{14}\cdot\frac{9}{59} \end{bmatrix} = \begin{bmatrix} \frac{5}{14} \\ \frac{9}{14} \end{bmatrix}.$$

Ruth's optimal strategy is

$$\begin{bmatrix} \frac{59}{14}\cdot\frac{9}{59} & \frac{59}{14}\cdot\frac{5}{59} \end{bmatrix} = \begin{bmatrix} \frac{9}{14} & \frac{5}{14} \end{bmatrix}.$$

b. Since the value is positive, the game favors Ruth (the row player). The value of the game is $\frac{59}{14} - 4 = \frac{3}{14}$.

14. a.

$$\begin{array}{c}\quad\quad\text{Strong} \quad \text{Avg.} \quad \text{Weak} \\ \begin{array}{c}A\\B\\C\end{array}\left[\begin{array}{ccc} 3000 & 2000 & 1000 \\ 6000 & 2000 & -3000 \\ 15,000 & 1000 & -10,000 \end{array}\right]\end{array}$$

b. Row 1, column 3 is a saddle point. The investors optimal strategy is to buy stock A.

15. For the given conditions, the matrices will not have a saddle point and therefore, will not be strictly determined.

16. The optimal strategy will remain the same as long as the row 1, column 1 value is less than the smallest value in the matrix or $2 + h < 6$ or $h < 4$.

17. a. Since both lines are straight lines, the intersection of the two lines would be the only solution to the system of equations. Moving from that intersection point would increase the value of one line while decreasing the value of the other.

b. You would use the equations
$y = a_{11}r + a_{12}(1 - r)$ and
$y = a_{21}r + a_{22}(1 - r)$ and find the point of intersection as in part a.

18.
$$a_{11}r + a_{21}(1-r) = a_{12}r + a_{22}(1-r)$$
$$a_{11}r + a_{21} - a_{21}r = a_{12}r + a_{22} - a_{22}r$$
$$a_{11}r - a_{21}r - a_{12}r + a_{22}r = a_{22} - a_{21}$$
$$r(a_{11} - a_{21} - a_{12} + a_{22}) = a_{22} - a_{21}$$
$$r = \frac{a_{22} - a_{21}}{a_{11} - a_{21} - a_{12} + a_{22}}$$

Chapter 9 Chapter Test

1. a. A move R_1 by R and C_3 by C results in a payoff of 6 to R.

b. A move R_3 by R and C_2 by C results in a payoff of 10 to C.

c. The least elements in the rows are
$$\begin{bmatrix} \underline{-5} & -1 & 6 \\ -2 & 5 & \underline{-8} \\ 8 & \underline{-10} & 5 \end{bmatrix}.$$
The maximum elements in the columns are
$$\begin{bmatrix} -5 & -1 & \underline{6} \\ -2 & \underline{5} & -8 \\ \underline{8} & -10 & 5 \end{bmatrix}.$$
Since none of the least row elements are also maximum column elements this game is not strictly determined. No.

2. Answers may vary. *Sample answer*: In a pure strategy the player always chooses the same row (or column). In a mixed strategy, the player probabilistically chooses a row (or column).

3. a. R: $\begin{bmatrix} \underline{-1} & 0 \\ 2 & \underline{-2} \\ 3 & \underline{2} \end{bmatrix}$, row 3

C: $\begin{bmatrix} -1 & 0 \\ 2 & -2 \\ \underline{3} & \underline{2} \end{bmatrix}$, column 2

value = 2

b. R: $\begin{bmatrix} \underline{0} & 2 & 4 \\ \underline{-4} & 0 & 3 \\ -2 & \underline{-3} & 0 \end{bmatrix}$, row 1

C: $\begin{bmatrix} 0 & 2 & 4 \\ -4 & 0 & 3 \\ -2 & -3 & 0 \end{bmatrix}$, column 1

value = 0

4. $\quad 1 \quad 2 \quad 3 \quad 4 \quad 5$; Yes; -1; C

$$\begin{array}{c}1\\2\\3\\4\\5\end{array}\begin{bmatrix} -1 & -1 & -1 & -1 & -1 \\ 2 & -2 & -2 & -2 & -2 \\ 3 & 3 & -3 & -3 & -3 \\ 4 & 4 & 4 & -4 & -4 \\ 5 & 5 & 5 & 5 & -5 \end{bmatrix}$$

The last element in the first row is a saddle point. The game is strictly determined, and the value is -1. It is better to be player C.

5. a. $[.4 \quad .6]\begin{bmatrix} 2 & -2 & 3 \\ -1 & 1 & -3 \end{bmatrix}\begin{bmatrix} .3 \\ .5 \\ .2 \end{bmatrix}$

$= [-.16]$

On average, C gains .16 every time the game is played.

b. $[.4 \quad .6]\begin{bmatrix} 2 & -2 & 3 \\ -1 & 1 & -3 \end{bmatrix}\begin{bmatrix} .4 \\ .1 \\ .5 \end{bmatrix}$

$= [-.24]$

The column player is better off using the mixed strategy $C = \begin{bmatrix} .4 \\ .1 \\ .5 \end{bmatrix}$.

6. a. Add 5 to each entry to make all the entries positive. We get $\begin{bmatrix} 5 & 4 \\ 1 & 10 \end{bmatrix}$.

Then maximize $M = z_1 + z_2$ subject to the constraints $\begin{cases} 5z_1 + 4z_2 \leq 1 \\ z_1 + 10z_2 \leq 1 \\ z_1 \geq 0,\ z_2 \geq 0 \end{cases}$

$$
\begin{array}{c}
\begin{array}{ccccc} z_1 & z_2 & t & u & M \end{array} \\
\begin{array}{c} t \\ u \\ M \end{array}
\left[
\begin{array}{ccccc|c}
1 & 8 & 1 & 0 & 0 & 1 \\
\underline{6} & 2 & 0 & 1 & 0 & 1 \\
-1 & -1 & 0 & 0 & 1 & 0
\end{array}
\right]
\end{array}
$$

$$
\begin{array}{c}
\begin{array}{ccccc} z_1 & z_2 & t & u & M \end{array} \\
\begin{array}{c} t \\ z_1 \\ M \end{array}
\left[
\begin{array}{ccccc|c}
0 & \frac{23}{3} & 1 & -\frac{1}{6} & 0 & \frac{5}{7} \\
1 & \frac{1}{3} & 0 & \frac{1}{6} & 0 & \frac{1}{6} \\
0 & -\frac{2}{3} & 0 & \frac{1}{6} & 1 & \frac{1}{6}
\end{array}
\right]
\end{array}
$$

$$
\begin{array}{c}
\begin{array}{ccccc} z_1 & z_2 & t & u & M \end{array} \\
\begin{array}{c} t \\ z_1 \\ M \end{array}
\left[
\begin{array}{ccccc|c}
1 & 0 & \frac{5}{23} & -\frac{2}{23} & 0 & \frac{3}{23} \\
0 & 1 & -\frac{1}{46} & \frac{5}{46} & 0 & \frac{2}{23} \\
0 & 0 & \frac{9}{46} & \frac{1}{46} & 1 & \frac{5}{23}
\end{array}
\right]
\end{array}
$$

$z_1 = \dfrac{3}{23},\ z_2 = \dfrac{2}{23},\ M = \dfrac{5}{23},\ v = \dfrac{23}{5}$

The optimal strategy for C is

$\begin{bmatrix} vz_1 \\ vz_2 \end{bmatrix} = \begin{bmatrix} \frac{3}{5} \\ \frac{2}{5} \end{bmatrix}$.

$y_1 = t = \dfrac{9}{46},\ y_2 = u = \dfrac{1}{46},\ M = \dfrac{5}{23},\ v = \dfrac{23}{5}$

The optimal strategy for R is

$\begin{bmatrix} vy_1 & vy_2 \end{bmatrix} = \begin{bmatrix} \frac{9}{10} & \frac{1}{10} \end{bmatrix}$.

b. The value of the game is

$v - 5 = \dfrac{23}{5} - 5 = -\dfrac{2}{5}$. Since this is negative, the game favors C.

Chapter 10

1. a. $i = \dfrac{.12}{12} = .01$

$n = (12)(2) = 24$

b. $i = \dfrac{.08}{4} = .02$

$n = (4)(5) = 20$

c. $i = \dfrac{.10}{2} = .05$

$n = (2)(20) = 40$

3. a. $i = \dfrac{.06}{1} = .06$

$n = (1)(4) = 4$

$P = \$500$

$F = \$631.24$

b. $i = \dfrac{.06}{12} = .005$

$n = (12)(10) = 120$

$P = \$800$

$F = \$1455.52$

c. $i = \dfrac{.04}{2} = .02$

$n = (2)(9.5) = 19$

$P = \$6177.88$

$F = \$9000$

15. a. $\left[\dfrac{1}{\left(1 + \frac{.06}{12}\right)^{12 \times 2}}\right] (\$4000) = \$3548.74$

b. $i = F - P = \$4000 - \$3548.74 = \$451.26$

c.

Month	Interest	Balance
0		$3548.74
1	$17.74	$(1.005)(\$3548.74) = \$3566.48$
2	$17.83	$(1.005)^2(\$3548.74) = \$3584.31$
3	$17.92	$(1.005)^3(\$3548.74) = \$3602.23$

5. $\left(1 + \dfrac{.06}{12}\right)^{12 \times 2} (\$1000) = \$1127.16$

7. $\left[\dfrac{1}{\left(1 + \frac{.06}{12}\right)^{12 \times 25}}\right] (\$100,000) = \$22,396.57$

9. $F = \left(1 + \dfrac{.06}{12}\right)^{12 \times 3} (\$6000) = \$7180.08$

$i = F - P = \$7180.08 - \$6000 = \$1180.08$

11. $B_5 = \left(1 + \dfrac{.03}{4}\right)^{4 \times 5} (\$1000) = \$1161.18$

$B_4 = \left(1 + \dfrac{.03}{4}\right)^{4 \times 4} (\$1000) = \$1126.99$

$i = B_5 - B_4 = \$1161.18 - \$1126.99 = \$34.19$

13. $P = \left[\dfrac{1}{\left(1 + \frac{.04}{4}\right)^{4 \times 1} - 1}\right] (\$406.04) = \$10,000$

17. $\left(1+\dfrac{.04}{4}\right)^{4\times6.25}(\$10,000) = \$12,824.32$

19. $\left[\dfrac{1}{\left(1+\frac{.04}{4}\right)^{4\times3}}\right](\$10,000) = \$8874.49$

21. a. $\dfrac{r}{12}(\$1000.00) = \$5.00$, so $r = .06 = 6\%$

    b. $(1.005)^3(\$1000.00) = \$1015.08$
        $\$1015.08 - \$1010.03 = \$5.05$

    c. $(1.005)^{24}(\$1000.00) = \$1127.16$
        $[(1.005)^{24} - (1.005)^{23}](\$1000.00) = \$5.61$

23. For $P = \$1000$,

$$F = \left(1+\dfrac{.06}{1}\right)^9(\$1000) = \$1689.48.$$

$\$1700$ in 9 years is more profitable.

25. $r_{\text{eff}} = \left(1+\dfrac{.622}{52}\right)^{52} - 1 \approx .8558$

This interest rate is better than 85%.

27. $\dfrac{r}{4}(\$10,000) = \$100$, so $r = .04$ and $i = .01$

$$\dfrac{1}{(1.01)^{12}}(\$10,000) = \$8874.49$$

29. a. $r = .04$
      $n = \dfrac{6}{12} = \dfrac{1}{2}$
      $P = \$500$
      $F = \$510$

    b. $r = .05$
      $n = 2$
      $P = \$500$
      $F = \$550$

31. $F = (1+3\cdot0.05)(\$1000) = \$1150$

33. $P = \left[\dfrac{1}{(1+2\cdot0.10)}\right](\$3000) = \$2500$

35. $\left(1+\dfrac{6}{12}r\right)(\$980) = \$1000$

    $r \approx .0408 = 4.08\%$

37. $(1+n\cdot0.06)(\$500) = \$800$, $n = 10$ years

39. $(1+n\cdot0.05)P = 2P$, $n = 20$ years

41. $F = (1+nr)P$; $P = \dfrac{F}{1+nr}$

43. (a)

45. $\left(1+\dfrac{.04}{4}\right)^{4\times1}(\$100) = \$104.06$

    $\dfrac{\$4.06}{\$100} = .0406 = 4.06\%$

47. $r_{\text{eff}} = \left(1+\dfrac{.04}{2}\right)^2 - 1 = .0404$; 4.04%

49. $r_{\text{eff}} = \left(1+\dfrac{.044}{12}\right)^{12} - 1 \approx .0449$; 4.49%

51. $\left(1+\dfrac{r}{4}\right)^4 - 1 = .0406$; $r \approx .04$; 4%

53. Since we start with $1000 and this amount doubles every six years, we have $2000 at the end of six years, $4000 at the end of twelve years and $8000 at the end of eighteen years. So, it will take 18 years for the investment to grow to $8000.

55. Assume an initial investment of $100.00. Then over a 10 year period, the amount of growth would be $100(1+0.04)^{10} = 148.02$ about a 48% increase; so answer d) is correct.

57. $1500(1+r)^7 = 2100$

$$\left(1+r\right)^7 = 1.4$$

$$(1+r) = \sqrt[7]{1.4}$$

$$r \approx 4.9\%$$

59. Assume $100 invested initially. Then 100 would increase to 102.5 after the first year. The total amount of the investment after three years would then be $100(1.025)(1.03)(1.084) = \$114.44$. If the same amount was invested at an interest rate of r % compounded annually, the amount would be the same, so;

$$100(1+r)^3 = 114.44$$

$$(1+r)^3 = 1.1444$$

$$(1+r) = \sqrt[3]{1.1444}$$

$$r \approx 4.6\%$$

61. After 1 year: $1.065(\$1000) = \$1065.00$

After 2 years: $(1.065)^2(\$1000) = \$1134.23$

After 3 years: $(1.065)^3(\$1000) = \$1207.95$

After 8 years: $(1.065)^8(\$1000) = \$1655.00$

After 11 years: $(1.065)^{11}(\$1000) = \$1999.15$

$1065.00; \$1134.23; \$12076.95; 8; 12$

63. $(1.08)^n(\$100,000)$ passes $1,000,000 when $n = 30$ years.

65. After year 1, option A would have a value of $F = (1 + 1(.08))(\$1000) = \$1080$ and option B would have a value of $F = (1 + .06)^1(\$1000) = \$1060$. After year 2, option A would have a value of $F = (1 + 2(.08))(\$1000) = \$1160$ and option B would have a value of $F = (1 + .06)^2(\$1000) = \$1123.60$. Continue until year 11 when option A has a value of $F = (1 + 11(.08))(\$1000) = \$1880$ and option B has a value of $F = (1 + .06)^{11}(\$1000) = \$1898.30$.

Exercises 10.2

1. a. $i = \dfrac{.06}{12} = .005$

$n = (12)(10) = 120$

$R = \$50$

$F = \$8193.97$

b. $i = \dfrac{.04}{2} = .02$

$n = (2)(10) = 20$

$R = \$8231.34$

$F = \$200,000$

3. $\left[\dfrac{\left(1 + \frac{.06}{4}\right)^{4\times 5} - 1}{\frac{.06}{4}} \right](\$1000) = \$23,123.67$

5. $\left[\dfrac{\frac{.08}{4}}{1 - \left(1 + \frac{.08}{4}\right)^{-4\times 7}} \right](\$100,000) = \$4698.97$

7. a. $\left[\dfrac{\left(1+\frac{.06}{12}\right)^{12\times4}-1}{\frac{.06}{12}}\right](\$500) = \$27,048.92$

 b. $\$27,048.92 - 48(\$500) = \$3048.92$

 c.

Month	Interest	Balance
1		\$500.00
2	$.005 \times 500 = \$2.50$	$(1.005)(500) + 500 = \$1002.50$
3	$.005 \times 1002.50 = \$5.01$	$(1.005)(1.002.50) + 500 = \$1507.51$

9. $\left[\dfrac{\frac{.06}{12}}{\left(1+\frac{.06}{12}\right)^{12\times3}-1}\right](\$12,000) = \$305.06$

Deposited: $36(\$305.06) = \$10,982.16$
Interest: $\$12,000 - \$10,982.16 = \$1017.84$

13. $\left[\dfrac{\frac{.04}{2}}{\left(1+\frac{.04}{2}\right)^{2\times15}-1}\right](\$1,000,000) = \$24,649.92$

11. $\left(1+\dfrac{.06}{12}\right)^{12}(\$2000) = \$2123.36$

$\left[\dfrac{\left(1+\frac{.06}{12}\right)^{12}-1}{\frac{.06}{12}}\right](\$200) = \$2467.11$

\$200 each month is better, by \$343.75.

15. $\left[\dfrac{1-\left(1+\frac{.12}{12}\right)^{-12\times15}}{\frac{.12}{12}}\right](\$30\ million) = \$2499\ million$

or approximately \$2.5 billion

17. After 10 years, the fund will be worth:

$\left[\dfrac{\left(1+\frac{.12}{12}\right)^{12\times10}-1}{\frac{.12}{12}}\right](\$100,000) = \$23,003,868.95$

After 10 years, the cost of the equipment will be:
$(1.06)^{10}(13,000,000) = \$23,281,020.06$
Therefore, the annuity will be short by
$\$23,281,020.06 - \$23,003,868.95$ or
\$277,151.11.

21. Jack withdraws for $12(.75) = 9$ months.

$\left[\dfrac{1-\left(1+\frac{.06}{12}\right)^{-12\times0.75}}{\frac{.06}{12}}\right](\$100) = \$877.91$

23. $\left[\dfrac{\left(1+\frac{.06}{12}\right)^{12\times10}-1}{\frac{.06}{12}}\right](\$1000) = \$163,879.35$

\$1000 at the end of each month is better.

19. $\left[\dfrac{\frac{.08}{4}}{\left(1+\frac{.08}{4}\right)^{4\times15}-1}\right](\$5,000,000) = \$43,839.83$

25. $\left(1+\dfrac{.08}{4}\right)^{4\times9}(\$1000)+\left[\dfrac{\left(1+\frac{.08}{4}\right)^{4\times9}-1}{\frac{.08}{4}}\right](\$100)=\$7239.32$

27. $\left[\dfrac{\left(1+\frac{.06}{12}\right)^{12\times10}-1}{\frac{.06}{12}}\right](\$100)+\left(1+\dfrac{.06}{12}\right)^{12\times3}(\$1000)=\$17{,}584.62$

29. The future value of the annuity will be $10R + \$5725.43$, so:

$$10R+\$5725.43=\left[\dfrac{(1+.06)^{10}-1}{.06}\right]R$$

$$10R+\$5725.43=13.180795R$$

$$\$5725.43=3.180795R$$

$$\$1800=R$$

31. (a)

33. $.05P = \$1200$, so $P = \$24{,}000$.

35. $P=(1.06)^7(\$10{,}000)=\$15{,}036.30$

$$R=\left[\dfrac{.06}{1-(1.06)^{-4}}\right](\$15{,}036.30)=\$4339.35$$

37. $P=\dfrac{R}{i}=\dfrac{\$60{,}000}{.06}=\$1{,}000{,}000$

$$\dfrac{\$1{,}000{,}000}{(1.06)^9}=\$591{,}898.46$$

39. Present value: $\left[\dfrac{1-(1.015)^{-30}}{.015}\right]\left(\dfrac{.04}{2}\right)(\$5000)+\dfrac{\$5000}{(1.015)^{30}}$

$$=\$5600.40$$

41. a. $\left[\dfrac{\frac{.18}{12}}{1-\left(1+\frac{.18}{12}\right)^{-12\times5}}\right](\$50{,}000)$

$$=\$1269.67$$

b. Treat as two annuities. The first has a present value of

$$\left[\dfrac{1-(1.01)^{-60}}{.01}\right](.015)(\$50{,}000)$$

$$=\$33{,}716.28.$$

The second has a present value of

$$\left[\frac{1-(1.01)^{-60}}{.01}\right](\$1269.67) \div (1.01)^{60}$$

$$= \$31,418.60.$$

The total present value = $65,134.88

43. Formula for a decreasing annuity: $B_{new} = (1+i)B_{previous} - R$

45. Set up a table with $Y_1 = (1.05 \wedge X - 1)/.05*1000$. After 2, 3, and 4 years, Y_1 equals $2050, $3152.50 and $4310.13. Y_1 equals $30,539 after 19 years, and exceeds $50,000 after 26 years.

47. Use $Y_1 = (1.001 \wedge X - 1)/.001*15$. $Y_1 = 503$ after 33 weeks.

49. The original $5000 will go through 10 years of interest and 10 years of payments. Therefore, the original money will be worth $5000(1.06)^{10}(0.997)^{10} = \$8689.21$. The $5000 deposited in the second year will go through 9 years of interest and 9 years of payment, therefore it will be worth $5000(1.06)^9(0.997)^9 = \$8222.03$. Continue this pattern for the 10 years and adding all the values together will give you a balance of $68,617.21 at the 0.3% payment. Using the same method for the 1.5% payment will yield a balance of $63,882.62 after the 10 years. The difference would then be $4734.59.

Exercises 10.3

1. $\left[\dfrac{\frac{.06}{12}}{1-\left(1+\frac{.06}{12}\right)^{-12\times5}}\right](\$10,000) = \$193.33$

3. $\left[\dfrac{1-\left(1+\frac{.12}{2}\right)^{-2\times10}}{\frac{.12}{2}}\right](\$1000) = \$11,469.92$

5. **a.** $\dfrac{.12}{12}(\$58,331) = \$583.31$

   b. $600 - \$583.31 = \$16.69$

   c. $58,331 - \$16.69 = \$58,314.31$

   d. $\left[\dfrac{1-\left(1+\frac{.12}{12}\right)^{-12(5)}}{\frac{.12}{12}}\right](\$600) = \$26,973.02$

   e. $\left[\dfrac{1-\left(1+\frac{.12}{12}\right)^{-12(4)}}{\frac{.12}{12}}\right](\$600) = \$22,784.38$

   $26,973.02 - \$22,784.38 = \$4188.65$
   (with a $.01 discrepancy due to rounding errors)

   f. Use result of part (d): $.01(\$26,973.02) = \$269.73$

7. a. $\left[\dfrac{\frac{.12}{12}}{1-\left(1+\frac{.12}{12}\right)^{-12\times3}}\right]($8000) = 265.71

b. $($265.71)(3)(12) = 9565.56

c. $$9565.56 - $8000 = 1565.56

d. $\left[\dfrac{1-\left(1+\frac{.12}{12}\right)^{-12(2)}}{\frac{.12}{12}}\right]($265.71) = 5644.58

e. $\left[\dfrac{1-\left(1+\frac{.12}{12}\right)^{-12(1)}}{\frac{.12}{12}}\right]($265.71) = 2990.59

f. $12($265.71) - ($5644.58 - $2990.59) = 534.53

g.

Payment	Amount	Interest	Applied to Principal	Unpaid balance
1	$265.71	$80.00	$185.71	$7814.29
2	265.71	78.14	187.57	7626.72
3	265.71	76.27	189.44	7437.28
4	265.71	74.37	191.34	7245.94

9. $\left(1+\dfrac{.09}{12}\right)($10,000) - $1125 = 8950

11. $\left[\dfrac{1-\left(1+\frac{.12}{2}\right)^{-2\times8}}{\frac{.12}{2}}\right]($1000) + \left[\dfrac{1}{\left(1+\frac{.12}{2}\right)^{2\times8}}\right]($10,000) = $14,042.36$

13. $\left[\dfrac{\frac{.12}{12}}{1-\left(1+\frac{.12}{12}\right)^{-4}}\right]100 = 256.28$

Payment	Amount	Interest	Applied to Principal	Unpaid balance
1	$256.28	$10.00	$246.28	$753.72
2	256.28	7.54	248.74	504.98
3	256.28	5.05	251.23	253.74
4	256.28	2.54	253.74	0.00

15. $\left[\dfrac{\frac{.063}{12}}{1-\left(1+\frac{.063}{12}\right)^{-12\times30}}\right]($360,000 - $60,000) = 1856.92

17. $\left[\dfrac{1-\left(1+\frac{.09}{12}\right)^{-12\times25}}{\frac{.09}{12}}\right]$ ($1200 - $200) = $119,161.62

19. $\left[\dfrac{1-\left(1+\frac{.06}{12}\right)^{-12\times3}}{\frac{.06}{12}}\right]$ ($100) = $3287.10

$6287.10 - ($2000 + $3287.10) = $1000.00

$\left(1+\dfrac{.06}{12}\right)^{12\times3}$ ($1000) = $1196.68

21. (a)

23. **a.** $\left[\dfrac{(1.06)^{20}-1}{.06}\right]$ ($5000) = $183,927.96

    b. $\left[\dfrac{.06}{1-(1.06)^{-10}}\right]$ ($183,927.96) = $24,989.92

    c. $\left[\dfrac{1-(1.06)^{-(10-5)}}{.06}\right]$ ($24,989.92) = $105,266.63

25. **a.** $\dfrac{\$26000}{36}$ = $722.22 monthly payment

    b. amount of loan = $26,000 - $1000 = $25,000
Monthly payment

$\left[\dfrac{\frac{.06}{12}}{1-\left(1+\frac{.06}{12}\right)^{-12\times3}}\right]$ ($25,000) = $760.55

    c. Option a is more favorable.

27. Assume you have a $100,000 mortgage, find your payments, and your balance after 15 years. The payment will be

$\left[\dfrac{\frac{.068}{12}}{1-\left(1+\frac{.068}{12}\right)^{-12\times30}}\right]$ ($100,000) = $651.93.

The balance after 15 years will be

$\left[\dfrac{1-\left(1+\frac{.068}{12}\right)^{-12\times15}}{\frac{.068}{12}}\right]$ ($651.93) = $73,441.68.

Therefore, the percent paid is

$\dfrac{100,000-73,441.68}{100,000} = 0.2656 = 26.56\%.$

29. The loan amount (F) is 36 times the payment amount (R) minus the interest. The loan amount is also

$$F = \left[\frac{1 - \left(1 + \frac{.06}{12}\right)^{-12 \times 3}}{\frac{.06}{12}} \right] R$$

$F = 32.8710R$
Therefore,

$$36R - \$1085.16 = 32.8710R$$
$$3.1290R = \$1085.16$$
$$R = \$346.81$$

31. a. $I_n + Q_n = R$ and $I_n = iB_{n-1}$

$$iB_{n-1} + Q_n = R$$

$$B_{n-1} = \frac{R - Q_n}{i}$$

b. $(1+i)\dfrac{R-Q}{i} - R = \dfrac{R - Q_{n+1}}{i}$

$$(1+i)(R-Q) - iR = R - Q_{n+1}$$
$$(1+i)R - (1+i)Q - iR = R - Q_{n+1}$$
$$R + iR - (1+i)Q - iR = R - Q_{n+1}$$
$$-(1+i)Q = -Q_{n+1}$$
$$(1+i)Q = Q_{n+1}$$

c. The amount of the portion applied to the principle in the next month is equal to the amount of the portion applied to the principle in the previous month multiplied by 1 plus the interest rate.

d. $Q_{11} = (1+i)Q_{10}$ and

$$= (1 + 0.01)(\$100)$$
$$= (1.01)(\$100)$$
$$Q_{11} = \$101.00$$
$$Q_{12} = (1+i)Q_{11}$$
$$= (1 + 0.01)(\$101)$$
$$= (1.01)(\$101)$$
$$Q_{11} = \$102.01$$

33. After 1 month:

$$\left(1 + \frac{.09}{12}\right)(\$2188.91) - \$100 = \$2105.33$$

After 2 months:

$$\left(1 + \frac{.09}{12}\right)(\$2105.33) - \$100 = \$2021.12$$

After 3 months:

$$\left(1 + \frac{.09}{12}\right)(\$2021.12) - \$100 = \$1936.28$$

The loan will be paid off after 24 months.

35. Enter 10000, then run $1.0075 * Ans - 166.68$ repeatedly. After 40 iterations, the balance is \$5741.79, which means \$4258.21 has been paid off. The balance drops below \$5000 after 46 months.

37. The balance B must drop to where

$$\left(\frac{.085}{12}\right)B < \$250, \text{ or } B < \$35,294.12.$$

Let $Y_1 = Y_6(300 - X)*1000$ where

$Y_6 = ((1+I)\wedge X - 1)/(I(1+I)\wedge X).$

Make a table. Then $B \le \$35,294.12$ after 260 payments which means that the next payment, after 261 months, is the first one where at least 75% goes toward debt reduction.

39. a. $\left[\dfrac{\frac{.029}{12}}{1-\left(1+\frac{.029}{12}\right)^{-12\times 3}}\right]($30,000)=$871.11$

    b. $\left[\dfrac{\frac{.074}{12}}{1-\left(1+\frac{.074}{12}\right)^{-12\times 3}}\right]($28,000)=$869.69$

    c. Option b is better by $1.42 per month.

Exercises 10.4

1. deferred

3. [amount after taxes] $=(1-.45)(300,000)$
$$=\$165,000$$

5. $\left[\dfrac{\left(1+\frac{.06}{1}\right)^{1\times 52}-1}{\frac{.06}{1}}\right]($5000)=$1,641,407.11$

7. If we assume a marginal tax bracket of 20%,

$\left[\dfrac{\left(1+\frac{.06}{1}\right)^{1\times 52}-1}{\frac{.06}{1}}\right](0.8)($5000)=$1,313,125.69$

9. a. For Earl:
[earnings after income tax]
$=[1-\text{tax bracket}]\cdot[\text{amount}]$
$=[.60]\cdot[5000]$
$=3000$

$\left[\dfrac{\left(1+\frac{.06}{1}\right)^{1\times 12}-1}{\frac{.06}{1}}\right]($3000)=$50,609.82$

This money then earns interest compounded annually for 36 years and grows to
$$\$50,609.82\cdot(1.06)^{36}=\$50,609.82(8.147252)$$
$$=\$412,330.96$$

b. For Larry:

$\left[\dfrac{\left(1+\frac{.06}{1}\right)^{1\times 36}-1}{\frac{.06}{1}}\right]($3000)=$357,362.60$

c. Earl paid in $12\times $3000=$36,000$ while Larry paid in $36\times $3000=$108,000.$ Larry paid in more.

d. Earl has $54,968.36 more than Larry.

11. $R=\dfrac{P(1+rt)}{12t}=\dfrac{4000(1+.10\cdot 1)}{12(1)}$
$$=\$366.67$$

13. $R=\dfrac{P(1+rt)}{12t}=\dfrac{3000(1+.09\cdot 3)}{12(3)}$
$$=\$105.83$$

15. $r=\dfrac{12Rt-P}{Pt}=\dfrac{12(171.21)(1)-2000}{2000(1)}$
$$\approx .0273 \text{ or about } 2.73\%$$

17. $r = \dfrac{12Rt - P}{Pt} = \dfrac{12(608.44)(3) - 20{,}000}{20{,}000(3)}$

$\approx .0317$ or about 3.17%

19. a. $[\text{total repayment}] = \dfrac{[\text{loan amount}]}{1 - rt} = \dfrac{880}{1 - .06(2)}$

$= 1000$

$[\text{monthly payment}] = \dfrac{[\text{total payment}]}{12t} = \dfrac{1000}{12(2)}$

$= \$41.67$

b. $R = \dfrac{P(1 + rt)}{12t} = \dfrac{880(1 + .06 \cdot 2)}{12(2)}$

$= \$41.07$

The monthly payment is less.

21. False. The effective rate differs from the APR only when discount points are involved.

23. False. The longer the mortgage will be held, the lower the effective rate.

25. False. The up-front fees must change proportionally for there to be no effect on the APR.

27. $Payment = \left[\dfrac{\frac{.09}{12}}{1 - \left(1 + \frac{.09}{12}\right)^{-12 \times 25}}\right](\$250{,}000) = \$209799$

New $P = 250{,}000 - 5000 = 245{,}000$
Using the Excel function 12*RATE(300, –2097.99, 245000, 0) gives .09249 or 9.25%; (d).

29. $Payment = \left[\dfrac{\frac{.055}{12}}{1 - \left(1 + \frac{.055}{12}\right)^{-12 \times 20}}\right](\$250{,}000) = \$1719.72$

New $P = 250{,}000 - 10{,}000 = 240{,}000$
Using the Excel function 12*RATE(240, –1719.72, 240000, 0) gives .06002 or about 6%; (a).

31. $[\text{monthly payment}] = \$581.03$
Using the Excel function 12*RATE(48, –581.03, 99000, –94342.20) gives .05999 or about 6%; (a).

33. $Payment = \left[\dfrac{\frac{.06}{12}}{1 - \left(1 + \frac{.06}{12}\right)^{-12 \times 30}}\right](\$100{,}000) = \$599.55$

New $P = 100{,}000 - 3000 = 97{,}000$
The mortgage has $360 - 84 = 276$ months to go. Therefore,

$balance = \left[\dfrac{1 - \left(1 + \frac{.06}{12}\right)^{-276}}{\frac{.06}{12}}\right](\$599.55) = \$89{,}639.31$

Using the Excel function 12*RATE(84, –599.55, 97000, –89639.39) gives .06560 or about 6.56%; (c).

35. APR: $n = 20 \cdot 12 = 240$

$$R = \frac{.005(1.005)^{240}}{1.005^{240} - 1} \cdot 80{,}000 = 573.14$$

$P = 80{,}000 - .03(80{,}000) = 77{,}600$

$12*\text{RATE}(n, R, P, 0) = 12*\text{RATE}(240, -573.14, 77{,}600, 0) = 6.38\%$

effective rate: $m = 10 \cdot 12 = 120$

$R = 573.14, P = 77{,}600$

$$B = \frac{1.005^{120} - 1}{.005(1.005)^{120}} \cdot 573.14 = 51{,}624.70$$

$12*\text{RATE}(m, R, P, B) = 12*\text{RATE}(120, -573.14, 77{,}600, -51{,}624.70) = 6.47\%$

37. APR: $n = 15 \cdot 12 = 180$

$$R = \frac{.0075(1.0075)^{180}}{1.0075^{180} - 1} \cdot 120{,}000 = 1217.12$$

$P = 120{,}000 - .01(120{,}000) = 118{,}800$

$\dfrac{(1+i)^{180} - 1}{i(1+i)^{180}} \cdot 1217.12 = 118{,}800$ yields

9.17%.

effective rate: $m = 5 \cdot 12 = 60$

$R = 1217.12, P = 118{,}800$

$$B = \frac{1.0075^{120} - 1}{.0075(1.0075)^{120}} \cdot 1217.12 = 96{,}081.51$$

$1217.12 - 96{,}081.51i = (1217.12 - 118{,}000i) \cdot (1+i)^{60}$

yields 9.27%.

39. The salesman is comparing the future value of the savings account to the sum of the present values of the loan payments at time of payment. A proper comparison would be the future value of the savings account, $1083.14, to the future value of the series of payments (assuming 4% interest)

$$\left[\frac{\left(1 + \frac{.04}{12}\right)^{24} - 1}{\frac{.04}{12}}\right]\$43.87) = \$1094.24$$

41. a. $P = \$200{,}000$ and $i = \dfrac{0.069}{12} = 0.00575$

$\text{Payment} = iP = (0.00575)(\$200{,}000) = \$1150$

b. $\text{Payment} = \left[\dfrac{\frac{.069}{12}}{1 - \left(1 + \frac{.069}{12}\right)^{-12 \times 10}}\right](\$200{,}000) = \$2311.87$

43. a. For the first 5 years; $\text{Payment} = \left[\dfrac{\frac{.06}{12}}{1 - \left(1 + \frac{.06}{12}\right)^{-12 \times 25}}\right](\$250{,}000) = \$1610.75$

b. The balance after 5 years; $\text{balance} = \left[\dfrac{1 - \left(1 + \frac{.06}{12}\right)^{-12 \times 20}}{\frac{.06}{12}}\right](\$1610.75) = \$224{,}829.73$

c. For the sixth year, $P = \$224{,}829.73$, $n = 240$, and $i = 0.044 + 0.025 = 0.069$.

$\text{Payment} = \left[\dfrac{\frac{.069}{12}}{1 - \left(1 + \frac{.069}{12}\right)^{-240}}\right](\$224{,}829.73) = \$1729.63$

45. a. The balance after 7 years; $balance = \left[\dfrac{1-\left(1+\frac{.069}{12}\right)^{-228}}{\frac{.069}{12}}\right]($1729.63) = $219,418.04$

b. Without the cap, $P = $219,418.04$, $n = 228$, and $i = 0.077 + 0.025 = 0.102$.

$Payment = \left[\dfrac{\frac{.102}{12}}{1-\left(1+\frac{.102}{12}\right)^{-228}}\right]($219,418.04) = 2181.80

c. Without the cap, the percentage increase from the sixth year to the seventh year would be
$\dfrac{2181.80-1729.63}{1729.63} = 0.2614 = 26.14\%$
Since this percentage is greater than the 7 % cap, the monthly payment will be
$(1.07)($1729.63) = 1850.70.

d. The interest due in the 73rd month would be $(0.0085)($219,418.04) = 1865.05.

e. Since the interest owed is more than the payment made, the balance will increase by
$$1865.05 - $1850.70 = 14.35. Therefore the new balance will be
$$219,418.04 + $14.35 = $219,432.39$.

47.
```
N=36
I%=11.08292218
PV=10000
PMT=-327.78
FV=0
P/Y=12
C/Y=12
PMT:BEGIN BEGIN
```

The APR is about 11.08%.

49.

	A	B	C	D
1	n =	300		
2	i =	0.00542		
3	loan amount =	300,000.00		
4	points:	3.00		
5				
6	$1/a_n$ =	0.0067520716		
7	monthly payment =	2,025.62		
8	APR =	0.06831867		
9				
10	formula in B6:	=(B2*(1+B2)^B1)/(-1+(1+B2)^B1)		
11	formula in B7:	=B6*B3		
12	formula in B8:	=12*RATE(B1,-B7,B3-0.01*B4*B3,0)		

The APR is about 6.83.%.

51

	A	B	C	D
1	n =	300		
2	i =	0.00542		
3	loan amount =	140,000.00		
4	points:	3		
5				
6	$1/a_n$ =	0.0067520716		
7	monthly payment =	945.29		
8	m =	84		
9	a_m =	127.1356748173		
10	unpaid balance =	120180.0853		
11	APR =	0.070787924		
12				
13				
14	formula in B6:	=(B2*(1+B2)^B1)/(-1+(1+B2)^B1)		
15	formula in B7:	=B6*B3		
16	formula in B9:	=(-1+(1+B2)^(B1-B8))/(B2*(1+B2)^(B1-B8))		
17	formula in B10:	=B9*B7		
18	formula in B11:	=12*RATE(B8,-B7,B3-0.01*B4*B3,-B10)		

The APR is about 7.08%.

53. Mortgage *A* costs an extra $1750 up front.
[difference in monthly payments] = $1181.61 − $1159.84 = $21.77
Using the Excel function NPER(.002, 21.77, −1750) gives 87.72 months.

55. Mortgage *A* costs an extra $2000 up front.
[difference in monthly payments] = $1303.85 − $1283.93 = $19.92
Using the Excel function NPER(.004, 19.92, −2000) gives 128.63 months.

Chapter 10 Supplementary Exercises

1. (d)

2. $\left[\dfrac{\frac{.06}{12}}{\left(1+\frac{.06}{12}\right)^{12\times10}-1} \right]($240,000) = 1464.49

3. The monthly mortgage payment should not exceed $\left(\dfrac{39,200}{12} \right)(.25) = $816.67.$

$\left[\dfrac{1-\left(1+\frac{.09}{12}\right)^{-12\times30}}{\frac{.09}{12}} \right]($816.67) = $101,497$

4. $50\left(1+\dfrac{.073}{365}\right)^{365} = 53.79

5. 9% compounded daily yields $\left(1+\dfrac{.09}{365}\right)^{365} -1 = .0942 = 9.42\%$ annually. 10% compounded

annually is better.

6. $\left[\dfrac{\left(1+\frac{.06}{12}\right)^{12\times5}-1}{\frac{.06}{12}}\right]$ ($200) = $13,954.01

7. a. $\left[\dfrac{\frac{.12}{12}}{1-\left(1+\frac{.12}{12}\right)^{-12\times15}}\right]$ ($200,000) = $2400.34

   b. $\left[\dfrac{1-\left(1+\frac{.12}{12}\right)^{-12(15-5)}}{\frac{.12}{12}}\right]$ ($2400.34) = $167,304.95

8. ($24,000)(1.005)^{120} = $43,665.52

9. $\dfrac{\$50,000}{\left(1+\frac{.06}{12}\right)^{12\times10}}$ = $27,481.64

10. $\dfrac{\$10,000}{\left(1+\frac{.06}{12}\right)^{12\times2}} + \dfrac{\$5000}{\left(1+\frac{.06}{12}\right)^{12\times3}}$ = $13,050.08

11. $\left[\dfrac{\frac{.06}{12}}{1-\left(1+\frac{.06}{12}\right)^{-12\times4}}\right]$ ($12,000 - $3,000) = $211.37

12. $\left[\dfrac{\frac{.04}{2}}{1-\left(1+\frac{.04}{2}\right)^{-2\times5}}\right]\left(1+\frac{.04}{2}\right)^{2\times2}$ ($100,000) = $12,050.34

13. $\dfrac{\$30,000}{\left(1+\frac{.06}{12}\right)^{12\times15}}$ = $12,224.47

    $105,003.50 - $12,224.47 = $92,779.03

    $\left[\dfrac{\frac{.06}{12}}{1-\left(1+\frac{.06}{12}\right)^{-12\times15}}\right]$ ($92,779.03) = $782.92

14. $\dfrac{\$100,000}{\left(1+\frac{.12}{12}\right)^{12\times10}}$ = $30,299.48

    $\left[\dfrac{\frac{.12}{12}}{1-\left(1+\frac{.12}{12}\right)^{-12\times10}}\right]$ ($509,289.22 - $30,299.48) = $6872.11

15. $\left[\dfrac{\left(1+\frac{.06}{12}\right)^{12\times30}-1}{\frac{.06}{12}}\right]$ ($100) = $100,451.50

16. $\left[\dfrac{1-\left(1+\frac{.12}{12}\right)^{-12\times10}}{\frac{.12}{12}}\right]$ ($2000) = \$139,401.04$

17. Investment A: $\left[\dfrac{(1+.06)^{10}-1}{.06}\right]1000 = \$13,180.79$

Investment B: $5000(1+.06)^5 + 5000 = \$11,691.13$

Thus Investment A is the better investment.

18. Present value of annuity is $\left[\dfrac{1-\left(1+\frac{.09}{12}\right)^{-12\times5}}{\frac{.09}{12}}\right]$ ($5) = \$240.87$

The present value of $1000 is $\dfrac{\$1000}{\left(1+\frac{.09}{12}\right)^{12\times5}} = \$638.70$

Yes, it is a bargain, since the present value is $240.87 + 638.70 = \$879.57.$

19. $\left(1+\dfrac{.10}{2}\right)^2 - 1 = .1025 = 10.25\%$

20. $\left(1+\dfrac{.18}{12}\right)^{12} - 1 = .1956 = 19.56\%$

21. $\left(1+\dfrac{.08}{4}\right)^{4\times15}$ ($10,000) + $\left[\dfrac{\left(1+\frac{.08}{4}\right)^{4\times15}-1}{\frac{.08}{4}}\right]$ ($1000) = \$146,861.85$

22. $R = \left[\dfrac{\frac{.06}{12}}{1-\left(1+\frac{.06}{12}\right)^{-36}}\right]$ ($10,000) = \$304.22$

Paym	Amount	Interest	Applied to Principal	Unpaid balance
1	$304.22	$50.00	$254.22	$9745.78
2	304.22	48.73	255.49	9490.29
3	304.22	47.45	256.77	9233.52
4	304.22	46.17	258.05	8975.47
5	304.22	44.88	259.34	8716.13
6	304.22	43.58	260.64	8455.49

23. $\left[\dfrac{(1.01)^{120}-1}{.01}\right]$ ($200)(1.01)^{120} = \$151,843.34$

24. $\left[\dfrac{\frac{.06}{12}}{1-\left(1+\frac{.06}{12}\right)^{-12\times5}}\right]$ ($300,000) = \$5799.84$

25. $\left[\dfrac{\frac{.09}{12}}{1-\left(1+\frac{.09}{12}\right)^{-12\times 30}} \right]($150,000) = \$1206.93$

26. **a.** $[\text{amount after taxes}] = (1-.30)(30,000)$
$$= \$21,000$$

     b. $30,000 \cdot (1.06)^5 = 40,146.77$
       $[\text{amount after taxes}] = (1-.35)(40,146.77) = \$26,095.40$

27. **a.** $[\text{amount after taxes}] = (1-0)(30,000) = \$30,000$

     b. $30,000 \cdot (1.06)^5 = \$40,146.77$

28. $r = \dfrac{12Rt - P}{Pt} = \dfrac{12(228.42)(2) - 5000}{5000(2)} = .048208 = 4.82\%$

29. Loan A is better, because the monthly payments for Loan B will be $\dfrac{3000(1.06)}{12} = \$265$.

30. $P = \$90,000 - .02(\$90,000) = \$88,200$
    $n = 15 \cdot 12 = 180$
    $R = \dfrac{.005}{1-(1.005)^{-80}} \cdot \$90,000 = \$759.47$

31. $m = 6 \cdot 12 = 72$
    $R = \$716.43$
    $P = \$100,000 - 0.3(\$100,000) = \$97,000$
    $B = \$81,298.32$

32. Mortgage A costs an extra \$2000 up front.
    $[\text{difference in monthly payments}] = \$1413.56 - \$1350.41 = \$63.15$
    $i = .00291667$, $R = \$63.15$, $P = -\$2000$
    Using the Excel function NPER(.00291667, 63.15, −2000) gives 33.3 months.

33. Through technology, you can find $x = .06$ on a graphing calculator by finding the intersection of
    $Y_1 = 245000$ and $Y_2 = \left((1+X/12)^{\wedge}240 - 1\right)/(X(1+X)^{\wedge}240) * 1755.21$.

34. Solve $567.79 - i(86,837.98) = \left[567.79 - i(97,000)\right](1+i)^{96}$, using technology, to find $i = .005$;
    the interest rate is $12i = .06$. On a graphing calculator, find the intersection of
    $Y_1 = 567.79 - X(86837.98)$ and $Y_2 = (567.79 - X * 97000)(1+X)^{\wedge}96$.

35. **a.** $P = \$380,000$ and $i = \dfrac{0.069}{12} = 0.00575$
     $Payment = iP = (0.00575)(\$380,000) = \$2185$

b. $Payment = \left[\dfrac{\frac{.069}{12}}{1 - \left(1 + \frac{.069}{12}\right)^{-12 \times 15}}\right]($380,000) = \$3394.34$

36. a.   For the first 5 years; $Payment = \left[\dfrac{\frac{.063}{12}}{1 - \left(1 + \frac{.063}{12}\right)^{-12 \times 25}}\right]($220,000) = \$1458.08$

b.   The balance after 5 years; $balance = \left[\dfrac{1 - \left(1 + \frac{.063}{12}\right)^{-12 \times 20}}{\frac{.063}{12}}\right]($1458.08) = \$198,690.34$

c.   For the sixth year, $P = \$198,690.34$, $n = 240$, and $i = 0.0455 + 0.028 = 0.0735$

$Payment = \left[\dfrac{\frac{.0735}{12}}{1 - \left(1 + \frac{.0735}{12}\right)^{-240}}\right]($198,690.34) = \$1582.46$

Conceptual Exercises

37. The effective rate will be slightly higher than the nominal rate.

38. No, not necessarily. It depends upon the number of compounding periods and the time.

39. No. Much more. For example, a house payment is a decreasing annuity. If you pay an additional 5% on the loan each month, the duration of the loan will decrease significantly more than just 5%.

40. The payment will decrease because the amount being applied to the principle will decrease. The total amount of interest paid will increase due to the length of time being added to the loan.

41. When you have successive payments, the interest on the loan is re-calculated more frequently. The interest on the loan is always the interest on the unpaid balance. If more frequent payments are made, the interest will decrease faster.

Chapter 10 Chapter Test

1. $\left(1 + \dfrac{9}{12} \times .065\right)($500) = \$524.38$

2. $\left[\dfrac{1}{(1 + 3 \times .04)}\right]($5000) = \$4464.29$

3. $F = \left(1 + \dfrac{.05}{4}\right)^{12}($1025) = \$1189.77$

$\$1189.77 - \$1025 = \$164.77$

4. $\left[\dfrac{1}{\left(1 + \frac{.08}{12}\right)^{120}}\right]($25,000) = \$11,263.09$

5. $\left[\dfrac{\left(1 + \frac{.06}{12}\right)^{180} - 1}{\frac{.06}{12}}\right]($250) = \$72,704.68$

6. $\left[\dfrac{1 - \left(1 + \frac{.06}{12}\right)^{-48}}{\frac{.06}{12}}\right]($500) = \$21,290.16$

7. $\left[\dfrac{\frac{.04}{2}}{\left(1 + \frac{.04}{2}\right)^{10} - 1}\right]($12,000) = \$1095.92$

8. $\left[\dfrac{\frac{.09}{12}}{1-\left(1+\frac{.09}{12}\right)^{-120}}\right]($200{,}000) = 2533.52

9. $\left(1+\dfrac{.06}{12}\right)^{48}($300) = 381.15

$\left[\dfrac{\left(1+\frac{.06}{12}\right)^{48}-1}{\frac{.06}{12}}\right]($50) = 2704.89

$$381.15 + $2704.89 = 3086.04

10. $\dfrac{.03}{2}($2499.93) = 37.50 interest

$350 - $37.50 = 312.50 paid on principal
$2499.93 - $312.50 = 2187.43

11. a. $\left[\dfrac{\frac{.09}{12}}{1-\left(1+\frac{.09}{12}\right)^{-60}}\right]($8000) = 166.07

    b. $\dfrac{.09}{12}($8000) = 60

    c. $\left(1+\dfrac{.09}{12}\right)($8000) - $166.07 = 7893.93

    d. $\left[\dfrac{1-\left(1+\frac{.09}{12}\right)^{-12}}{\frac{.09}{12}}\right]($166.07) = 1899.00

12. $\left[\dfrac{1-\left(1+\frac{.06}{4}\right)^{-100}}{\frac{.06}{4}}\right]($3500) = $180{,}686.46$

$180{,}686.46 + $45{,}000 = $225{,}686.46$

13. a. $F = \left[\dfrac{\left(1+\frac{.06}{1}\right)^{11}-1}{\frac{.06}{1}}\right]($5000) = $74{,}858.21$

This money then earns interest compounded annually for 29 years and grows to

$Balance = ($74{,}858.21)(1.06)^{29} = $405{,}610.84$

    b. $F = \left[\dfrac{\left(1+\frac{.06}{1}\right)^{29}-1}{\frac{.06}{1}}\right]($5000) = $368{,}198.99$

14. $R = \dfrac{P(1+rt)}{12t} = \dfrac{4000(1+.10\cdot 3)}{12(3)}$

$\quad = \$144.44$

15. $P = 250{,}000 - 2500 = 247{,}500$
Using the Excel function 12*RATE(180, –2510.31, 247500, 0) gives .09 or 9%.

16. a. It will decrease because the cost of the points will be spread over a longer period of time.

  b. Increasing the number of points will increase the effective cost of the loan, therefore, increasing its interest rate.

17. If there are no points, or the mortgage is not terminated early.

18. a. $P = \$500{,}000$ and $i = \dfrac{0.078}{12} = 0.0065$

$\quad Payment = iP = (0.0065)(\$500{,}000) = \$3250$

  b. $Payment = \left[\dfrac{\frac{.078}{12}}{1-\left(1+\frac{.078}{12}\right)^{-12\times 20}}\right](\$500{,}000) = \$4120.18$

Chapter 11

Exercises 11.1

1. $a = 4, b = -6, \dfrac{b}{1-a} = \dfrac{-6}{1-4} = 2$

3. $a = -\dfrac{1}{2}, \ b = 0, \ \dfrac{b}{1-a} = \dfrac{0}{1+\frac{1}{2}} = 0$

5. $a = -\dfrac{2}{3}, \ b = 15, \ \dfrac{b}{1-a} = \dfrac{15}{1+\frac{2}{3}} = 9$

7. a. $y_0 = 10, \ y_1 = \dfrac{1}{2}(10) - 1 = 4,$

$y_2 = \dfrac{1}{2}(4) - 1 = 1, \ y_3 = \dfrac{1}{2}(1) - 1 = -\dfrac{1}{2},$

$y_4 = \dfrac{1}{2}\left(-\dfrac{1}{2}\right) - 1 = -\dfrac{5}{4}$

b.

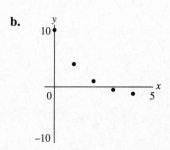

c. $y_n = \dfrac{-1}{1-\frac{1}{2}} + \left(10 - \dfrac{-1}{1-\frac{1}{2}}\right)\left(\dfrac{1}{2}\right)^n$

$= -2 + 12\left(\dfrac{1}{2}\right)^n$

9. a. $y_0 = 3.5, \ y_1 = 2(3.5) - 3 = 4,$

$y_2 = 2(4) - 3 = 5, \ y_3 = 2(5) - 3 = 7,$

$y_4 = 2(7) - 3 = 11$

b.

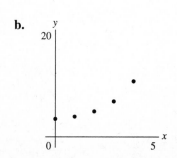

c. $y_n = \dfrac{-3}{1-2} + \left(3.5 - \dfrac{-3}{1-2}\right)(2)^n$

$= 3 + (.5)2^n$

11. a. $y_0 = 17.5, \ y_1 = -.4(17.5) + 7 = 0,$

$y_2 = -.4(0) + 7 = 7,$

$y_3 = -.4(7) + 7 = 4.2,$

$y_4 = -.4(4.2) + 7 = 5.32$

b.

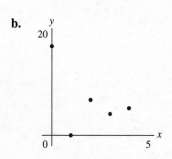

c. $y_n = \dfrac{7}{1+.4} + \left(17.5 - \dfrac{7}{1+.4}\right)(-.4)^n$

$= 5 + 12.5(-.4)^n$

13. a. $y_0 = 15, \ y_1 = 2(15) - 16 = 14,$

$y_2 = 2(14) - 16 = 12,$

$y_3 = 2(12) - 16 = 8,$

$y_4 = 2(8) - 16 = 0$

b.

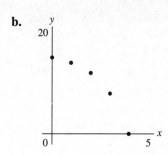

c. $y_n = \dfrac{-16}{1-2} + \left(15 - \dfrac{-16}{1-2}\right)(2)^n$

$= 16 - 2^n$

15. $y_0 = 6 - 5(.2)^0 = 1$

$y_1 = 6 - 5(.2)^1 = 5$

$y_2 = 6 - 5(.2)^2 = 5.8$

$y_3 = 6 - 5(.2)^3 = 5.96$

$y_4 = 6 - 5(.2)^4 = 5.992$

17. $y_n = 1.05 y_{n-1}, \ y_0 = 1000$

19. $y_n = .99 y_{n-1} - 1,000,000, \ \ y_0 = 70,000,000$

21. a. $y_0 = 1, \ y_1 = 1 + 2 = 3, \ y_2 = 3 + 2 = 5,$

$y_3 = 5 + 2 = 7, \ y_4 = 7 + 2 = 9$

b.

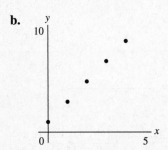

The points lie on a straight line.

c. $a = 1$, so the denominator of $\dfrac{b}{1-a}$ is zero.

23. $1.20(55) - 36 = \$30$

25. a. $y_n = 1.04 y_{n-1} + 250, \ y_0 = 800$

b. $y_n = \dfrac{250}{1 - 1.04} + \left(800 - \dfrac{250}{1 - 1.04} \right)(1.04)^n$

$y_n = -6250 + 7050(1.04)^n$

c. $y_7 = -6250 + 7050(1.04)^7 \approx 3027.32$

about \$3027.32

27. a. $y_n = 0.85 y_{n-1} + 0$

$a = 0.85, b = 0$

Use the formula: $y_n = \dfrac{b}{1-a} + \left(y_0 - \dfrac{b}{1-a} \right) a^n$

$y_n = \dfrac{0}{1 - 0.85} + \left(20,000 - \dfrac{0}{1 - 0.85} \right) 0.85^n$

b. $y_n = 20,000 \cdot 0.85^n$

c. After 5 years, $n = 5$, so $y_5 = 20{,}000 \cdot 0.85^5 = 8874.11$.

29. a.–b.

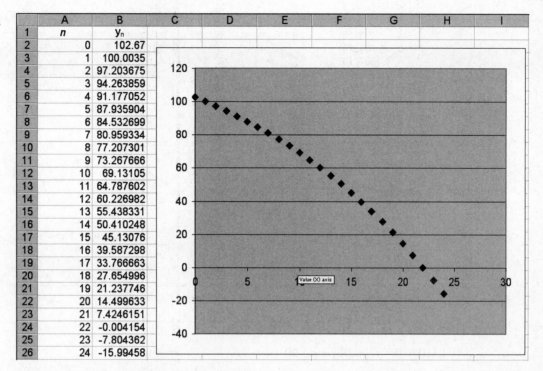

	A	B
1	*n*	y_n
2	0	102.67
3	1	100.0035
4	2	97.203675
5	3	94.263859
6	4	91.177052
7	5	87.935904
8	6	84.532699
9	7	80.959334
10	8	77.207301
11	9	73.267666
12	10	69.13105
13	11	64.787602
14	12	60.226982
15	13	55.438331
16	14	50.410248
17	15	45.13076
18	16	39.587298
19	17	33.766663
20	18	27.654996
21	19	21.237746
22	20	14.499633
23	21	7.4246151
24	22	-0.004154
25	23	-7.804362
26	24	-15.99458

c. $y_{14} = 50.41$; $y_n \approx 0$ for $n = 22$

31. a.–b.

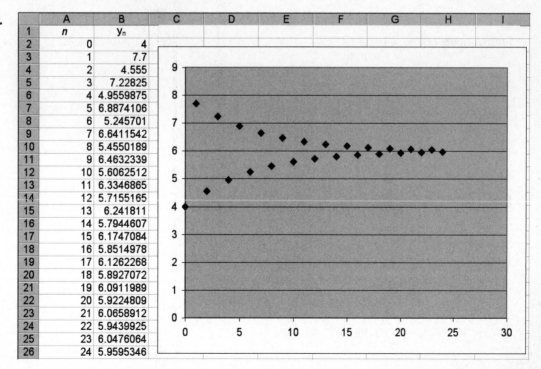

	A	B
1	*n*	y_n
2	0	4
3	1	7.7
4	2	4.555
5	3	7.22825
6	4	4.9559875
7	5	6.8874106
8	6	5.245701
9	7	6.6411542
10	8	5.4550189
11	9	6.4632339
12	10	5.6062512
13	11	6.3346865
14	12	5.7155165
15	13	6.241811
16	14	5.7944607
17	15	6.1747084
18	16	5.8514978
19	17	6.1262268
20	18	5.8927072
21	19	6.0911989
22	20	5.9224809
23	21	6.0658912
24	22	5.9439925
25	23	6.0476064
26	24	5.9595346

c. $y_{12} = 5.715516$; $|6 - y_n| < .1$ for $n \ge 19$

33. a.–b.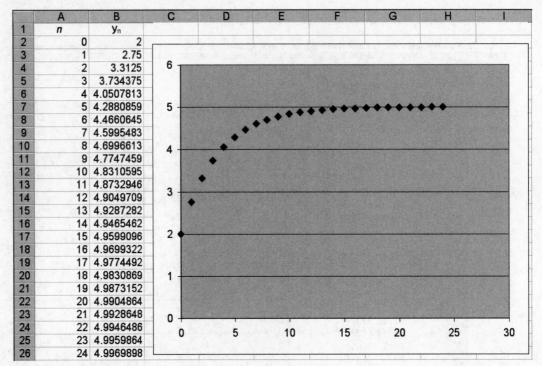

	A	B
1	*n*	y_n
2	0	2
3	1	2.75
4	2	3.3125
5	3	3.734375
6	4	4.0507813
7	5	4.2880859
8	6	4.4660645
9	7	4.5995483
10	8	4.6996613
11	9	4.7747459
12	10	4.8310595
13	11	4.8732946
14	12	4.9049709
15	13	4.9287282
16	14	4.9465462
17	15	4.9599096
18	16	4.9699322
19	17	4.9774492
20	18	4.9830869
21	19	4.9873152
22	20	4.9904864
23	21	4.9928648
24	22	4.9946486
25	23	4.9959864
26	24	4.9969898

c. $y_{10} = 4.831059; |5 - y_n| < .01$ for $n \geq 20$

Exercises 11.2

1. $b = 5, \ y_0 = 1$

From formula (2), $y_n = 1 + 5n$.

3. $y_0 = 80, i = \dfrac{.09}{12} = .0075, \ n = 5 \times 12 = 60$

From formula (4), $y_n = 80(1.0075)^{60}$

5. $y_0 = 80, i = \dfrac{1}{365}, \ n = 5 \times 365 = 1825$

From formula (4), $y_n = 80\left(1 + \dfrac{1}{365}\right)^{1825}$.

7. $y_0 = 80, \ i = .07, n = 5$

From formula (3), $y_n = 80 + .07 \times 80 \times 5 = 108$

9. $y_0 = 1$

a. $i = .40, n = 1, \ y_n = 1(1.40)^1 = \$1.40$

b. $i = \dfrac{.40}{2} = .20, \ n = 2 \times 1 = 2,$

$y_n = 1(1.20)^2 = \$1.44$

c. $i = \dfrac{.40}{4} = .10,\ n = 4 \times 1 = 4,$

     $y_n = 1(1.10)^4 \approx \$1.46$

11. a. $y_0 = 10;\ y_1 = 2(10) - 10 = 10;$

      $y_2 = 10;\ y_3 = 10;\ y_4 = 10$

      points lie on a horizontal line

b. $y_0 = 11;\ y_1 = 2(11) - 10 = 12;$

      $y_2 = 2(12) - 10 = 14;$

      $y_3 = 2(14) - 10 = 18;\ y_4 = 2(18) - 10 = 26;$

      points curve upward

c. $y_0 = 9;\ y_1 = 2(9) - 10 = 8;$

      $y_2 = 2(8) - 10 = 6;\ y_3 = 2(6) - 10 = 2;$

      $y_4 = 2(2) - 10 = -6;$ points curve downward

13. $a = .4,\ b = 3;\ y_n = \dfrac{3}{1 - .4} + \left(7 - \dfrac{3}{1 - .4}\right)(.4)^n$

           $= 5 + 2(.4)^n;$

   as n gets large, y_n approaches 5.

15. $a = -5,\ b = 0;$

     $y_n = \dfrac{0}{1 + 5} + \left(2 - \dfrac{0}{1 + 5}\right)(-5)^n = 2(-5)^n;$

   as n gets large, y_n gets arbitrarily large,

   alternating between being positive and negative.

17. $a = 1 + \dfrac{.066}{12} = 1.0055,\ b = -1600;$

     $y_n = 1.0055 y_{n-1} - 1600,\ y_0 = 250,525$

19. $a = 1,\ b = \dfrac{50,000}{25} = 2000;$

     $y_n = y_{n-1} - 2000,\ y_0 = 50,000;$

     $y_n = 50,000 - 2000n$

21. $a = 2,\ b = 0;\ y_n = 2y_{n-1},\ y_0 = 500;$

     $y_n = 500(2)^n;$

     $500,000 = 500(2)^n$

     $n \approx 9.97$

     $9.97 \times 5 \approx 50$ min

     Answer (b) is correct.

23.   Rate = \$50/day

     $14 \times 50 = \$700$

     Answer (c) is correct.

25. $a = 2,\ b = 0;\ y_n = 2y_{n-1};\ y_n = y_0(2)^n$

     $y_5 = 100,000 = y_0(2)^5$

     $y_0 = \dfrac{100,000}{2^5} = 3125$

     Then $y_2 = 3125(2)^2 = 12,500.$

     Answer (c) is correct.

27. $a = 2,\ b = 0;\ y_n = 2y_{n-1};\ y_0 = 10^5$

     $y_n = 10^5(2)^n;\ n = 4$

     $y_4 = 10^5(2)^4$

     Answer (c) is correct.

29. $y_0 = 1000,\ i = \dfrac{.06}{4} = .015;\ y_n = 1.015 y_{n-1}$

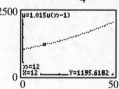

   For $n = 3 \times 4 = 12,\ y_{12} = \$1195.62;$

   $y_n = 1659$ for $n = 34,$ or $8\frac{1}{2}$ years;

   $y_n \geq 2000$ for $n \geq 47,$ so it will double in

   47 quarters.

31. $y_0 = 200,\ i = .045;\ y_n = y_{n-1} + 9$

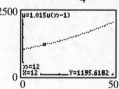

   For $n = 5,\ y_5 = \$245;$

   $y_n \geq 308$ for $n \geq 12$ years;

   $y_n \geq 400$ for $n \geq 23$ years

33. $y_n = 1.015y_{n-1} - 105$, $y_0 = 2000$

Set the calculator to sequence mode and enter the following in the "Y=" screen.

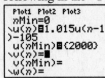

Table setup:

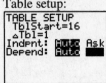

Table:

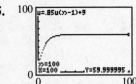

35.

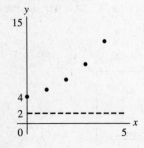

y_n approaches 60.

Exercises 11.3

1. (a), (b), (d), (f), (h)

3. (b), (d), (e), (f)

5. (b), (d), (e), (f)

7. (a), (c), (h), and possibly (g)

9. Possible answer:

11. Possible answer:

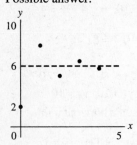

13. Possible answer:

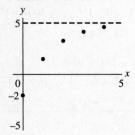

15. Draw $y = \dfrac{b}{1-a} = -2$ as a dashed line.

$a = 3 > 0$, so the graph is monotonic.

$|a| = 3 > 1$, so the graph is repelled from $y = -2$.

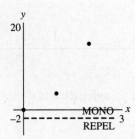

17. Draw $y = \dfrac{b}{1-a} = 6$ as a dashed line.

$y_0 = 6$, so the graph is constant.

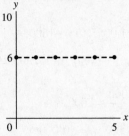

19. Draw $y = \dfrac{b}{1-a} = 4$ as a dashed line.

$a = -2 < 0$, so the graph is oscillating. $|a| = 2 > 1$, so the graph is repelled from $y = 4$.

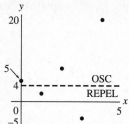

21. Draw $y = \dfrac{b}{1-a} = 10{,}000$ as a dashed line.

$a = .7 > 0$, so the graph is monotonic. $|a| = .7 < 1$, so the graph is attracted to $y = 10{,}000$.

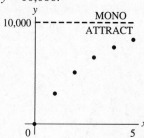

23. Draw $y = \dfrac{b}{1-a} = 1$ as a dashed line.

$a = -.6 < 0$, so the graph is oscillating. $|a| = .6 < 1$, so the graph is attracted to $y = 1$.

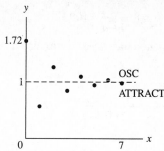

25. $a = 1 + i = 1.0075, b = -450$

The loan, y_0, must be less than

$$\frac{b}{1-a} = \$60{,}000.$$

27. $a = 1 + i = 1.06, b = -120$

   a. $y_n = 1.06 y_{n-1} - 120$

   b. The deposit, y_0, must be at least

$$\frac{b}{1-a} = \$2000.$$

29. $a = 1 + i = 1.05, b = -1500$

$y_n = 1.05 y_{n-1} - 1500$

The loan, y_0, must be less than

$\dfrac{b}{1-a} = \$30{,}000$. If it is greater than or equal to $\$30{,}000$, the loan will never be paid off.

For Exercises 31–35, choose y_0 as the beginning value; choose $a > 0$ or < 0 depending on whether the graph is monotonic or oscillating, also $|a| > 1$ or < 1 depending on whether the graph is unbounded or approaches a value; and if it approaches or is repelled from a value, choose b so that $\dfrac{b}{1-a}$ equals that value.

31. Possible answer: $y_n = .5 y_{n-1} + 4,\ y_0 = 1$

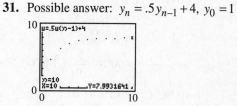

33. Possible answer: $y_n = 2 y_{n-1},\ y_0 = 1$

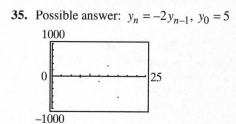

35. Possible answer: $y_n = -2 y_{n-1},\ y_0 = 5$

Exercises 11.4

1. $a = 1 + i = 1.0075, b = -261.50$;
$y_n = 1.0075 y_{n-1} - 261.50,\ y_0 = 32{,}500$

3. $a = 1 + i = 1.015$, $b = 200$;
$$y_n = 1.015 y_{n-1} + 200, \quad y_0 = 4000$$

5. $a = 1 + i = 1.01$, $b = -660$, $\dfrac{b}{1-a} = 66,000$

$$y_{120} = 0 = 66,000 + (y_0 - 66,000)(1.01)^{120}$$

$$y_0 = \frac{-66,000}{(1.01)^{120}} + 66,000 \approx \$46,002.34$$

7. $a = 1 + i = 1.06$, $b = 300$, $\dfrac{b}{1-a} = -5000$, $y_0 = 0$

$$y_{20} = -5000 + [0 - (-5000)](1.06)^{20}$$
$$\approx \$11,035.68$$

9. $a = 1 + i = 1.005$, $y_0 = 0$

$$y_{144} = 6000 = \frac{b}{-.005} + \left(0 - \frac{b}{-.005}\right)(1.005)^{144}$$
$$= \frac{(1.005)^{144} - 1}{.005} b$$

$$b = \frac{6000 \times .005}{(1.005)^{144} - 1} \approx \$28.55$$

11. $a = 1 + i = 1.01$, $y_0 = 4000$

$$y_{36} = 0 = \frac{b}{-.01} + \left(4000 - \frac{b}{-.01}\right)(1.01)^{36}$$
$$= 4000(1.01)^{36} + 100[(1.01)^{36} - 1]b$$

$$b = \frac{-4000(1.01)^{36}}{100[(1.01)^{36} - 1]} \approx -132.86$$
$$\$132.86$$

13. $a = 1 + i = 1.08$, $b = -4$
Use $n\text{Min} = 0$, $u(n) = 1.08u(n-1) - 4$,
$u(n\text{Min}) = \{45\}$
$u(10) \approx \$39.205$ million
$u(n) \le 0$ for $n = 30$ years

15. $a = 1 + i = 1.005$, $b = 100$
Use $n\text{Min} = 0$, $u(n) = 1.005u(n-1) + 100$,
$u(n\text{Min}) = \{0\}$
$u(5) \approx \$505.03$
$u(10) \approx \$1022.80$
$u(15) \approx \$1553.65$
$u(n) \ge \$3228$ for $n = 30$
$u(n) > \$4000$ for $n = 37$

Exercises 11.5

1. $a = 1 + .03 - .01 = 1.02$, $b = 0$, $\dfrac{b}{1-a} = 0$

$y_n = 1.02 y_{n-1}$, $y_0 = 100$ million
$a > 0$: monotonic; $|a| > 1$: repelled from
$y = 0$

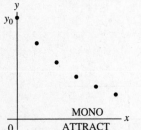

3. $a = 1 - .25 = .75$, $b = 0$, $\dfrac{b}{1-a} = 0$

$y_n = .75 y_{n-1}$
$a > 0$: monotonic; $|a| < 1$: attracted to $y = 0$

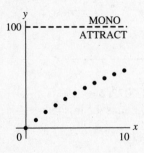

5. $y_n = y_{n-1} + .08(100 - y_{n-1}) = .92 y_{n-1} + 8$,
$y_0 = 0$

$a = .92$, $b = 8$, $\dfrac{b}{1-a} = 100$

$a > 0$: monotonic; $|a| < 1$: attracted to
$y = 100$

7. $y_n = y_{n-1} + .30(12 - y_{n-1})$
$\quad = .7y_{n-1} + 3.6, \ y_0 = 0$

$a = .7, b = 3.6, \dfrac{b}{1-a} = 12$

$a > 0$: monotonic; $|a| < 1$: attracted to $y = 12$

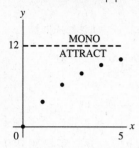

9. $a = 1 + i = 1.05, b = -1000, \dfrac{b}{1-a} = 20{,}000$

$\quad y_n = 1.05y_{n-1} - 1000, \ y_0 = 30{,}000$

$a > 0$: monotonic; $|a| > 1$: repelled from $y = 20{,}000$

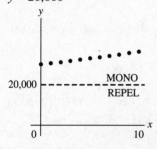

11. $y_n = y_{n-1} + .20(70 - y_{n-1}) = .8y_{n-1} + 14,$
$\quad y_0 = 40$

$a = .8, b = 14, \dfrac{b}{1-a} = 70$

$a > 0$: monotonic; $|a| < 1$: attracted to $y = 70$

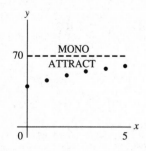

13. $p_n = 20 - .1q_n = 20 - .1(5p_{n-1} - 10)$
$\quad = -.5p_{n-1} + 21, \ p_0 = 4.54$

$a = -.5, b = 21, \dfrac{b}{1-a} = 14$

$a < 0$: oscillating, $|a| < 1$: attracted to $y = 14$

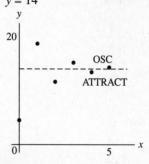

15. After 2 hours, 0.5 is left. After four hours, $(0.5)(0.5) = 0.25$ is left. Answer is b).

17. Use $n\text{Min} = 0$,
$u(n) = (1 + .035 - .02)u(n-1) - .0003,$
$u(n\text{Min}) = \{5\}$
$u(5) \approx 5.38$ million
$u(n) > 6$ for $n = 13$, or in the year 2022
$u(n) \geq 10$ for $n = 47$, or in the year 2056

19. $y_n = y_{n-1} - .13y_{n-1}$
$\quad y_n = (.87)y_{n-1}, \ y_0 = 130$

$a = .87, b = 0, \dfrac{b}{1-a} = 0$

$y_n = 130(.87)^n$

After 5 hours, $y_n = 130(0.87)^5 = 64.79$ (This problem can be solved through trial and error or by the use of logarithms)

After 24 hours, $y_n = 130(0.87)^{24} = 4.6$ milligrams will remain.

Chapter 11 Supplementary Exercises

1. a. $y_1 = -3(1) + 8 = 5; \quad y_2 = -3(5) + 8 = -7;$
$\qquad y_3 = -3(-7) + 8 = 29$

b. $a = -3, b = 8$

$\qquad y_n = \dfrac{b}{1-a} + \left(y_0 - \dfrac{b}{1-a}\right)a^n$

$\qquad = 2 + (1 - 2)(-3)^n$

$\qquad = 2 - (-3)^n$

   c. $2 - (-3)^4 = -79$

2. a. $y_1 = 10 - \dfrac{3}{2} = \dfrac{17}{2};\ y_2 = \dfrac{17}{2} - \dfrac{3}{2} = 7;$

   $y_3 = 7 - \dfrac{3}{2} = \dfrac{11}{2}$

   b. $a = 1,\ b = -\dfrac{3}{2}$

   $y_n = 10 - \dfrac{3}{2}n$

   c. $10 - \dfrac{3}{2}(6) = 1$

3. $a = 1 + i = 1 + \dfrac{.0305}{4} = 1.007625,\ b = 0,\ \dfrac{b}{1-a} = 0$

   $y_n = y_0(1.007625)^n$
   Solve.
   $2474 = y_0(1.007625)^{28}$
   $y_0 = \dfrac{2474}{1.2370} \approx \$2000$

4. $a = 1 + i = 1.001,\ b = 0,\ \dfrac{b}{1-a} = 0,\ y_0 = 1000$

   $y_{104} = 0 + (1000 - 0)(1.001)^{104}$
   $= \$1109.54$

5. Draw $y = \dfrac{b}{1-a} = 6$ as a dashed line.

   $a = -\dfrac{1}{3} < 0,$ so the graph is oscillating.

   $|a| = \dfrac{1}{3} < 1,$ so the graph is attracted to

   $y = 6.$

6. Draw $y = \dfrac{b}{1-a} = 4$ as a dashed line.

   $a = 1.5 > 0,$ so the graph is monotonic.

   $|a| = 1.5 > 1,$ so the graph is repelled from $y = 4$.

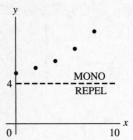

7. a. $y_n = 1.03y_{n-1} - 600,\ y_0 = 120,000$

   b. $\dfrac{b}{1-a} = 20,000;$

   $y_{20} = 20,000 + (120,000 - 20,000)(1.03)^{20}$
   $\approx 200,611$

8. a. $y_n = 1.01y_{n-1} - 360,\ y_0 = 35,000$

   b. $\dfrac{b}{1-a} = 36,000;$

   $y_{84} = 36,000 + (35,000 - 36,000)(1.01)^{84}$
   $\approx \$33,693.28$

9. $a = 1 + i = 1.001,\ y_0 = 0,\ 21 \times 52 = 1092$

   $y_{1092} = 40,000$
   $= \dfrac{b}{-.001} + \left(0 - \dfrac{b}{-.001}\right)(1.001)^{1092}$
   $= 1000[(1.001)^{1092} - 1]b$
   $b = \dfrac{40,000}{1000[(1.001)^{1092} - 1]} \approx \$20.22$

10. $a = 1 + i = 1.005,\ y_0 = 33,100$

   $y_{240} = 0$
   $= \dfrac{b}{-.005} + \left(33,100 - \dfrac{b}{-.005}\right)(1.005)^{240}$
   $= 33,100(1.005)^{240} + 200[(1.005)^{240} - 1]b$
   $b = \dfrac{-33,100(1.005)^{240}}{200[(1.005)^{240} - 1]} \approx -237.14$
   $\$237.14$

11. $a = 1 + i = 1.08$, $b = -2400$, $\dfrac{b}{1-a} = \$30,000$

$$y_{18} = 0 = 30,000 + (y_0 - 30,000)(1.08)^{18}$$

$$y_0 = \dfrac{-30,000}{(1.08)^{18}} + 30,000 \approx \$22,492.53$$

12. $y_n = 1.005 y_{n-1} + 50$, $y_0 = 0$

$a = 1.005$, $b = 50$, $\dfrac{b}{1-a} = -10000$

$$y_{48} = -10000 + [0 - (-10000)](1.005)^{48}$$
$$\approx \$2704.89$$

13. $y_n = y_{n-1} + .10(1,000,000 - y_{n-1})$
$\quad = .9 y_{n-1} + 100,000$, $y_0 = 0$

$a = .9$, $b = 100,000$, $\dfrac{b}{1-a} = 1,000,000$

$a > 0$: monotonic, $|a| < 1$: attracted to
$y = 1,000,000$

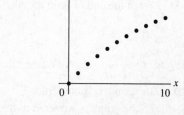

14. $y_n = y_{n-1} - .08 y_{n-1} = .92 y_{n-1}$, $y_0 = 100$

$a = .92$, $b = 0$, $\dfrac{b}{1-a} = 0$

$a > 0$: monotonic; $|a| < 1$: attracted to $y = 0$

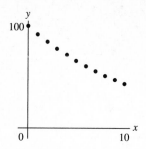

Chapter 11 Chapter Test

1. **a.** $y_1 = \dfrac{1}{2} y_0 + 3 = \dfrac{1}{2}(-2) + 3 = 2$

$\quad y_2 = \dfrac{1}{2} y_1 + 3 = \dfrac{1}{2}(2) + 3 = 4$

$\quad y_3 = \dfrac{1}{2} y_2 + 3 = \dfrac{1}{2}(4) + 3 = 5$

b. $y_n = \dfrac{3}{1 - \frac{1}{2}} + \left(-2 - \dfrac{3}{1 - \frac{1}{2}}\right)\left(\dfrac{1}{2}\right)^n$

$\quad y_n = 6 - 8\left(\dfrac{1}{2}\right)^n$

c. $y_4 = 6 - 8\left(\dfrac{1}{2}\right)^4 = 5.5$

2. **a.** $y_1 = y_0 + 50 = 60 + 50 = 110$
$\quad y_2 = y_1 + 50 = 110 + 50 = 160$
$\quad y_3 = y_2 + 50 = 160 + 50 = 210$

b. $y_n = y_0 + bn$
$\quad y_n = 60 + 50n$

c. $y_4 = 60 + 50(4) = 260$

3. **a.** Draw the line $y = \dfrac{b}{1-a} = 50$ as a dashed
line. $a = .6 > 0$, so the graph is monotonic.
$|a| = .6 < 1$, so the graph is attracted to the
line $y = 50$.

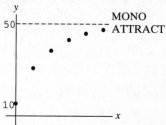

b. Draw the line $y = \dfrac{b}{1-a} = 4$ as a dashed line.

$a = -1.25 < 0$, so the graph is oscillating.

$|a| = 1.25 > 1$ so the graph is repelled from the line $y = 4$.

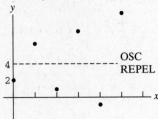

4. $y_n = .8y_{n-1} + b; \; y_0 = 1$

$y = \dfrac{b}{1-a} = \dfrac{b}{1-.8} = 1.5$

$b = .3$

Add .3 ppm each day.

5. a. $31,726

b. $2 \times 4 = 8\%$

c. $3000

d. $y_1 = 1.02(31,726) - 3000$
$= \$29,360.52$

6. a. $100 + 354.15 = 454.15$

b. $8,888.01i = 92.58$
$\quad\quad i \approx .010417$
$12i = .125$
12.5%

c. $A = 8888.01 + 357.84 = \$9245.85$
$C = 454.15 - 92.58 = \$361.57$
$B = 8888.01 - 361.57 = \$8526.44$

7. $i = \dfrac{.09}{12} = .0075;$

$y_n = 1.0075y_{n-1} - 1000; \; y_{300} = 0;$

$\dfrac{b}{1-a} = \dfrac{-1000}{1-1.0075} \approx 133,333.33;$

$y_{300} = 133,333.33 + (y_0 - 133,333.33)(1.0075)^{300}$

$0 = 133,333.33 + (y_0 - 133,333.33)(1.0075)^{300}$

$y_0 = 119,161.62$

The amount that can be borrowed is $119,161.62.

8. a. $y_n = 1.06y_{n-1} + 500; \; y_0 = 0$

b. $\dfrac{b}{1-a} = \dfrac{500}{1-1.06} \approx -8333.333;$

$y_n = -8333.333 + [0 - (-8333.333)](1.06)^n$

$\quad = -8333.333 + 8333.333(1.06)^n$

$y_{17} = -8333.333 + 8333.333(1.06)^{17}$

$\quad = 14,106.44$

After 17 years the account contains $14,106.44.

9. $y_n = 1.005y_{n-1} + D; \; y_0 = 0;$

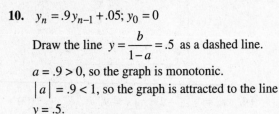

$y_n = \dfrac{D}{1-1.005} + \left(0 - \dfrac{D}{1-1.005}\right)(1.005)^n$

$\quad = -200D + 200D(1.005)^n;$

$y_{120} = -200D + 200D(1.005)^{120}$

$20,000 = -200D + 200D(1.005)^{120}$

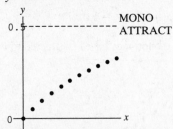

$D = \dfrac{20,000}{-200 + 200(1.005)^{120}} \approx 122.041$

Deposit $122.04 at the end of each month.

10. $y_n = .9y_{n-1} + .05; \; y_0 = 0$

Draw the line $y = \dfrac{b}{1-a} = .5$ as a dashed line.

$a = .9 > 0$, so the graph is monotonic.

$|a| = .9 < 1$, so the graph is attracted to the line $y = .5$.

Chapter 12

Exercises 12.1

1. Statement

3. Statement

5. Not a statement—not a declarative sentence.

7. Not a statement—not a declarative sentence.

9. Statement

11. Not a statement—x is not specified.

13. Not a statement—not a declarative sentence.

15. Statement

17. p: The Phelps library is in New York.
    q: The Phelps library is in Dallas.
    Then we have $p \lor q$

19. p: The Smithsonian Museum of Natural History has displays of rocks.
    q: The Smithsonian Museum of Natural History has displays of bugs.
    Then we have $p \land q$

21. p: Amtrak trains go to Chicago.
    q: Amtrak trains go to Cincinnati.
    Then we have $\sim p \land \sim q$ or $\sim(p \lor q)$

23. **a.** Ozone is opaque to ultraviolet light, and life on earth requires ozone.

    b. Ozone is not opaque to ultraviolet light, or life on earth requires ozone.

    c. Ozone is not opaque to ultraviolet light, or else life on earth does not require ozone.

    d. It is not the case that life on earth does not require ozone.

25. **a.** Florida borders Alabama or Florida borders Mississippi: $p \lor q$

    b. Florida borders Alabama and Florida does not border Mississippi $p \land \sim q$

    c. Florida borders Mississippi and Florida does not border Alabama $q \land \sim p$

    d. Florida does not border Alabama and Florida does not border Mississippi $\sim p \land \sim q$

Exercises 12.2

1. Since r is a statement form, so is $\sim r$. Then $p \land \sim r$ is a statement form, and so is $\sim(p \land \sim r)$. Since q is a statement form, $\sim(p \land \sim r) \lor q$ is a statement form.

3. Since p is a statement form, so is $\sim p$. Since q and r are statement forms, so are $\sim p \vee r$ and $q \wedge r$, and hence also $(\sim p \vee r) \rightarrow (q \wedge r)$.

5.

p	q	p	$\wedge$	$\sim$	q
T	T	T	**F**	F	T
T	F	T	**T**	T	F
F	T	F	**F**	F	T
F	F	F	**F**	T	F
(1)	(2)		(4)	(3)	

7.

p	q		$(p$	$\vee$	$\sim q)$	$\vee$	$\sim p)$
T	T		T	F	**T**		F
T	F		T	T	**T**		F
F	T		F	F	**T**		T
F	F		T	T	**T**		T
(1)	(2)			(5)	(3)	(6)	(4)

9.

p	q	$\sim$	$((p \vee q)$	$\wedge$	$(p \wedge q))$
T	T	**F**	T	T	T
T	F	**T**	T	F	F
F	T	**T**	T	F	F
F	F	**T**	F	F	F
(1)	(2)	(6)	(3)	(5)	(4)

11.

p	q		p	$\oplus$	$(\sim p$	$\vee$	$q)$
T	T		T	**F**	F	T	T
T	F		T	**T**	F	F	F
F	T		F	**T**	T	T	T
F	F		F	**T**	T	T	F
(1)	(2)		(5)	(3)	(4)		

13.

p	q	r		$(p$	$\wedge$	$\sim$	$r)$	$\vee$	q
T	T	T		T	F	F	T	**T**	T
T	T	F		T	T	T	F	**T**	T
T	F	T		T	F	F	T	**F**	F
T	F	F		T	T	T	F	**T**	F
F	T	T		F	F	F	T	**T**	T
F	T	F		F	F	T	F	**T**	T
F	F	T		F	F	F	T	**F**	F
F	F	F		F	F	T	F	**F**	F
(1)	(2)	(3)			(5)	(4)		(6)	

15.

p	*q*	*r*	~	[(*p*	∧	*r*)	∨	*q*]
T	T	T	**F**	T	T	T	T	T
T	T	F	**F**	T	F	F	T	T
T	F	T	**F**	T	T	T	T	F
T	F	F	**T**	T	F	F	F	F
F	T	T	**F**	F	F	T	T	T
F	T	F	**F**	F	F	F	T	T
F	F	T	**T**	F	F	T	F	F
F	F	F	**T**	F	F	F	F	F
(1)	(2)	(3)	(6)		(4)		(5)	

17.

p	*p*	∨	~	*p*
T	T	**T**	F	T
F	F	**T**	T	F
(1)		(3)	(2)	

19.

p	*q*	*r*	*p*	⊕	(*q*	∨	*r*)
T	T	T	T	**F**	T	T	T
T	T	F	T	**F**	T	T	F
T	F	T	T	**F**	F	T	T
T	F	F	T	**T**	F	F	F
F	T	T	F	**T**	T	T	T
F	T	F	F	**T**	T	T	F
F	F	T	F	**T**	F	T	T
F	F	F	F	**F**	F	F	F
(1)	(2)	(3)		(5)		(4)	

21.

p	*q*	*r*	(*p*	∨	*q*)	∧	(*p*	∨	*r*)
T	T	T	T	T	T	**T**	T	T	T
T	T	F	T	T	T	**T**	T	T	F
T	F	T	T	T	F	**T**	T	T	T
T	F	F	T	T	F	**T**	T	T	F
F	T	T	F	T	T	**T**	F	T	T
F	T	F	F	T	T	**F**	F	F	F
F	F	T	F	F	F	**F**	F	T	T
F	F	F	F	F	F	**F**	F	F	F
(1)	(2)	(3)		(4)		(6)		(5)	

23.

p	q	(p	∨	q)	∧	~	(p	∨	q)
T	T	T	T	T	**F**	F	T	T	T
T	F	T	T	F	**F**	F	T	T	F
F	T	F	T	T	**F**	F	F	T	T
F	F	F	F	F	**F**	T	F	F	F
(1)	(2)		(3)		(6)	(5)		(4)	

25.

p	q	r	~	(p	∨	q)	∧	r
T	T	T	F	T	T	T	**F**	T
T	T	F	F	T	T	T	**F**	F
T	F	T	F	T	T	F	**F**	T
T	F	F	F	T	T	F	**F**	F
F	T	T	F	F	T	T	**F**	T
F	T	F	F	F	T	T	**F**	F
F	F	T	T	F	F	F	**T**	T
F	F	F	T	F	F	F	**F**	F
(1)	(2)	(3)	(5)		(4)		(6)	

27.

p	q	r	~	p	∨	(q	∧	r)
T	T	T	F	T	**T**	T	T	T
T	T	F	F	T	**F**	T	F	F
T	F	T	F	T	**F**	F	F	T
T	F	F	F	T	**F**	F	F	F
F	T	T	T	F	**T**	T	T	T
F	T	F	T	F	**T**	T	F	F
F	F	T	T	F	**T**	F	F	T
F	F	F	T	F	**T**	F	F	F
(1)	(2)	(3)	(4)		(6)		(5)	

29.

p	q	~	p	∨	~	q		~	(p	∧	q)
T	T	F	T	**F**	F	T		**F**	T	T	T
T	F	F	T	**T**	T	F		**T**	T	F	F
F	T	T	F	**T**	F	T		**T**	F	F	T
F	F	T	F	**T**	T	F		**T**	F	F	F
(1)	(2)	(3)		(5)	(4)			(4)		(3)	

They are identical.

31.

p	q	p	⊕	q		(p	∨	q)	∧	~	(p	∧	q)
T	T	T	**F**	T		T	T	T	**F**	F	T	T	T
T	F	T	**T**	F		T	T	F	**T**	T	T	F	F
F	T	F	**T**	T		F	T	T	**T**	T	F	F	T
F	F	F	**F**	F		F	F	F	**F**	T	F	F	F
(1)	(2)		(3)				(4)		(6)	(5)		(3)	

They are identical.

33.

p	q	r	(p	∧	q)	∨	r	p	∧	(q	∨	r)
T	T	T	T	T	T	**T**	T	T	**T**	T	T	T
T	T	F	T	T	T	**T**	F	T	**T**	T	T	F
T	F	T	T	F	F	**T**	T	T	**T**	F	T	T
T	F	F	T	F	F	**F**	F	T	**F**	F	F	F
F	T	T	F	F	T	**T**	T	F	**F**	T	T	T
F	T	F	F	F	T	**F**	F	F	**F**	T	T	F
F	F	T	F	F	F	**T**	T	F	**F**	F	T	T
F	F	F	F	F	F	**F**	F	F	**F**	F	F	F
(1)	(2)	(3)		(4)		(6)			(5)		(4)	

$(p \wedge q) \vee r$ is T and $p \wedge (q \vee r)$ is F when p is F and r is T. Otherwise the tables are identical.

35. a.

p	p		p
T	T	**F**	T
F	F	**T**	F
(1)		(2)	

b.

p	q	(p		p)		(q		q)
T	T	T	F	T	**T**	T	F	T
T	F	T	F	T	**T**	F	T	F
F	T	F	T	F	**T**	T	F	T
F	F	F	T	F	**F**	F	T	F
(1)	(2)		(3)		(5)		(4)	

c.

p	q	(p		q)		(p		q)
T	T	T	F	T	**T**	T	F	T
T	F	T	T	F	**F**	T	T	F
F	T	F	T	T	**F**	F	T	T
F	F	F	T	F	**F**	F	T	F
(1)	(2)		(3)		(5)		(4)	

d.

p	q	p		((p		q)		q)
T	T	T	**F**	T	F	T	T	T
T	F	T	**F**	T	T	F	T	F
F	T	F	**T**	F	T	T	F	T
F	F	F	**T**	F	T	F	T	F
(1)	(2)		(5)		(3)		(4)	

37. p has truth value T and q has truth value F.

    a. $p \quad \vee \quad \sim \quad q$

            T   **T**   T   F

    b. $\sim \quad p \quad \wedge \quad q$

            F   T   **F**   F

    c. $p \quad \oplus \quad q$

            T   **T**   F

    d. $\sim \quad p \quad \oplus \quad q$

            F   T   **F**   F

    e. $\sim \quad (p \quad \oplus \quad q)$

           F   T   T   F

    f. $(p \quad \vee \quad q) \quad \oplus \quad \sim \quad q$

          T   T   F   **F**   T   F

39. p has truth value F and q has truth value T.

    a. $p \quad \wedge \quad \sim \quad q$

            F   **F**   F   T

    b. $\sim \quad (p \quad \oplus \quad q)$

           F   F   T   T

    c. $p \quad \wedge \quad q$

            F   **F**   T

    d. $\sim \quad p \quad \wedge \quad \sim \quad q$

           T   F   **F**   F   T

41. $(p \oplus q) \wedge r$

$$
\begin{array}{c}
(p \oplus q) \wedge r \\
\text{F} \\
\diagdown \\
p \oplus q \qquad r \\
\text{T} \qquad \text{F} \\
\diagdown \\
p \qquad q \\
\text{T} \qquad \text{F}
\end{array}
$$

43. $(p \vee q) \wedge (p \vee \sim r)$

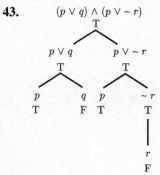

45.
$(p$	$\wedge$	$q)$	$\vee$	$\sim$	p
T	F	F	**F**	F	T

0 will be displayed.

47. a.

p	q		$(p$	$\vee$	$q)$	$\oplus$	$\sim$	p
T	T		T	T	T	**T**	F	T
(1)	(2)			(3)		(5)	(4)	

Calculator will display 1.

b.

p	q		$(p$	$\vee$	$q)$	$\oplus$	$\sim$	p
T	F		T	T	F	**T**	F	T
(1)	(2)			(3)		(5)	(4)	

Calculator will display 1.

49. a.

p	q		$(p$	$\wedge$	$\sim$	$q)$	$\oplus$	p
T	T		T	F	F	T	**T**	T
(1)	(2)		(4)		(3)		(5)	

Calculator will display 1.

b.

p	q		$(p$	$\wedge$	$\sim$	$q)$	$\oplus$	p
T	F		T	T	T	F	**F**	T
(1)	(2)			(4)	(3)		(5)	

Calculator will display 0.

51.

P	Q		♦ 3
1	1	1	
1	0	0	
0	1	1	
0	0	1	
------	------	------	

L1 ="not(LP xor L

Exercises 12.3

1.

p	q		p	→	~	q
T	T		T	**F**	F	T
T	F		T	**T**	T	F
F	T		F	**T**	F	T
F	F		F	**T**	T	F
(1)	(2)			(4)	(3)	

3.

p	q		(p	⊕	q)	→	q
T	T		T	F	T	**T**	T
T	F		T	T	F	**F**	F
F	T		F	T	T	**T**	T
F	F		F	F	F	**T**	F
(1)	(2)			(3)		(4)	

5.

p	q	r		(~	p	∧	q)	→	r
T	T	T		F	T	F	T	**T**	T
T	T	F		F	T	F	T	**T**	F
T	F	T		F	T	F	F	**T**	T
T	F	F		F	T	F	F	**T**	F
F	T	T		T	F	T	T	**T**	T
F	T	F		T	F	T	T	**F**	F
F	F	T		T	F	F	F	**T**	T
F	F	F		T	F	F	F	**T**	F
(1)	(2)	(3)			(4)		(5)		(6)

7.

p	q		(p	→	q)	↔	(~	p	∨	q)
T	T		T	T	T	**T**	F	T	T	T
T	F		T	F	F	**T**	F	T	F	F
F	T		F	T	T	**T**	T	F	T	T
F	F		F	T	F	**T**	T	F	T	F
(1)	(2)			(4)		(6)	(3)		(5)	

9.

p	q	r		(p	→	q)	→	r
T	T	T		T	T	T	**T**	T
T	T	F		T	T	T	**F**	F
T	F	T		T	F	F	**T**	T
T	F	F		T	F	F	**T**	F
F	T	T		F	T	T	**T**	T
F	T	F		F	T	T	**F**	F
F	F	T		F	T	F	**T**	T
F	F	F		F	T	F	**F**	F
(1)	(2)	(3)			(4)		(5)	

11.

p	q	r	~	(p	∨	q)	→	(~	p	∧	r)
T	T	T	F	T	T	T	**T**	F	T	F	T
T	T	F	F	T	T	T	**T**	F	T	F	F
T	F	T	F	T	T	F	**T**	F	T	F	T
T	F	F	F	T	T	F	**T**	F	T	F	F
F	T	T	F	F	T	T	**T**	T	F	T	T
F	T	F	F	F	T	T	**T**	T	F	T	F
F	F	T	T	F	F	F	**T**	T	F	T	T
F	F	F	T	F	F	F	**F**	T	F	F	F
(1)	(2)	(3)	(5)		(4)		(8)	(6)			(7)

13.

p	q	(p	∨	q)	↔	(p	∧	q)
T	T	T	T	T	**T**	T	T	T
T	F	T	T	F	**F**	T	F	F
F	T	F	T	T	**F**	F	F	T
F	F	F	F	F	**T**	F	F	F
(1)	(2)		(3)		(5)		(4)	

15.

p	q	r	[p	∧	(q	∨	r)]	↔	[(p	∧	q)	∨	(p	∧	r)]
T	T	T	T	T	T	T	T	**T**	T	T	T	T	T	T	T
T	T	F	T	T	T	T	F	**T**	T	T	T	T	T	F	F
T	F	T	T	T	F	T	T	**T**	T	F	F	T	T	T	T
T	F	F	T	F	F	F	F	**T**	T	F	F	F	T	F	F
F	T	T	F	F	T	T	T	**T**	F	F	T	F	F	F	T
F	T	F	F	F	T	T	F	**T**	F	F	T	F	F	F	F
F	F	T	F	F	F	T	T	**T**	F	F	F	F	F	F	T
F	F	F	F	F	F	F	F	**T**	F	F	F	F	F	F	F
(1)	(2)	(3)		(7)		(4)		(9)		(5)		(8)		(6)	

17. $((\sim p) \wedge (\sim q)) \rightarrow ((\sim p) \wedge q)$

19. $(((\sim p) \wedge (\sim q)) \vee r) \rightarrow ((\sim q) \wedge r)$

21.

~	p	→	q
F	T	**T**	F

23.

q	→	p
F	**T**	T

25.

(p	⊕	q)	→	p
T	T	F	**T**	T

27.

(p	∧	~	q)	→	(~	p	⊕	q)
T	T	T	F	**F**	F	T	F	F

29. $p \quad \rightarrow \quad [p \quad \wedge \quad (p \quad \oplus \quad q)]$
     T   **T**   T   T   T   T   F

31. $p \leftrightarrow q$

33. $q \rightarrow p$; hypothesis: q; conclusion: p

35. $q \rightarrow p$; hypothesis: q; conclusion: p

37. $\sim p \rightarrow \sim q$; hypothesis: $\sim p$; conclusion: $\sim q$

39. $\sim p \rightarrow \sim q$; hypothesis: $\sim p$; conclusion: $\sim q$;
     F T **T** F T   TRUE

41. $\sim q \rightarrow \sim p$; hypothesis: $\sim q$; conclusion: $\sim p$;
                                       TRUE

43. **a.** hyp: A person is healthy.
             con: The person lives a long life.

     b. hyp: The train stops at the station.
             con: A passenger requests the stop.

     c. hyp: The azalea grows.
             con: The azalea is exposed to sunlight.

     d. hyp: I will go to the store.
             con: Jane goes to the store.

45. **a.** If City Sanitation collects the garbage, then
             the mayor calls.

     b. The price of beans goes down if there is no
             drought.

     c. Goldfish swim in Lake Erie if Lake Erie is
             fresh water.

     d. Tap water is not salted if it boils slowly.

47. **a.** $Z = 0 + 0 = 0$, so the condition fails and
             $A = 4$.

     b. $Z = 8 + (-8) = 0$, so the condition fails and
             $A = 4$.

     c. $Z = -3 + 3 = 0$, so the condition fails and
             $A = 4$.

     d. $X = -3 \leq 0$, so the condition fails and
             $A = 4$.

     e. $X = 8 > 0$ and $Z = 8 + (-3) = 5 \neq 0$, so the
             condition is met and $A = 6$.

f. $X = 3 > 0$ and $Z = 3 + (-8) = -5 \neq 0$, so the
        condition is met and $A = 6$.

49. **a.** $B = -6 < 0$, so the condition is met and
             $Y = 7$.

     b. $B = -6 < 0$, so the condition is met and
             $Y = 7$.

     c. $C = (-2)(6) = -12 < 10$ and $B \geq 0$, so the
             condition fails and $Y = 0$.

     d. $B = -1 < 0$, so the condition is met and
             $Y = 7$.

     e. $A = 4 \geq 0$ and $B = 3 \geq 0$, so the condition
             fails and $Y = 0$.

     f. $A = 3 \geq 0$ and $B = 1 \geq 0$, so the condition
             fails and $Y = 0$.

51. **a.** $A = -1 < 0$ and $B = -2 < 0$, so the condition
             is met and
             $C = (-1)(-2) + 4 = 6$.

     b. $B = 8 \geq 6$, so the condition is met and
             $C = (-2)(8) + 4 = -12$

     c. $B = 3 \geq 0$ and $B = 3 < 6$, so the condition
             fails and $C = 0$.

     d. $A = 3 \geq 0$ and $B = -2 < 6$, so the condition
             fails and $C = 0$.

     e. $B = 8 \geq 6$, so the condition is met and
             $C = (3)(8) + 4 = 28$.

     f. $A = 3 \geq 0$ and $B = -3 < 6$, so the condition
             fails and $C = 0$.

53. **a.** $C = 0 - 0 = 0$, $B = 0$, so the condition fails
             and $D = 0$, $X = 3$.

     b. $C = 6 - 3 = 3 > 0$, $B = 3 > 0$, so the
             condition fails and $D = 0$, $X = 3$.

     c. $C = -5 - 3 = -8 < 0$, so the condition is met
             and $D = -40$, $X = -37$.

     d. $C = 3 - 5 = -2 < 0$, so the condition is met
             and $D = -10$, $X = -7$.

     e. $B = -3 < 0$, so the condition is met. $C = 8$
             and $D = 40$, $X = 43$.

     f. $B = -3 < 0$, so the condition is met. $C = -2$
             and $D = -10$, $X = -7$.

Exercises 12.4

1.

[(p	→	q)	∧	q]	→	p
F	T	T	T	T	**F**	F

When p is false and q is true, the statement is FALSE.

3. Show that the corresponding bi-conditional is a tautology.

p	q	(p	→	q)	↔	(~	(p	∧	~	q))
T	T	T	T	T	**T**	T	T	F	F	T
T	F	T	F	F	**T**	F	T	T	T	F
F	T	F	T	T	**T**	T	F	F	F	T
F	F	F	T	F	**T**	T	F	F	T	F
(1)	(2)		(3)		(7)	(6)		(5)	(4)	

5. a.

p	~	p	↔	p	|	p
T	F	T	**T**	T	F	T
F	T	F	**T**	F	T	F
(1)	(2)		(4)		(5)	

b.

p	q	(p	∨	q)	↔	(p	|	p)	|	(q	|	q)
T	T	T	T	T	**T**	T	F	T	T	T	F	T
T	F	T	T	F	**T**	T	F	T	T	F	T	F
F	T	F	T	T	**T**	F	T	F	T	T	F	T
F	F	F	F	F	**T**	F	T	F	F	F	T	F
(1)	(2)		(5)		(7)		(3)		(6)		(4)	

c.

p	q	(p	∧	q)	↔	(p	|	q)	|	(p	|	q)
T	T	T	T	T	**T**	T	F	T	T	T	F	T
T	F	T	F	F	**T**	T	T	F	F	T	T	F
F	T	F	F	T	**T**	F	T	T	F	F	T	T
F	F	F	F	F	**T**	F	T	F	F	F	T	F
(1)	(2)		(5)		(7)		(3)		(6)		(4)	

d. $p \to q$

$\Leftrightarrow \sim p \lor q$ Implication (10a)

$\Leftrightarrow (\sim p \mid \sim p) \mid (q \mid q)$ Part (b) above

$\Leftrightarrow p \mid (q \mid q)$ Part (a) above

e. $p \mid q \Leftrightarrow \sim(p \land q)$

(Compare truth table for $\sim(p \land q)$ with truth table for $p \mid q$.)

7.

p	q	c	(p	→	q)	↔	[(p	∧	~	q)	→	c)]
T	T	F	T	T	T	**T**	T	T	F	F	T	T	F
T	F	F	T	F	F	**T**	T	T	T	T	F	F	F
F	T	F	F	T	T	**T**	F	F	F	F	T	T	F
F	F	F	F	T	F	**T**	F	F	F	T	F	T	F
(1)	(2)	(3)		(6)		(8)		(5)	(4)		(7)	

9. False: consider p FALSE and q TRUE.

11. $p \oplus q \Leftrightarrow (p \vee q) \wedge \sim(p \wedge q) \Leftrightarrow \sim[\sim(p \vee q) \vee \sim(\sim p \vee \sim q)]$

13. $(p \vee q) \to (q \wedge \sim r)$
$\Leftrightarrow \sim(p \vee q) \vee (q \wedge \sim r)$ Implication (10a)

15. $\sim(p \wedge \sim q) \to (p \vee \sim r)$
$\Leftrightarrow (p \wedge \sim q) \vee (p \vee \sim r)$ Implication (10a) and double negation (1)

17. a. $\sim(p \wedge q) \Leftrightarrow \sim p \vee \sim q$
   Arizona does not border California, or Arizona does not border Nevada.

   b. $\sim(p \vee q) \Leftrightarrow \sim p \wedge \sim q$
   There are no tickets available, and the agency cannot get tickets.

   c. $\sim(p \vee q) \Leftrightarrow \sim p \wedge \sim q$
   The killer's hat was neither white nor gray.

19. $\sim(p \vee \sim q \vee r) \Leftrightarrow \sim p \wedge q \wedge \sim r$

21. a. Jeremy does not take 12 credits and Jeremy does not take 15 credits.

   b. Sandra does not receive a gift from Sally or Sandra does not receive a gift from Sacha.

23. a. "The Old Man and the Sea" was written by Ernest Hemingway or "The Old Man and the Sea" was written by Jack London.
   Negation: "The Old Man and the Sea" was not written by Ernest Hemingway and "The Old Man and the Sea" was not written by Jack London.

   b. "H.M.S. Pinafore" was written by Gilbert and "H.M.S. Pinafore" was written by Sullivan.
   Negation: "H.M.S. Pinafore" was not written by Gilbert or "H.M.S. Pinafore" was not written by Sullivan.

25. a. I have a ticket to the theater, and I did not spend a lot of money.

   b. Basketball is played on an indoor court, and the players do not wear sneakers.

   c. The stock market is going up, and interest rates are not going down.

   d. Humans have enough water, and humans are not staying healthy.

27. a. If the number of odd numbers in a sum is even, then the sum is even.
   True
   For example: the number of odd numbers in the sum is even: $1+ 3 + 5+ 7$.
   The sum of the odd numbers is 16 (even)

   b. If K is next to W, then the computer keyboard is not standard.
True.

29. a. Contrapositive: If a bird is not a hummingbird, then it is not small (F).
Converse: If a bird is a hummingbird, then it is small (T).

   b. Contrapositive: If two nonvertical lines are not parallel, they do not have the same slope (T).
Converse: If two nonvertical lines are parallel, they have the same slope (T).

   c. Contrapositive: If we are not in France, then we are not in Paris (T).
Converse: If we are in France, we must be in Paris (F).

   d. Contrapositive: If you can legally make a U-turn, then the road is not one-way (T).
Converse: If you cannot legally make a U-turn, then the road is one-way (F).

31. Ask either guard, "If I asked you whether your door was the door to freedom, would you say yes?" Phrasing the question this way forces the guard that always lies to tell the truth as to whether his door is the door to freedom.

33.

p	q	r	$($	p	$\to$	$q)$	$\wedge$	$($	q	$\to$	$r)$	$\to$	$($	p	$\to$	$r)$
T	T	T		T	T	T	T		T	T	T	**T**		T	T	T
T	T	F		T	T	T	F		T	F	F	**T**		T	F	F
T	F	T		T	F	F	F		F	T	T	**T**		T	T	T
T	F	F		T	F	F	F		F	T	F	**T**		T	F	F
F	T	T		F	T	T	T		T	T	T	**T**		F	T	T
F	T	F		F	T	T	F		T	F	F	**T**		F	T	F
F	F	T		F	T	F	T		F	T	T	**T**		F	T	T
F	F	F		F	T	F	T		F	T	F	**T**		F	T	F
(1)	(2)	(3)			(4)		(7)			(5)		(8)			(6)	

Exercises 12.5

1. m = "Sue goes to the movies."
r = "Sue reads."

   1. $m \vee r$ hyp.
   2. $\sim m$ hyp.
   3. r disj. syll. (1, 2)

3. a = "My allowance comes this week."
p = "I pay the rent."
b = "My bank account will be in the black."
e = "I will be evicted."

   1. $(a \wedge p) \to b$ hyp.
   2. $\sim p \to e$ hyp.
   3. $\sim e \wedge a$ hyp.
   4. $\sim e$ subtr. (3)
   5. p mod. tollens (2, 4)
   6. a subtr. (3)
   7. b mod. ponens (5, 6, 1)

5. p = "The price of oil increases."
   a = "The OPEC countries are in agreement."
   d = "There is a U.N. debate."
   1. $p \to a$ hyp.
   2. $\sim d \to p$ hyp.
   3. $\sim a$ hyp.
   4. $\sim p$ mod. tollens (1, 3)
   5. d mod. tollens (2, 4)

7. g = "The germ is present."
   r = "The rash is present."
   f = "The fever is present."
   1. $g \to (r \land f)$ hyp.
   2. f hyp.
   3. $\sim r$ hyp.
   4. $\sim r \lor \sim f$ addition (3)
   5. $\sim(r \land f)$ DeMorgan (4)
   6. $\sim g$ mod tollens (1, 5)

9. c = "The material is cotton."
   r = "The material is rayon."
   d = "The material can be made into a dress."
   1. $(c \lor r) \to d$ hyp.
   2. $\sim d$ hyp.
   3. $\sim(c \lor r)$ mod. tollens (1, 2)
   4. $\sim c \land \sim r$ DeMorgan (3)
   5. $\sim r$ subtraction (4)

11. s = "Salaries go up."
    m = "More people apply."
    If s is false and m is true, then
    $(s \to m) \land (m \lor s)$ is true but s is false.
    $(s \to m) \land (m \lor s) \Rightarrow s$. The argument is invalid.

13. y = "The balloon is yellow."
    p = "The ribbon is pink."
    h = "The balloon is filled with helium."
    1. $y \lor p$ hyp.
    2. $h \to \sim y$ hyp.
    3. h hyp.
    4. $\sim y$ mod. ponens (2, 3)
    5. p disj. syllogism (1, 4)
    The argument is valid.

15. p = "The papa bear sits."
    m = "The mama bear stands."
    b = "The baby bear crawls on the floor."
    1. $p \to m$ hyp.
    2. $m \to b$ hyp.
    3. $\sim b$ hyp.
    4. $\sim m$ mod. tollens (2, 3)
    5. $\sim p$ mod. tollens (1, 4)
    The argument is valid.

17. w = "Wheat prices are steady."
    e = "Exports will increase."
    s = "The GNP will be steady."
    If w and s are true and e is false, then
    $[w \to (e \lor s)] \land (w \lor s)$ is true but e is false.
    $[w \to (e \lor s)] \land (w \land s) \Rightarrow e$. The argument is invalid.

19. i = "Tim is industrious."
    p = "Tim is in line for a promotion."
    l = "Tim is thinking of leaving."
    If l and p are true and i is false, then
    $(i \to p) \land (p \lor l)$ is true but $l \to i$ is false.
    $(i \to p) \land (p \lor l) \Rightarrow l \to i$
    The argument is valid.

21. s = "Sam goes to the store."
    m = "Sam needs milk."
    $H_1 = s \to m$
    $H_2 = \sim m$
    $C = \sim s$
    1. s $\sim C$
    2. $s \to m$ H_1
    3. m $\sim H_2$;
      mod. ponens (1, 2)

23. n = "The newspaper reports the crime."
    t = "Television reports the crime."
    s = "The crime is serious."
    k = "A person is killed."
    $H_1 = (n \land t) \to s$
    $H_2 = k \to n$
    $H_3 = k$
    $H_4 = t$
    $C = s$
    1. $\sim s$ $\sim C$
    2. $(n \land t) \to s$ H_1
    3. $\sim(n \land t)$ mod. tollens (1, 2)
    4. $\sim n \lor \sim t$ DeMorgan (3)
    5. t H_4
    6. $\sim n$ disj. syllogism (4, 5)
    7. $k \to n$ H_2
    8. $\sim k$ $\sim H_3$; mod. tollens (6, 7)

25. j = "Jimmy finds his keys."
    h = "He does his homework."
    $H_1 = \sim j \to h$
    $H_2 = \sim h$

Direct proof

1. $\sim j \to h$ H_1
2. $\sim h$ H_2
3. j mod. tollens (1, 2)

Indirect proof

1. $\sim j$ $\sim C$
2. $\sim j \to h$ H_1
3. h mod. ponens (1, 2) $\Big\}$ contradiction
4. $\sim h$ H_2

27. m = "Marissa goes to the movies."
k = "Marissa is in a knitting class."
i = "Marissa is idle."
$H_1 = \sim m \to \sim i$
$H_2 = \sim k \to i$
$H_3 = \sim k$

Direct proof

1. $\sim m \to \sim i$ H_1
2. $\sim k \to i$ H_2
3. $\sim k$ H_3
4. i mod. ponens (2, 3)
5. m mod. tollens (1, 4)

Indirect proof

1. $\sim m$ $\sim C$
2. $\sim m \to \sim i$ H_1
3. $\sim i$ mod. ponens (1, 2)
4. $\sim k \to i$ H_2
5. k mod. tollens (3, 4) $\Big\}$ contradiction
6. $\sim k$ H_3

Exercises 12.6

1. a. "1 is even or 1 is divisible by 3" is FALSE.

b. "4 is even or 4 is divisible by 3" is TRUE.

c. "3 is even or 3 is divisible by 3" is TRUE.

d. "6 is even or 6 is divisible by 3" is TRUE.

e. "5 is even or 5 is divisible by 3" is FALSE.

3. a. $\forall x\, p(x)$

b. $\sim [\forall x\, p(x)]$

c. $\forall x \sim p(x)$

d. (c) implies (b), since if nobody is taking a writing course, it follows that not everybody is.

5. Abby's statement: $\forall x \sim p(x)$, or $\sim [\exists x\, p(x)]$. This is surely false. Abby meant to say, "not all men cheat on their wives" $\sim [\forall x\, p(x)]$, or $\exists x \sim p(x)$.

7. a. $\forall x\, p(x)$

b. $\exists x \sim p(x)$

c. $\exists x\, p(x)$

d. $\sim [\forall x\, p(x)]$

e. $\forall x \sim p(x)$

f. $\sim [\exists x\, p(x)]$

g. (b) and (d); (e) and (f) · (b) and (d) both say that there are some university professors who don't like poetry; (e) and (f) both say that no university professors like it.

9. a. TRUE;

$p(4) = (4$ is prime$) \to (4^2 + 1$ is even$)$ is TRUE, because the hypothesis is FALSE.

b. FALSE;

$p(2) = ($if 2 is prime$) \to (2^2 + 1$ is even$)$ is FALSE.

11. a. T; every x is either even or odd.

b. $[\forall x\, p(x)] \bigvee [\forall x\, q(x)]$
    F   **F**   F

c. T; 5 is odd, hence even or odd, for instance.

d. F; no x is both even or odd.

e. F; 5 is not both even and odd, for instance.

f. $[\exists x\, p(x)] \bigwedge [\exists x\, q(x)]$
    F   **T**   F

g. F; (4 is even) $\to$ (4 is odd) is false, for instance.

h. $[\forall x\, p(x) \to [\forall x\, g(x)]$
    F   **T**   F

13. a. Not every dog has his day.

 b. No men fight wars.

 c. Some mothers are unmarried.

 d. There exists a pot without a cover.

 e. All children have pets.

 f. Every month has 30 days.

15. a. "The sum of any two nonnegative integers is greater than 12." FALSE: consider $x = 1$, $y = 2$. "There exist two nonnegative integers whose sum is not greater than 12."

 b. "For any nonnegative integer, there is another nonnegative integer that, added to the first, makes a sum greater than 12." TRUE

 c. "There is a nonnegative integer that, added to any other nonnegative integer, makes a sum greater than 12." TRUE (Try $x = 13$.)

 d. "There are two nonnegative integers, the sum of which is greater than 12." TRUE (Try $x = 6$, $y = 7$.)

17. a. FALSE: let $x = 2$, $y = 3$.

 b. TRUE: for any x, let $y = x$.

 c. TRUE: let $x = 1$.

 d. FALSE: no y is divisible by every x.

 e. TRUE: for any y, let $x = y$.

 f. TRUE: any x divides itself.

19. a. $S \subseteq T$ translates as $\forall x[x \geq 8 \rightarrow x \leq 10]$.

 b. No; consider $x = 11$.

21. $S = \{2, 4, 6, 8\}$, $T = \{1, 2, 3, 4, 6, 8\}$. So, $\forall x[x \in S \rightarrow x \in T)$.

23. The solutions to $(x - 8)(x - 3) = 0$ are 8 and 3. Only 3 is in U. The solutions to $x^2 = 9$ are -3 and 3. Only 3 is in U. So, $\forall x(x \in S \leftrightarrow x \in T)$. Therefore, $S = T = \{3\}$.

Exercises 12.7

 1. $(p \vee q) \wedge (p \vee {\sim}q)$

 3. $((p \wedge q) \wedge {\sim}r) \wedge ({\sim}q \vee r)$

 5.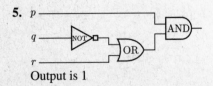

 Output is 1

7.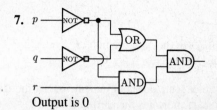

Output is 0

9. The circuit represents the logic statement

$(p \wedge q) \vee (p \wedge \sim q)$

$\Leftrightarrow p \wedge (q \vee \sim q)$ Distributive law (4b)

$\Leftrightarrow p \wedge t$ (7a)

$\Leftrightarrow p$ (6d)

11. The circuit represents the logic statement

$((p \wedge q) \wedge r) \vee \sim((p \vee q) \vee \sim r)$

$\Leftrightarrow (r \wedge (p \wedge q)) \vee (r \wedge \sim(p \vee q))$ Commutative law (2b) and DeMorgan's law (8a)

$\Leftrightarrow r \wedge [(p \wedge q) \vee \sim(p \vee q)]$ Distributive law (4b)

13.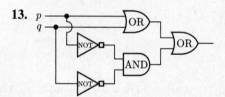

15.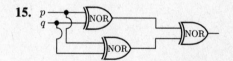

17.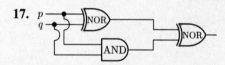

19.

Chapter 12 Supplementary Exercises

1. a. Statement

    b. Not a statement—not a declarative sentence.

    c. Statement

    d. Not a statement—"he" is not specified.

    e. Statement

2. a. If two lines are perpendicular, then their slopes are negative reciprocals of each other.

    b. If goldfish can live in a fishbowl, then the water is aerated.

    c. If it rains, then Jane uses her umbrella.

    d. If Sally gives Morris a treat, then he ate all his food.

3. a. Contrapositive: If the Yankees are not playing in Yankee Stadium, then they are not in New York City; converse: If the Yankees are playing in Yankee Stadium, then they are in New York City.

    b. Contrapositive: If the quake is not considered major, then the Richter scale does not indicate the earthquake is a 7; converse: If the quake is considered major, then the Richter scale indicates the earthquake is a 7.

    c. Contrapositive: If a coat is not warm then it is not made of fur; converse: If a coat is warm then it is made of fur.

    d. Conrapositive: If Jane is not in Moscow then she is not in Russia; converse: If Jane is in Moscow then she is in Russia.

4. a. $p =$ (two triangles are similar) and
$q =$ (their sides are equal).
$p \rightarrow q$ negated becomes $\sim(p \rightarrow q)$ or $p \wedge \sim q$, or "two triangles are similar but their sides are unequal."

    b. $U =$ {real numbers} and $p(x) = (x^2 = 5)$.

        $\exists x \, p(x)$ negated becomes $\forall x \sim q(x)$ or "For every real number x, $x^2 \neq 5$."

    c. $U =$ {positive integers},

        $p(n = (n$ is even), and $q(n) = (n^2$ is even).

        $\forall n[p(n) \rightarrow q(n)]$ negated becomes $\exists n \sim [p(n) \rightarrow q(n)]$ or $\exists n[p(n) \wedge q(n)]$, or "There exists a positive integer n such that n is even but n^2 is not even."

    d. $U =$ {real numbers} and $p(x) = (x^2 + 4 = 0)$.

        $\exists x \, p(x)$ negated becomes $\forall x \sim p(x)$ or "For every real number x, $x^2 + 4 \neq 0$."

5. a.

p		p	$\vee$	$\sim$	p
T		T	**T**	F	T
F		F	**T**	T	F
(1)			(3)	(2)	

Tautology

    b.

p	q		$(p$	$\rightarrow$	$q)$	$\leftrightarrow$	$(\sim$	p	$\vee$	$q)$
T	T		T	T	T	**T**	F	T	T	T
T	F		T	F	F	**T**	F	T	F	F
F	T		F	T	T	**T**	T	F	T	T
F	F		F	T	F	**T**	T	F	T	F
(1)	(2)			(4)		(6)	(3)		(5)	

Tautology

   c. Let p and q be TRUE.

     $(p \wedge \sim q) \leftrightarrow \sim(\sim p \wedge q)$

     T FFT **F** TFT FT

     Not a tautology

   d. Let p, q, and r be FALSE.

     $[p \rightarrow (q \rightarrow r)] \leftrightarrow [(p \rightarrow q) \rightarrow r]$

     F T F T F **F** F T F F F

     Not a tautology

6. a.

p	q	r	p	$\rightarrow$	$(\sim$	q	$\vee$	$r)$
T	T	T	T	**T**	F	T	T	T
T	T	F	T	**F**	F	T	F	F
T	F	T	T	**T**	T	F	T	T
T	F	F	T	**T**	T	F	T	F
F	T	T	F	**T**	F	T	T	T
F	T	F	F	**T**	F	T	F	F
F	F	T	F	**T**	T	F	T	T
F	F	F	F	**T**	T	F	T	F
(1)	(2)	(3)		(6)	(4)		(5)	

b.

p	q	r	p	$\wedge$	$(q$	$\leftrightarrow$	$(r$	$\wedge$	$p))$
T	T	T	T	**T**	T	T	T	T	T
T	T	F	T	**F**	T	F	F	F	T
T	F	T	T	**F**	F	F	T	T	T
T	F	F	T	**T**	F	T	F	F	T
F	T	T	F	**F**	T	F	T	F	F
F	T	F	F	**F**	T	F	F	F	F
F	F	T	F	**F**	F	T	T	F	F
F	F	F	F	**F**	F	T	F	F	F
(1)	(2)	(3)		(6)		(5)		(4)	

7. a. True—a version of disjunctive syllogism

   b. False—consider p false and q true.

8. a. True—contrapositive

   b. False—consider p, q, and r false.

9. a. False—consider p false and q true.

   b. True—modus tollens

10. a. $C = 3(4) + 5 = 17 > 0$ and $B = 5 > 0$, so the condition is met and $Z = 17$.

   b. $B = 2 \not> 3$, so the condition fails and $Z = 100$.

   c. $C = 3(-4) + 5 = -7 \not> $, so the condition fails and $Z = 100$.

   d. $B = -2 \not> 3$, so the condition fails and $Z = 100$.

11. a. $C = 10$, so the condition is met and $Z = 5 \times 10 = 50$.

   b. $C = 10$, so the condition is met and $Z = (5)(-5) = -25$.

   c. $X = -10 \not> 0$, $C = 2 \not\geq 10$, so the condition fails and $Z = (-10) + (-5) = -15$.

   d. $X = 2 > 0$, $Y = 5 > 0$, so the condition is met and $Z = 2 \times 5 = 10$.

12. a. Cannot be determined

   b. TRUE, by the contrapositive

   c. TRUE

   d. Cannot be determined

   e. Cannot be determined

13. a. Cannot be determined

   b. Cannot be determined

   c. TRUE, by contraposition and DeMorgan

14. a. TRUE (given the additional assumption that at least two mathematicians exist)

   b. FALSE; let $p(x) = $ "like rap music" and $U = \{$mathematicians$\}$.
$\forall x\, p(x) \quad \sim\exists x\, p(x)$

   c. TRUE

15. a. Cannot be determined

   b. Cannot be determined

   c. True; $\exists x[\sim r(x)] \Leftrightarrow \sim\forall x[r(x)]$

16. $t = $ "Taxes go up."
$s = $ "I sell the house."
$m = $ "I move to India."

   1. $t \rightarrow (s \wedge m)$ hyp.
   2. $\sim m$ hyp.
   3. $\sim s \vee \sim m$ addition (2)
   4. $\sim(s \wedge m)$ DeMorgan (3)
   5. $\sim t$ mod. tollens (1, 4)

17. m = "I study mathematics."
b = "I study business."
p = "I can write poetry."

1. $m \wedge b$	hyp.
2. $b \rightarrow (\sim p \vee \sim m)$	hyp.
3. b	subtraction (1)
4. $\sim p \vee \sim m$	mod. ponens (2, 3)
5. m	subtraction (1)
6. $\sim p$	disj. syllogism (4, 5)

18. d = "I shop for a dress."
h = "I wear high heels."
s = "I have a sore foot."

1. $d \rightarrow h$	hyp.
2. $s \rightarrow \sim h$	hyp.
3. d	hyp.
4. h	mod. ponens (1, 3)
5. $\sim s$	mod. tollens (2, 4)

19. a = "Asters grow in the garden."
d = "Dahlias grow in the garden."
s = "It is spring."

1. $a \vee d$	hyp.
2. $s \rightarrow \sim a$	hyp.
3. s	hyp.
4. $\sim a$	mod. ponens (2, 3)
5. d	disj. syllogism (1, 4)

20. t = "The professor gives a test."
h = "Nancy studies hard."
d = "Nancy has a date."
s = "Nancy takes a shower."

$H_1 = t \rightarrow h$
$H_2 = d \rightarrow s$
$H_3 = \sim t \rightarrow \sim s$
$H_4 = d$
$C = h$

1. $\sim h$	$\sim C$
2. $t \rightarrow h$	H_1
3. $\sim t$	mod. tollens (1, 2)
4. $\sim t \rightarrow \sim s$	H_3
5. $\sim s$	mod. ponens (3, 4)
6. $d \rightarrow s$	H_2
7. $\sim d$	$\sim H_4$; mod. tollens (5, 6)

Contradiction

21.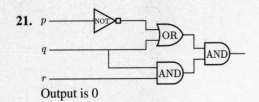

Output is 0

22. $((p \wedge \sim q) \wedge \sim (q \vee \sim r)) \wedge r$

$\Leftrightarrow ((p \wedge \sim q) \wedge (\sim q \wedge \sim r)) \wedge r$ DeMorgan's law (8a)

$\Leftrightarrow ((p \wedge r) \wedge \sim q) \wedge r$ Distributive law (4b) and commutative law (2b)

$\Leftrightarrow (p \wedge (r \wedge r)) \wedge (\sim q \wedge r)$ Distributive law (4b)

$\Leftrightarrow p \wedge (\sim q \wedge r)$ Idempotent law (5b)

Conceptual Exercises

23. $(p$ NOR $q)$ NOR $(p$ AND $q)$

p	q	$(p$	NOR	$q)$	NOR	$(p$	AND	$q)$	$\leftrightarrow$	$(p$	XOR	$q)$
T	T	T	F	T	F	T	T	T	**T**	T	F	T
T	F	T	F	F	T	T	F	F	**T**	T	T	F
F	T	F	F	T	T	F	F	T	**T**	F	T	T
F	F	F	T	F	F	F	F	F	**T**	F	F	F
(1)	(2)		(3)		(6)		(4)		(7)		(5)	

24.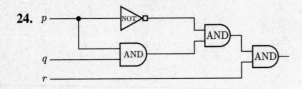

25.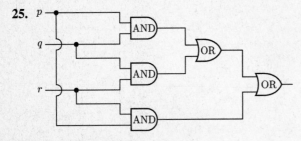

Chapter 12 Chapter Test

1.

p	q	$(p$	$\wedge$	$\sim$	$q)$	$\rightarrow$	q
T	T	T	F	F	T	**T**	T
T	F	T	T	T	F	**F**	F
F	T	F	F	F	T	**T**	T
F	F	F	F	T	F	**T**	F
(1)	(2)		(4)	(3)		(5)	

2. a. Statement

b. Not a statement—not a declarative sentence.

    c. Statement

    d. Statement

    e. Not a statement—not a declarative sentence.

3. If the coach does not buy Bob ice cream, then Bob did not hit a triple or a home run.

4. Every integer is either even or greater than 8.

5. $\forall x[p(x)]$ is FALSE; let $x = 2$, so $x^2 = 4$ is even; but $x^3 + 1 = 9$, which is not prime. $\exists x[\sim p(x)]$ is TRUE; as just shown, $\sim p(x)$ is TRUE for $x = 2$.

6. a.

p	q	(p	$\wedge$	$\sim$	q)	$\vee$	(q	$\rightarrow$	$\sim$	p)
T	T	T	F	F	T	**F**	T	F	F	T
T	F	T	T	T	F	**T**	F	T	F	T
F	T	F	F	F	T	**T**	T	T	T	F
F	F	F	F	T	F	**T**	F	T	T	F
(1)	(2)		(5)	(3)		(7)		(6)	(4)	

Neither a tautology nor a contradiction.

b.

p	q	(p	$\oplus$	$\sim$	q)	$\leftrightarrow$	(p	$\leftrightarrow$	q)
T	T	T	T	F	T	**T**	T	T	T
T	F	T	F	T	F	**T**	T	F	F
F	T	F	F	F	T	**T**	F	F	T
F	F	F	T	T	F	**T**	F	T	F
(1)	(2)		(4)	(3)		(6)		(5)	

A tautology.

c.

p	q	$\sim$	(p	$\rightarrow$	q)	$\wedge$	q
T	T	F	T	T	T	**F**	T
T	F	T	T	F	F	**F**	F
F	T	F	F	T	T	**F**	T
F	F	F	F	T	F	**F**	F
(1)	(2)	(4)		(3)		(5)	

A contradiction.

7. a. TRUE; both parts of the statement are TRUE.

    b. FALSE; the hypothesis is TRUE but the conclusion is FALSE.

    c. TRUE; the hypothesis is FALSE so the statement is TRUE.

    d. FALSE; the hypothesis is TRUE but the conclusion is FALSE.

8. w = "You win the lottery."
   g = "You have a good job."
   m = "You will have a lot of money."
   r = "You get robbed."

1. $(w \lor g) \to m$	hyp.
2. $r \to \sim m$	hyp.
3. r	hyp.
4. $\sim m$	mod. ponens (2, 3)
5. $\sim(w \lor g)$	mod. tollens (1, 4)
6. $\sim w \land \sim g$	DeMorgan (5)
7. $\sim g$	subtraction (6)

   The argument is valid.

9. **a.** Every English dictionary contains the word "Internet."

   b. There is a student at the university who does not listen to jazz.

   c. There is a floor at which the elevator does not stop.

10. **a.** $p \to q$

    b. $q \to p$

    c. $p \to q$

    d. $p \to q$

    e. $q \to p$

11. **a.** $D = 4 - 9 \cdot 3 = -23$
       $G = D + A = -19$

    b. $D = 185 - 30 \cdot 6 = 5$
       $G = 0$

    c. $D = 0 - (30(-6)) = 180$
       $G = D + A = 180$

    d. $G = 0$

12. **a.** False; setting r TRUE and p FALSE makes $p \land (q \lor r)$ FALSE and $(p \land q) \lor r$ TRUE.

    b. True; if $p \land (q \lor r)$ is TRUE, then so is $q \lor r$, which means that either r is TRUE or else q is TRUE and $\sim q$ is FALSE. Either way, $\sim q \to r$ is TRUE.

13. $((p \land q) \land r) \lor (p \land r))$
    $\Leftrightarrow ((p \land q) \lor p) \land r$ Distributive law (4b) and commutative law (2b)
    $\Leftrightarrow p \land r$ Subtraction

14.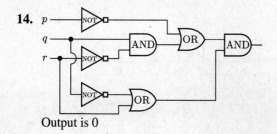

Output is 0

EXPLORATIONS IN FINITE MATHEMATICS

The student oriented software package *Explorations in Finite Mathematics*, which runs on both Macintosh and PC computers, was developed to accompany the textbook. *Explorations in Finite Mathematics* is easy to use and focuses on the learning of concepts. The software is accompanied by a 67-page manual in Adobe Reader format.

To obtain a CD containing *Explorations in Finite Mathematics*, send an email to David Schneider at dis@math.umd.edu. (You may freely make copies of the CD.) The software also is available for download from the website www.pearsoned.com/goldstein.

Explorations in Finite Mathematics consists of two programs called FINITE1 (contains the matrix, Markov processes, systems of linear equations and inequalities, and linear programming routines) and FINITE2 (contains the counting, probability, statistics, financial, difference equation, and truth table routines).

FINITE1

Graphical Solution of Linear Programming Problems uses animation to find the solution of a linear programming problem.

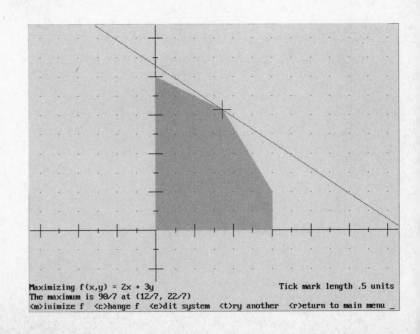

```
Maximizing f(x,y) = 2x + 3y                        Tick mark length .5 units
The maximum is 90/7 at (12/7, 22/7)
<m>inimize f  <c>hange f  <e>dit system  <t>ry another  <r>eturn to main menu _
```

Elimination uses Gauss-Jordan elimination to row-reduce a matrix.

$$\left[\begin{array}{cc|c} 2 & -\frac{1}{2} & 0 \\ \\ 1 & 3 & \frac{7}{2} \end{array} \right]$$

```
<m>ultiply a row by a constant          Do <n>ext step automatically
E<x>change two rows                      Do <e>very step automatically
<a>dd a multiple of one row to another   Get <h>elp
<p>rint matrix      <s>ave matrix        <r>eturn to main menu
Press letter of your choice (m x a p n e h s  or r):
```

Inverse of a Matrix uses elementary row operations to invert a matrix via the Gauss-Jordan method.

$$\left[\begin{array}{ccc|ccc} \frac{2}{7} & -3 & 5 & 1 & 0 & 0 \\ \\ 0 & 2 & -4 & 0 & 1 & 0 \\ \\ 1 & 6 & \frac{5}{2} & 0 & 0 & 1 \end{array} \right]$$

```
<m>ultiply a row by a constant          Do <n>ext step automatically
E<x>change two rows                      Do <e>very step automatically
<a>dd a multiple of one row to another   Get <h>elp
<p>rint matrix       <s>ave matrix       <r>eturn to main menu
Press letter of your choice (m x a p n e h s  or r):
```

Simplex Method carries out the simplex method on an initial simplex tableau.

	x	y	u	v	w	M	
u	6	3	1	0	0	0	96
v	1	1	0	1	0	0	18
w	2	6	0	0	1	0	72
M	-80	-70	0	0	0	1	0

```
<c>hoose pivot element        Do <n>ext step automatically
Get <h>elp                    Do <e>very step automatically
<p>rint matrix   <s>ave to disk   <t>ry a new matrix   <r>eturn to main menu
Press letter of your choice (c h p n e s t or r):
```

Regular Stochastic Matrices

calculates the stable matrix of a regular stochastic matrix and displays successive generations of the matrix and an initial distribution.

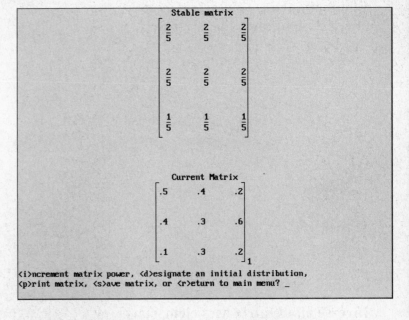

Least Squares Line

Least Squares Line calculates the line of best fit and also allows the user to move a line around the screen to find the best fit.

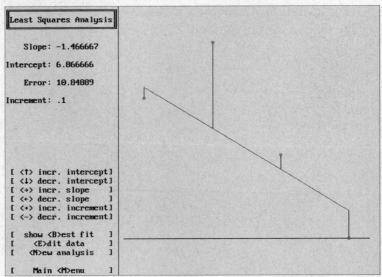

Absorbing Stochastic Matrices

calculates the stable matrix of an absorbing stochastic matrix and displays successive generations of the matrix. The routine also calculates the fundamental matrix.

Matrix Operations evaluates a matrix expression.

```
This routine evaluates a matrix expression.  Some examples are:

A + B           A*B  (A times B)          A^2 (same as A²)
2A - B          A^-1 (A inverse)          I(n) (nxn identity matrix)
A^-1*B          det(A) (determinant)      trans(A) (transpose)
I(3) - A        A*(B+C)                   det(A)*B

Saved matrices may be loaded from the disk.

Note: This routine can be used without matrices to evaluate numerical
      expressions such as (2 + sqr(3))/4, 5*exp(2), and 8.2*sin(1).
```

Expression to evaluate: _

Graphing Linear Equations graphs one or more straight lines.

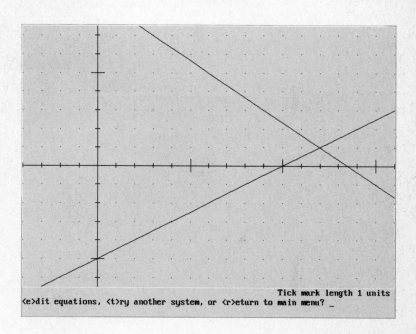

```
                                        Tick mark length 1 units
<e>dit equations, <t>ry another system, or <r>eturn to main menu? _
```

Graphing Linear Inequalities graphs a system of linear inequalities.

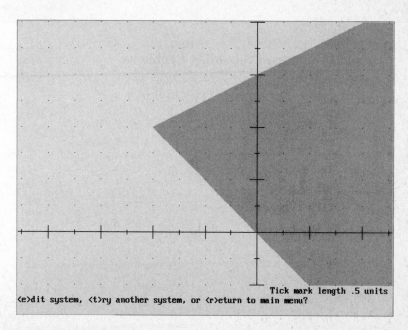

```
                                        Tick mark length .5 units
<e>dit system, <t>ry another system, or <r>eturn to main menu?
```

FINITE2

Venn Diagrams draws a Venn diagram for a set-theoretic expression. Alternately, the routine asks the user to find a set-theoretic expression for a randomly chosen Venn diagram.

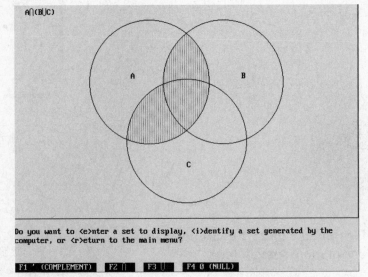

Venn Diagram Counting Problems solves two-circle or three-circle Venn diagram counting problems.

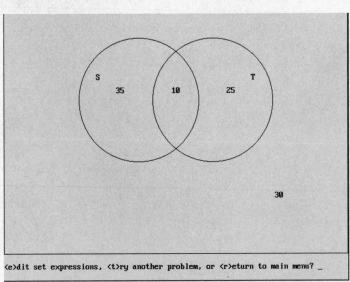

Venn Diagram Probability Problems solves two-circle or three-circle Venn diagram probability problems.

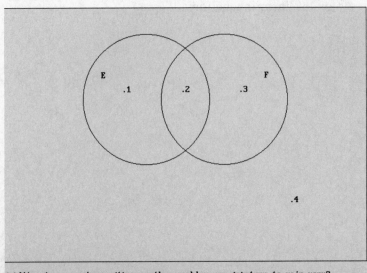

Counting Routines The four routines Factorial Computation, Combinations, Permutations, and Multiplication Principle are used to compute *n!, C(n, k), P(n, k)* and any product consisting of these numbers and ordinary integers.

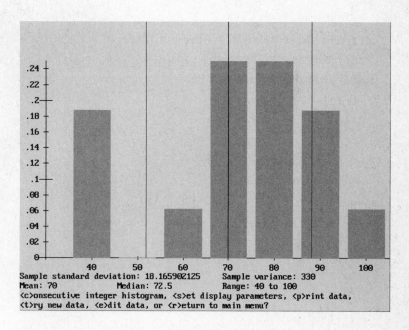

```
Press <Enter> after each value.  Press <Enter> again once you have entered
all values.  Each value must either be an integer or be of the form n!, C(n,r),
or P(n,r), for factorial, combinations, or permutations. Press <Esc> to return
to the main menu.

Value: 100!
Value: C(100,50)
Value: P(40,10)
The product is 28962934207504440416800055071730392618848179469631419666309940255
45015860812251214124641983342055987082678826674839572736338308286518506099276001
221128899294930012343291881717760000000000000000000000000000000

Value:
```

Statistics calculates range, median, mean, variance, and standard deviation for a collection of data and displays histograms.

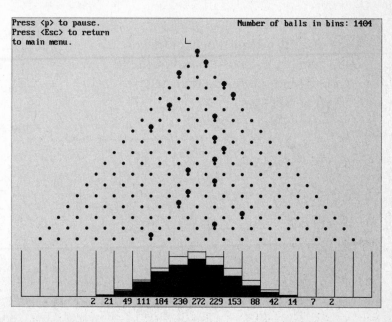

```
Sample standard deviation: 18.165902125          Sample variance: 330
Mean: 70                  Median: 72.5            Range: 40 to 100
<c>onsecutive integer histogram, <s>et display parameters, <p>rint data,
<t>ry new data, <e>dit data, or <r>eturn to main menu?
```

Galton Board uses animation to simulate a binomial distribution. The probability of a ball bouncing to the right is specified by the user.

```
Press <p> to pause.                    Number of balls in bins: 1404
Press <Esc> to return
to main menu.
```

```
2  21  49  111  184  230  272  229  153  88  42  14  7  2
```

Binomial Distribution calculates binomial probabilities for a range of numbers or a single number.

```
Probability that, of n binomial trials each with a p probability of success,
from m1 to m2 of the trials will result in success.
Press <Esc> to return to main menu.

n = 40
p = .25
m1 = 16
m2 = 40
There will be from 16 to 40 successes with probability .0262448840837
n = 100
p = .5
m1 = 50
m2 = 50
There will be exactly 50 successes with probability .0795892373872
n =
```

Areas Under Normal Curve calculates and illustrates the area of a region under a normal curve.

```
The region between 11.25 and 15 has area .6687123
mu = 12
sigma = 1.5
<t>ry again or
<r>eturn to main menu? _
```

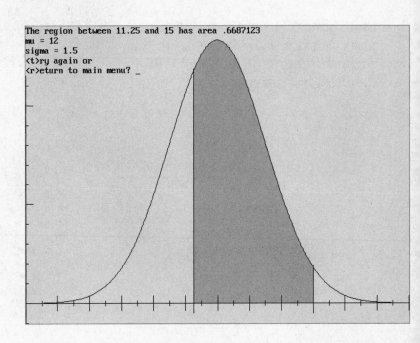

Financial Routines Four routines analyze a bank deposit earning simple or compound interest, a mortgage, and an annuity.

```
              A M O R T I Z A T I O N   O F   A   L O A N

                       Principal:    $300,000.00
                                                          Enter new values to see
         Annual interest rate (percent):      6.500%     effect on computed item.
         # of times compounded per year:         12
                                                          Press <PgUp>, <PgDn>, or
         Duration of loan (# of periods):        360     <S> to change starting
                                                          period of chart. Press
         Computed Amount of Payment ----->   $1,896.20   <P> to print full chart.

PAY                        DEBT                           Press <Tab> to make the
NUM.   PAYMENT   INTEREST  REDUCTION    BALANCE           currently selected item
  1    1896.20   1625.00    271.20    299,728.80          the computed item.
  2    1896.20   1623.53    272.67    299,456.12
  3    1896.20   1622.05    274.15    299,181.97          Press <N> to start a new
  4    1896.20   1620.57    275.64    298,906.34          problem or <Esc> to
  5    1896.20   1619.08    277.13    298,629.21          return to the main menu.
  6    1896.20   1617.57    278.63    298,350.58
  7    1896.20   1616.07    280.14    298,070.44          -------- Interest --------
  8    1896.20   1614.55    281.66    297,788.79          This page:      $19,401.27
  9    1896.20   1613.02    283.18    297,505.60          Total:         $382,633.47
 10    1896.20   1611.49    284.72    297,220.89          ----- Debt Reduction -----
 11    1896.20   1609.95    286.26    296,934.63          This page:       $3,353.18
 12    1896.20   1608.40    287.81    296,646.82
```

Difference Equations draws the graph of a difference equation. The routine also displays a table giving the values of the first 81 terms.

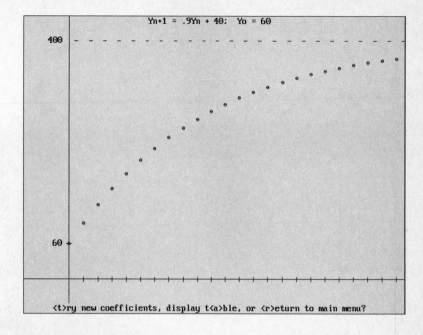

Truth Tables constructs a truth table for a logical expression with up to 24 statements.

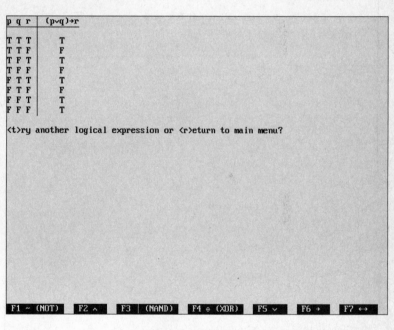